Extracting Innovations

Mining, Energy, and Technological Change in the Digital Age

Extracting Innovations

Mining, Energy, and Technological Change in the Digital Age

Edited by

Martin J. Clifford

Robert K. Perrons

Saleem H. Ali

Tim A. Grice

CRC Press
Taylor & Francis Group
Boca Raton London New York

CRC Press is an imprint of the
Taylor & Francis Group, an **informa** business

MATLAB® is a trademark of The MathWorks, Inc. and is used with permission. The MathWorks does not warrant the accuracy of the text or exercises in this book. This book's use or discussion of MATLAB® software or related products does not constitute endorsement or sponsorship by The MathWorks of a particular pedagogical approach or particular use of the MATLAB® software.

CRC Press
Taylor & Francis Group
6000 Broken Sound Parkway NW, Suite 300
Boca Raton, FL 33487-2742

First issued in paperback 2020

© 2018 by Taylor & Francis Group, LLC
CRC Press is an imprint of Taylor & Francis Group, an Informa business

No claim to original U.S. Government works

ISBN 13: 978-0-367-65724-6 (pbk)
ISBN 13: 978-1-138-04082-3 (hbk)

Visit the Taylor & Francis Web site at
http://www.taylorandfrancis.com

and the CRC Press Web site at
http://www.crcpress.com

Contents

Section III   Social Responsibility and Environmental Stewardship

Editors

Martin J. Clifford is a postdoctoral researcher in Energy and Environmental Policy at the Department of Geography, University of Delaware, Newark, Delaware. His work focuses on the socioeconomic and environmental dynamics of large and small-scale extractive industries, particularly in lower income countries. Much of his empirical work in this regard took place in Ghana and Guyana. This ties into his broader interests in international development and sustainable natural resource management. Much of his recent scholarship has focused on the Minamata Convention on mercury reduction and its impact on the gold mining sector. He has also completed a study on pathways for socioecological resilience in small-island developing states with extractive resources, focusing on the island nation of Nauru. This work was supported by the Pavetta Foundation. He earned a PhD and MSc in International Development from the University of Reading, Reading, England and the University of Manchester, Manchester, England, respectively.

Prior to joining the Queensland University of Technology (QUT) in Australia as an associate professor in 2011, **Robert K. Perrons** worked in a wide variety of roles and locations for Shell International's Exploration and Production division. He started his career in Shell's Strategy and Economics team in 1997, and then worked for several years as a production engineer in the company's overseas operations (offshore and onshore). He left Shell for 3 years to work as an Industrial Research Fellow at the University of Cambridge in the United Kingdom, and then rejoined Shell again in 2004 to become the company's Executive Coordinator of Research and Development. He earned a Bachelor of Engineering in mechanical engineering from McMaster University in Canada, a Master's degree in Technology and Policy from the Massachusetts Institute of Technology (MIT), and a PhD in engineering from the University of Cambridge, where he was a Gates Cambridge Scholar. He is a Fellow of the UK's Institution of Mechanical Engineers (IMechE) and Engineers Australia, and is chartered as a professional engineer (Eur. Ing.)

in Europe. He continues to stay connected to the University of Cambridge as an Affiliated Researcher, and is an adviser to MIT's Mining, Oil and Gas Club. In addition to his QUT position, Dr. Perrons serves the Australian government on an *ad hoc* basis as a member of their "Expert Network." His duties in this role include providing advice about the energy industry and emerging innovations in that sector, with the overarching objective of accelerating the commercialization of promising new energy technologies.

Saleem H. Ali holds the Blue and Gold Distinguished Professorship in Energy and the Environment at the University of Delaware, Newark, Delaware where he also directs the Gemstones and Sustainable Development Knowledge Hub, supported by the Tiffany & Co. Foundation. He is also a Senior Fellow at Columbia University's Center on Sustainable Investment. Professor Ali has held the Chair in Sustainable Resources Development at the University of Queensland's Sustainable Minerals Institute in Brisbane, Australia (where he retains professorial affiliation). His books include *Treasures of the Earth: Need, Greed and a Sustainable Future* (Yale University Press); *Environmental Diplomacy* (with Lawrence Susskind, Oxford University Press) and *Mining, the Environment and Indigenous Development Conflicts* (University of Arizona Press). Corporate and government experience includes employment in General Electric's Technical Leadership Program; a Baker Foundation Fellowship at Harvard Business School and a Research Internship at the UK House of Commons. He is a member of the United Nations International Resource Panel, was chosen as a Young Global Leader by the World Economic Forum in 2011, and received an Emerging Explorer award from the National Geographic Society in 2010. He earned his doctorate in Environmental Planning from Massachusetts Institute of Technology (MIT), Cambridge, Massachusetts; a Master's degree in Environmental Studies from Yale University, New Haven, Connecticut; and a Bachelor's degree in Chemistry from Tufts University, Boston, Massachusetts (summa cum laude).

 **Tim A. Grice** is an honorary senior research fellow at The University of Queensland's Centre for Social Responsibility in Mining and the Founding Director of Leapfrog International. Grice's academic and applied work focuses on extractive governance and inclusive development in Asia–Pacific, Africa, and Latin America. Dr. Grice has led resource governance projects for The World Bank, The International Finance Corporation, The United Nations Development Programme, Extractive Industries Transparency Initiative, Transparency International, The Natural Resource Governance Institute, and the Australian Department of Foreign Affairs and Trade. Dr. Grice earned a PhD from the University of Queensland, and is a member of Transparency International and the World Future Society.

Contributors

Saleem H. Ali
Department of Geography
University of Delaware
Newark, Delaware

Sara Bice
Crawford School of Public Policy
Australian National University
and
School of Public Policy and
    Management
Tsinghua University
Beijing, China

Martin J. Clifford
Department of Geography
University of Delaware
Newark, Delaware

Andrew Cooke
Levin Sources
Cambridge, United Kingdom

Richard Mohan David
ADCO
Abu Dhabi, United Arab Emirates

Ian Dover
METS Ignited
Brisbane, Queensland, Australia

Colette Einfeld
Melbourne School of Government
University of Melbourne
Melbourne, Australia

Guillaume Echevarria
Laboratoire Sols et Environnement
Université de Lorraine–INRA
Nancy, Lorraine, France

Peter D. Erskine
Centre for Mined Land
    Rehabilitation
Sustainable Minerals Institute
The University of Queensland
Brisbane, Queensland, Australia

Energistics Consortium
Houston, Texas

Brian J. Evans
Petroleum Engineering
Faculty of Science and Engineering
Curtin University
Bentley, Western Australia,
    Australia

Sharon Flynn
Oceana Gold
Melbourne, Victoria, Australia

Victor Gerardo Ortiz Gallardo
Instituto Mexicano del Petróleo
Mexico City, Mexico

Tim A. Grice
Centre for Social Responsibility in
    Mining
The University of Queensland
Brisbane, Queensland, Australia

David W. Hainsworth
CSIRO Energy
Canberra, Australian Capital
    Territory, Australia

Jill Harris
Sustainable Minerals Institute
The University of Queensland
Brisbane, Queensland, Australia

Lindsey Harris
University of Hawaii
Honolulu, Hawaii

Bruce Harvey
Sustainable Minerals Institute
The University of Queensland
Brisbane, Queensland, Australia

P.J. Hollenbeck
ARANZ Geo
Christchurch, New Zealand

John Jackson
JKTech
The University of Queensland
Brisbane, Queensland, Australia

Craig A.R. James
CSIRO Energy
Canberra, Australian Capital
    Territory, Australia

Valentina Kaman
ExxonMobil PNG
Port Moresby, Papua New Guinea

Deanna Kemp
Sustainable Minerals Institute
The University of Queensland
Brisbane, Queensland, Australia

Mehmet Kizil
The University of Queensland
Brisbane, Queensland, Australia

Peter Knights
The University of Queensland
Brisbane, Queensland, Australia

Gary Krieger
Newfields LLC
Atlanta, Georgia

and

ExxonMobil PNG
Port Moresby, Papua New Guinea

Nadja Kunz
Norman B. Keevil Mining School
    and the Liu Institute
University of British
    Columbia
Vancouver, British Columbia,
    Canada

Yolande Kyngdon-McKay
Levin Sources
Cambridge, United Kingdom

Estelle Levin-Nally
Levin Sources
Cambridge, United Kingdom

Chen Li
University of Melbourne
Melbourne, Australia

Derek McAuley
Horizon Digital Economy
    Research
University of Nottingham
Nottingham, United Kingdom

Sam Macaulay
University of Technology Sydney
Sydney, New South Wales,
    Australia

John McGagh
Snowy Hydro
Rio Tinto
Huelva, Spain

Justin Montgomery
Massachusetts Institute of Technology
Cambridge, Massachusetts

John D. Moore
Senior Advisor
Public and Government Affairs
ExxonMobil
Alaska

Philip N. Nkrumah
Centre for Mined Land
    Rehabilitation
Sustainable Minerals Institute
The University of Queensland
Brisbane, Queensland, Australia

Andry Nowosiwsky
ExxonMobil Corporation
Esso Australia Pty. Ltd.
Melbourne, Victoria, Australia

Francis O'Sullivan
Massachusetts Institute of
    Technology
Cambridge, Massachusetts

John R. Owen
Sustainable Minerals Institute
The University of Queensland
Brisbane, Queensland, Australia

Robert K. Perrons
Queensland University of
    Technology
Brisbane, Queensland, Australia

and

Centre for Strategy and
    Performance
University of Cambridge
Cambridge, United Kingdom

Jonathon C. Ralston
CSIRO Energy
Canberra, Australian Capital
    Territory, Australia

Serkan Saydam
School of Mining Engineering
UNSW Sydney
Sydney, New South Wales,
    Australia

Ali Soofastaei
The University of Queensland
Brisbane, Queensland, Australia

Jonathan Stacey
Levin Sources
Cambridge, United Kingdom

John Steen
UQ Business School
The University of Queensland
Brisbane, Queensland, Australia

Miranda Taylor
CEO
National Energy Resources
    Australia (NERA)
Kensington, Western Australia

Kane Usher
METS Ignited
Brisbane, Queensland, Australia

Anthony van der Ent
Centre for Mined Land
    Rehabilitation
Sustainable Minerals Institute
The University of Queensland
Brisbane, Queensland, Australia

Introduction—Extracting Innovations

Martin J. Clifford and Saleem H. Ali

> The key challenge is not so much globalization. It is actually, what I call the Fourth Industrial Revolution. Because it's technology which creates the major changes in our daily lives. It's technology that creates the fears. What we want to do is make the world much more aware. On the one hand is the opportunity of the new technology but on the other hand is the risks and dangers we encounter.
>
> —Klaus Schwab

Extraction of mineral resources marks temporal milestones in human civilization. From the Stone Age to the Bronze and Iron Ages, we have developed as a species through a remarkable ability to innovate in our methods of extracting the most useful mined elements. Fossil fuels also have catalyzed innovation through the advent of the Industrial Revolution from coal to oil to natural gas harkening back to similar subterranean innovations that also characterized other earthen resources.

In this book we consider the most contemporary innovations that are now propelling the extractive industries forward while also creating new environmental and social challenges. Since the extractive industries are considered nonrenewable on human time scales in terms of their extraction from the Earth's crust, their extraction is becoming more technically arduous. Economic geologists often have been reluctant to use the term "nonrenewable" in this context because of the relentless ability of technologies to extract more inaccessible ores. Yet, the specter of exercising innovations on conventional geological ore bodies is narrowing and new frontiers of resource bases are being explored from deep sea beds to asteroid deposits.

We consider such new frontiers of mineral extraction and the innovations they require, but are most concerned about the ways in which industry is having to adapt to imminent structural changes in our economic and social systems. The founder of the World Economic Forum, Klaus Schwab, heralded this as the dawn of the "Fourth Industrial Revolution," characterized by a highly networked digital society, and carrying important implications for the extractive industries. This has led to opportunities for industry to save costs and reduce occupational risk through automation and remote operations. At the same time, communities have demanded greater environmental protection from the negative impacts of mining which has necessitated a range of ecological innovations.

The socioecological fabric of innovation in the extractive industries is thus considered in this volume through an integrative approach that brings together engineers, natural scientists, and social scientists. The book had its genesis in a symposium that was organized by coeditor Robert K. Perrons at the Queensland University of Technology in Brisbane, Australia, in partnership with the Oil, Gas, and Mining Club of the Massachusetts Institute of Technology. This collaborative effort between two major research universities alongside numerous external participants galvanized interest in such a volume from practitioners as well. The result is this book which brings together academic voices in concert with practitioners to give an empirically grounded and realistic evaluation of the innovations in this sector.

The goal of this book is also to consider the limits of innovation in various arenas. With concerns such as climate change and tailings waste management perennially on the horizon, there is also a need to consider broader paradigm shifts in the sector in terms of reuse and recycling of materials and a gradual movement toward what is being termed a "circular economy." We hope that this book will pave the way toward such a long-term trajectory for more efficient and sustainable resource usage.

Book Structure

This book is divided into three sections. Section I of the book provides more contextual depth to the impetus for the (thus far presumed) need for greater integration of technological innovation in the resources sector and the processes and organizational designs that might facilitate this. It opens with an analysis by John Steen, Sam Macaulay, Nadja Kunz, and John Jackson on the "innovation ecosystem" in mining (Chapter 1). The authors weave a contextual examination of the nature, drivers, trends, and challenges of innovation within the sector. Lessons that can be drawn from the adoption of innovation in other, comparative industries, namely agriculture and aerospace, are cited as encouraging examples for the mining industry to follow suit. Engagingly, the narrative here is interspersed with quotes derived from interviews with mining managers and executives gathered over the course of 2 years.

Following this, Kane Usher and Ian Dover offer their take on the coexistent "barriers and imperatives" to innovation in the mining sector (Chapter 2). They reassert the contradictory tale outlined in the previous chapter of global conditions necessitating change but the specific nature of mining as an activity making this inherently challenging. This, they argue, results in a propensity to adopt and preserve low risk methods that "work" rather than being involved in the "creative destructionist" arms race that has characterized the manufacturing, electronics, and software sectors. They propose that effective leadership from senior levels in mining organizations can be used to drive companies through "three horizons of growth" and into a more innovation-friendly environment.

Chapter 3 draws the same line of enquiry for the oil and gas sector, which has traditionally had a reputation for being slow to develop and adopt innovation. One of the editors of this book, Robert K. Perrons, offers an analysis of his survey, conducted in collaboration with the Society of Petroleum Engineers that sought to gather insights directly from those involved in the industry into exactly how innovations happen in their sector and which networks prove most valuable in this regard. While the results might make gloomy reading for academic and independent research bodies, they do confirm the trend noted at several points within this book of a shift toward externally produced innovation.

Indeed, relying upon a closed system has had very real impacts for the Mexican hydrocarbon sector, as Victor Gerardo Ortiz Gallardo outlines in his brief appraisal that constitutes the first of the book's shorter "vignette" contributions (Chapter 4). Due to total monopoly held by the state-owned oil company, PEMEX, his organization, the Mexican Petroleum Institute, and a small team of other suppliers, has worked solely with one entity for many years. He highlights how this has slowed progress in technological developments, despite some interesting projects.

In the case of Australia, the government has, in fact, set out to ensure that its most important economic contributor is "future-proofed" with the establishment of the National Energy Resources Australia (NERA), a not-for-profit with a mandate to grow collaboration and innovation in the country's energy resources industry. In Chapter 5, Miranda Taylor, NERA's chief executive officer, briefly highlights the context under which the organization came about, their identified goals, and the approaches to ensuring continued competitiveness and innovation in the coming decades.

Closing off the first section and setting the scene for the next is John McGagh, former Head of Innovation at Rio Tinto, who gives his viewpoint on what makes "now the time" for kickstarting a new wave of advances in the mining sector and what represents the most promising areas for technological improvements (Chapter 6). Specifically, he picks out effective use of "big data," much wider use of automated systems, and the need for a renewed look at process methods as fruitful areas for future developments.

Section II of the volume looks at principles, practicalities, and potential of such developments in action, highlighting how the use of technological innovation is beginning to permeate the extractive industries in a myriad of ways. An engaging way to start this collection is through Jonathon Ralston, Craig James, and David Hainsworth's longer term overview of the emergence and evolution of digital technologies in mining (Chapter 7). They use the example of longwall coal operations, stretching back 60 years and working through to the modern era. They round the chapter off by highlighting emerging areas of technological innovation as well as important areas for future consideration and development of the digital mining ecosystem.

Brian J. Evans, Head of the Petroleum Engineering Department at Curtin University in Perth, Australia, also takes a broad view of what the implementation of technology might look like on a pragmatic level and some of the (often less considered) knock-on effects for a daily operation in Chapter 8. He considers the example of an operating room for a liquefied natural gas (LNG) gas train, and what the full integration of data sources, analytics, and automation might look like for an individual using these systems.

Moving to more specific areas of focus and research, coauthors Robert K. Perrons and Derek McAuley provide another valuable contribution on this topic in Chapter 9, which explains how the "big data" revolution will probably unfold differently in the mining sector than it has elsewhere throughout the marketplace. The authors point to important differences in the costs associated with acquiring large volumes of data in a mining environment, and make the case for a more focused data collection and processing strategy than what is being evangelized these days by many vendors and consultants in the data analytics domain.

And there are signs that the oil & gas industry is also refining and re-shaping its data management philosophies in the face of "big data," too. In Chapter 10, Richard Mohan David explores how companies in that sector are more effectively using data and technology as parts of the evolution of "digital oilfield" technologies. As well as pointing out some of the more prominent "strategic initiatives" for integrating next generation technologies, he narrows in on what he terms "data driven technologies," which are beginning to harness the depths of previously underutilized data, as being particularly promising.

Chapters 11 and 12 provide examples of just this type of data use in action. Justin Montgomery and Francis O'Sullivan's modeling of the productivity of unconventional oil and gas wells, something which has proven to be extremely unreliable, is based upon predictive analytics. This form of analytics is reliant upon capacity for data mining, modeling, machine learning, and artificial intelligence that previously were not technologically possible and have found increasing weight in natural resource research. Ali Soofastaei, Peter Knights, and Mehmet Kizil also employ data analytics in their proposed model to optimize the fuel efficiency of haulage trucks. Using artificial neural network modeling and genetic algorithms, which again rely on computing reams of data and running a huge number of complex functions, they arrive at an extremely high agreement between (their) modeled and observed fuel consumption. With material haulage consuming a significant proportion of energy consumption at mining sites, this proposed model has significant potential to improve efficiencies in active mines or before production even begins. The wider point of these two studies, of course, is demonstrating that the large amount of data collected daily by energy companies can be put to effective use with appropriate harnessing of increasingly advanced technological techniques.

As is highlighted in many of the overviews of the sector, in the modern context of the industry, the largest percentage of innovation arises not in the research and development departments of companies, but from firms that supply them with products and services. P.J. Hollenbeck of ARANZ Geo, a technology provider for the mining, energy, and environmental industries, provides us with an archetypal example of this in practice in Chapter 13. He takes us through the development and numerous uses of the company's Leapfrog Aspect Viewer, a geological modeling process that uses augmented reality to allow the user to visualize the model and physical space simultaneously using the camera software on a mobile device. This is a perfect illustration of the innovative work that companies that supply the industry are driving through. Encouragingly, extractive companies are increasingly alert to the potential of collaboration. Energistics Consortium, a not-for-profit industry consortium working with extractive firms and suppliers, aims to bridge this process in Chapter 14. They provide standards, protocols, and tools not only to corporations but also to associated technology companies, with the aim of ensuring a consistency and interconnectedness across the board. In a brief vignette, they review the context behind and effectiveness of their work.

A total alternative to attempting to streamline efficiency and boost productivity in known energy resources, however, is to seek out and exploit entirely new frontiers, something that underpins the emerging deep-water and embryonic off-Earth mining industries, which round off our second section. Again, and centrally with the theme of this book, moving such ideas toward reality has been the result of substantial technological progress over recent decades. Lindsey Harris of the University of Hawaii summarizes the technical advances and the social, environmental, and regulatory dynamics surrounding the growing discussion of the deep-water mining industry in Chapter 15. Then, Serkam Saydam, who has collaborated with NASA and the Kennedy Space Center, reviews the literature relating to the prospects of mining on asteroids, the Moon, and Mars and summarizes the major remaining technological challenges that need to be overcome for the industry to realistically come into existence in Chapter 16.

In the third section of the book, elements of social responsibility and environmental stewardship that intersect with innovation and technological adaptation are introduced. As Sharon Flynn puts across in her responses at the outset of this portion of the volume (Chapter 17), the seminal World Bank report into the social and environmental performance of the extractive industries and the consequent introduction of Performance Standards by the International Finance Corporation (IFC) have, commendably, "rocked the boat," heralding a shift in corporate approaches from well-critiqued standard offerings toward more collaborative, sustainable, and socially embedded strategies. As she goes on to emphasize, a rapid adoption of such approaches will be essential as future mining operations are likely to increasingly have to look toward areas of higher population density or ecological diversity in the search for new ore bodies.

The proliferation of public-private partnerships (PPPs) between companies, host governments, and local communities are a prominent example of the "new breed" of social responsibility arrangements. The work of companies like ExxonMobil, whose LNG project in Papua New Guinea is the subject of John Moore, Andry Nowosiwsky, Valentina Kaman, and Gary Krieger's Chapter 18, seeks to go beyond traditional Corporate Social Responsibility (CSR) approaches to create "shared value" and a platform for more impacting and lasting social integrity. The authors provide an engaging and in-depth discussion of the context, critiques, and considerations regarding PPPs, and the company's experience in implementing such a partnership. It is a highly informative case study for academics and (governmental and corporate) policymakers alike.

A likely and seemingly emerging trend is for these partnerships to stretch across sectors, as evidenced by the study outlined in Chapter 19 by Tim Grice and Saleem Ali, two of the volume's editors. This initiative to extend mobile banking and financial inclusion in Papua New Guinea resource regions is anticipated to involve financial regulators, banks, and telephone companies alongside extractive operators and host communities in a "multistakeholder working group." It will be extremely interesting to see how these types of schemes develop.

Bruce Harvey's comments on the emerging field of "local level agreements" (LLAs) in Chapter 20 carries a similar sentiment to the two preceding chapters. The central tenet of these agreements is the sharing of benefits that result from resource exploitation between companies and land-connected communities. By highlighting the "ideal" content and processes in establishing of such agreements, the author attempts to emphasize this approach as an innovative way of ensuring much more direct, collaborative, and longstanding interaction between companies and host communities than can be said of pre-existing CSR initiatives.

If the subject of PPPs and LLAs deal with proactive social engagement, there are also significant challenges and substantial scope for improvement in the way companies react to social criticism and incidents. For example, previous approaches for dealing with social incidents (e.g., protests, strikes) involving resource developers have come under scrutiny. Simultaneously, there is also increasing external demand to deliver technically rigorous social due diligence in increasingly demanding circumstances, which many companies have struggled with. Deanna Kemp, John R. Owen and Jill Harris examine in Chapter 21 how social science methods and models, which have traditionally been kept "outside the fence" by many mining companies, could provide a much more nuanced and, as a result, informed understanding and mitigation toward dealing with social incidents and helping to stop them reoccurring.

Colette Einfeld, Sara Bice, and Chen Li's Chapter 22 on the challenges posed by social media for extractive industries' community relations practices is extremely interesting for its focus on less orthodoxly considered elements in modern Public Relations (PR), especially among industries with more traditional approaches, like in mining, oil, and gas. Using an analysis of Twitter in relation to proposed or existing coal seam gas ("fracking") operations in Australia, the authors point to how the rapid and geographically untethered spread of multiple streams of information to a huge disparate audience has given birth to new forms of "concerned communities" and activism. Despite not existing in an electronic dimension, these streams can have very real reputational and operational impacts and require a new, as of yet unconsolidated, approach by companies toward protecting their social license to operate.

Moving on to environmental innovations, in Chapter 23, Philip N. Nkrumah, Guillaume Echevarria, Peter D. Erskine, and Anthony van der Ent present the interesting case of phytomining, in which metallophytic plant species, especially those that are "hyperaccumulators," are used to draw metals from the ground which are then harvested from their biomass. This interesting idea could be used as a method to rehabilitate former mining lands by removing elevated metallic compounds in an economic and ecological fashion. It can also be used, the authors suggest, as a mining method on scattered, less economically viable deposits, possibly as smaller, "agromining" operations. Using the example of nickel accumulation, the authors outline suitable environs, selection of appropriate species, processing methods, and the potential economics of the practice.

Interestingly, the ecologically friendly methods outlined here may owe a partial debt to research carried out at the Eden Project, an equally conceptual and practical location in a former clay mine, as described by Saleem Ali in Chapter 24. The Eden Project is not merely a tourist attraction but a pragmatic exercise in post-mining rehabilitation, centered around an ethos of renewability. Part of this project is restoration research, including identifying species suitable for the phytomining methods, mentioned by Nkrumah and his coauthors, in the facility's greenhouses. This was in addition to stabilizing and construction of the site itself, using the pit and thousands of tons of topsoil, in both a feat of engineering and a utopian ideal of what can be done with former mining sites.

Finally, Jonathan Stacey, Yolande Kyngdon-McKay, and Estelle Levin-Nally introduce a method for an innovative rehabilitation approach tailored for the artisanal and small-scale mining (ASM) sector in Chapter 25. As a sector that is largely unregulated and informal, driven by poverty, and has proven resistant to change, this approach is a tough task. It is also a worthy one, particularly with ASM; the social and environmental impacts are very clearly felt by local communities, and the price of permanently debilitating formerly useful land is a heavy price to pay for the (typically) short-term gains of extracting what is underneath. The authors introduce a detailed, locally appropriate, and potentially replicable framework for rehabilitation in this context, the Frugal Rehabilitation Methodology, implemented in Mongolia. This approach is a very welcome addition to the ASM literature, which often attempts to quantify and reduce environmental impacts during operations, but has thus far had very little to say on what should be done afterwards.

The "Conclusion" chapter of the volume, by Robert K. Perrons, draws together the various narrative strands running through the text and speculates on what lies ahead. While obvious differences exist between each sector of the extractive industries, and indeed between different contexts and companies, there are shared challenges: increasingly tough conditions under which to maintain efficiency and productivity, coupled with burgeoning environmental and social demands. Perhaps most encouragingly—and something which this book hopes to demonstrate, through appropriate use of technology and innovation—the means are seemingly there to deal with these circumstances and to follow directions that were inconceivable in living memory. We hope that this book provides a conceptual and pragmatic taste of the current groundswell in new approaches to manage and advance the extractive industries, and acts as an inspiration for continued progress in future research and implementation of these ideals.

Section I

Processes and Organizational Designs Underpinning Innovation in the Resources Sector

1

Understanding the Innovation Ecosystem in Mining and What the Digital Revolution Means for It

John Steen, Sam Macaulay, Nadja Kunz, and John Jackson

CONTENTS

1.1 Introduction

One thing that is clear about economic development is that sustainable improvements in productivity come from technical change (Eslake and Walsh 2011). During the period of inflated commodity prices leading up to 2013, profitability in mining was high but productivity declined as mining companies rapidly expanded their operations. With a focus on production rather than efficiency, mining companies opened high-cost mines to meet demand (Eslake and Walsh 2011; Syed 2013).

Since the sustained decline in prices over the past years, productivity has returned to center stage as a concern for mine managers. Between 2013 and 2015, surveys of executives showed that productivity declines were identified as a key risk for the industry and cost reduction was prioritized (Mitchell and Steen 2014; Mitchell et al. 2017). Although the initial focus was on labor cost reduction and squeezing input prices, mining companies renewed an interest in innovation to address long-term challenges that could not be met by short-term cost cutting. Several reports noted that in the long term, mining productivity will decline due to exhaustion of high-quality resource deposits in easy-to-access locations (Syed 2013). Critical inputs for mines such as energy and water also will become more expensive. Increased scrutiny by community and governments also will challenge the social license to operate. Consequently, mining companies must find ways to minimize environmental impacts and show benefits to the regions in which they operate. This focus also will require the mining industry to innovate.

The mining industry has a long history of innovation. One study of copper mining shows the step change reductions in operating costs that come with successive introduction of new technology over the past century (Tilton and Landsberg 1997). Although the economies of scale that come with industry expansion can account for 30% of cost reduction, innovations such as solvent extraction electrowinning (SXEW; hydrometallurgical processing rather than smelting) and computerization of operations can account for 70% of the cost reduction during that time.

However, the pressure to improve the innovation performance of the mining industry to address these long-term challenges means that mining companies need to fundamentally change the way they innovate. We agree with recent commentary by the Rio Tinto Group that innovation in mining is different from other industries (Shook 2015), but we also suggest that important similarities with other industries provide examples of how mining companies can embrace a more innovative future. Furthermore, digital technology will be a key enabler of innovation in the mining industry and will have profound effects on the nature of competitive advantage in the mining industry sector and the relationship between mining companies and their technology suppliers.

Characterizing the nature of innovation in the mining industry sector is important because most mining innovation takes the form of process innovation and on-site problem solving, which is not well captured by traditional innovation measures such as research and development (R&D) expenditure and patents (Kastelle and Steen 2011). Mining companies are also net-consumers of innovation and rely heavily on the mining equipment, technology, and services (METS) sector for solutions and new technology. Understanding this relationship between mining companies and their supply chains is crucial for making sense of innovation in mining as a special case of open innovation (Dodgson and Steen 2008). Another feature of innovation in mining is that mines are systems of connected technologies known as complex capital goods (Acha et al. 2004). This point has two implications for mine innovation in that innovation can be hidden within these complex capital goods and not readily measured in surveys of mining innovation. The other implication is that new technology enters a mine as part of a deeply interdependent social, technical, and organizational system and this implication makes the introduction of innovation especially challenging in the mining industry sector.

Although mining does have obstacles to innovation, other industries also share these attributes of being innovation consumers and innovating in complex production systems. Mining executives are currently interested in digital technology for improving efficiency through cost reduction, but we can draw on experiences in adjacent industries to argue that digital technology also can be used to facilitate the introduction of more radical innovations in complex business environments.

We discuss how digital technology within the innovation process might allow mining companies to innovate in areas that will become critical in the future such as better use of water and managing the mine through the entire lifecycle from planning to decommissioning. Throughout the discussion, we illustrate key points with quotes from interviews that were conducted between 2014 and 2016 with mining managers and executives from Australia, North America, Africa, and South America as part of exploratory research into innovation and productivity in the mining industry. We conducted all these interviews with ethical research approval from the University of Queensland and recorded them with the permission of the interviewees for analysis.

1.2 Does Innovation Happen Differently in the Mining Sector?

The mining industry has a reputation for being reluctant to innovate (Hanson et al. 1997; Dodgson and Steen 2008; Mitchell et al. 2014). This reputation is not solely based on the views of outsiders. Rather, it is often heard from those in the industry. For example, the Anglo American plc chief executive officer (CEO), Mark Cutifani, declared that unless mining companies improve their innovation performance they will become subsidiaries of proven innovators like General Electric (GE) Mining (Mitchell et al. 2014). Several industry reports have commented on the low rate of R&D spending in mining, especially in comparison to similar industries such as oil and gas (Mitchell and Steen 2014). In our interview series we generally found that interviewees agreed that mining was not a very innovative industry and this lack was attributed to a range of factors. Some interviewees pointed to a general problem of a business culture that avoided innovation and preferred to use established technology and procedures:

> ...you can see the productivity gain would be enormous and the technical challenge is not difficult at all. But it's more of a cultural thing within the mining space; there seems to be a bit of push back on innovating and finding a new way of doing things. (Australian GM)

Several interviewees mentioned the short-term time horizon often applied to innovation investments. The underlying reason for this short-term horizon, though not clear, may be associated with the volatile nature of cash flows and profits that are most driven by commodity prices.

> ...gosh, how many times have I seen let's use the word innovation or something; people trying to launch initiatives, be it at the asset level or even at the group level. But the moment anything like innovation comes along, and anything that could be bucketed under innovation, it has a very short shelf life in the mining culture. (North American CFO)
>
> In one of my predecessor companies, our leaders set up a technology group and they were working on things like "Let's trans-levitate rock out of the pit using linear synchronous motors." And I am sorry, you can work on that for 10 years and you are not going to deliver any value. So, I am sorry to be this way, but this pie in the sky, "we are going to reinvent the world," I am not a big fan of that. (North American CEO)
>
> Through necessity, we kind of moved into "We need to generate value, and we need to generate value on a very short time horizon." So that sort of eliminates anything too clever and too thoughtful. (UK GM)

One way of thinking about this short-term focus is that mining companies have a limited appetite for business risk. In one discussion, an Australian mine manager explained that mining companies had a finite budget for risk and most of this was allocated to exploration and development. When innovation does happen, it tends to be in response to persistent challenges that threaten the business, such as sustained periods of low margins or logistics for a remote operation. Mining businesses tend to innovate when they must do so.

> ...many years ago we had to work with a lot of smaller margins in terms of operating a gold mine or a copper mine or a silver mine. So, we were a little more entrepreneurial and innovative about some of the things that we did. I think the last decade of high metal prices is taking some of that out of us. (North American CFO)
>
> Yeah, so, I think the mining industry is pretty good at R&D and innovation. Some of that actually would come out of necessity: if you're working in remote parts of the world you need to be able to somehow make the mine more independent. (North American R&D director)

This conservatism also means that the preferred approach to innovation is to be a fast follower rather than the leader who wears the cost of developing the technology and solving the problems associated with it.

> I think you have always got to be open-minded over technology. And particularly you have always got to be looking for disruptive technology and that doesn't mean you need to be the proving ground; you just need to be a fast follower and so if Rio Tinto wants to go and manage automated trucks and stuff like that, okay we will watch them closely, but will be prepared to implement that when they get it right. (Australian COO)

In addition to the broad explanation of culture, some interviewees offered more specific reasons for why mining companies were reluctant to be radical innovators. One of these reasons was due to the nature of competition. As long as the big mining companies compete on ownership of long-life, low-cost deposits, technology only needs to be fit for purpose rather than source of competitive advantage. In other words, mining companies only innovate when they are forced to.

> The definition of a world class resource is one that survives five generations of incompetent management. You end up with these big behemoth mining companies sitting on great geological resources, and it sort of crowds out the ability for innovative companies to come in and shake things about. (South American GM)

Beyond the nature of competition, the way that mining companies have been traditionally structured as portfolios of relatively independent mines also limits innovation to small-scale programs that fit the budget cycle of a mine manager (Mitchell et al. 2017). Radical innovation needs longer time frames and greater financial returns that can be achieved from implementing new processes across the business to achieve higher returns on investment.

> If you want to introduce disruptive innovation in mining, you cannot do it with the level of autonomy and empowerment that is currently given to site managers. There has to be some recentralization of the technology to make a major change. (North American R&D director)

Paradoxically, some executives identify the problem of justifying innovation in a large-scale operation with high operating costs unless the return was of a magnitude that was similar to the scale of the costs and profit. While pointing to the need for innovation to generate immediate returns at low risk, these returns also had to be sufficiently large to justify attention from senior managers. This is a paradox because high returns on investment must involve taking bigger risks. Clearly this mindset of wanting big rewards with minimal risk is going to be a barrier to any systematic innovation program due to the incompatibility of high returns and low risk.

> If you get a lot of automation into your mine, and the trucks, and the data, and – you will see the massive benefits, but counterfactual to that, as you say, there's actually a fairly massive cost, and if all you are doing

is realizing 10 percent on your cost, then it's pretty hard to justify that economically, and you know, if you are making 40 dollars a ton it's not shifting the dial, you know, for the amount of risk you're taking on. (South American Mining Executive)

While coming up with reasons for the mining industry's reputation for slow and incremental innovation, several respondents questioned how sustainable this lethargic pace of innovation was in the long term. The mining industry executives understand the importance of innovation but when it comes to committing resources to make it happen, mining companies have little resolve to move forward.

....but it is clear to me that we have got to ask the question, are there more incremental solutions available, or is it time to do something more radical? (African Chairman)

You get to a point where, without some sort of innovation or something like that, you would get to a point where you are just going to get diminishing returns on the effort you are trying to put into improve it. (Australian GM)

The evidence from these interviews reinforces the view that innovation in mining has become slow and incremental, especially in recent years. However, finding quantitative answers to the question of what type of innovation do mining companies do and how much innovation happens in the mining industry is difficult due to the lack of fine-grained data on the mining industry sector. Innovation surveys based on the European Union Community Innovation Survey have been around in many countries since the 1970s, but these cover a cross section of national economies and do not capture sufficient data from the mining industry sector to address the question of what mining innovation looks like (Dodgson and Steen 2008). Answering this question is important because different industries have different innovation signatures and comparing dissimilar industries may lead to erroneous conclusions about the future of mining and how it might be shaped by innovation.

1.3 Mining Companies as Process Innovators and Consumers of Innovation

If innovation does have different characteristics in the mining industry compared to other parts of the economy, then what might some of these differences look like? In considering this question it is useful to think about classifications of various forms of innovation. One of the founders of innovation economics, Schumpeter (1912), took a broad view of innovation that could include new products and services, new production processes, new markets, and new sources of supply, as well as changing the structure of the industry.

Discussing these different forms of innovation is important because in our interviews we encountered many differing interpretations of what constitutes innovation in mining. This variation often extends to differences in the same mine site. While one manager will explain that no innovation happens on site, another manager will give details on several innovations that were introduced to key production activities. This disconnect is not because the innovation is secret but rather because there is not a common language for innovation in mining.

Most mining companies tend not to produce new products although there are some exceptions in vertically integrated businesses that transform commodities into value-added products. It is also possible to argue that blending sources of commodities from different mines to meet customer needs, such as iron ore in the Pilbara region of Australia, is a form of product innovation. Most commonly, innovation in mining takes the form of improved processes. Although many of our interviewees do not call it innovation, it is apparent that process innovation is very important to the mining industry. Some examples from our interviews include improving efficiency in asset utilization and management:

> In the future there are huge improvements that people are looking at in terms of the planning and scheduling of fleets, the planning and scheduling of fleet maintenance as well, those sorts of things. (North American CEO)

Other examples of process integration included better connections between the steps in the ore processing pathway.

> …they improved the output from 13 million ton to 14.5 million, you know, that's 1.5 million ton over 13, but with the same fleet, with the same volume mined, but much improved. And it was a huge joint effort. It involved right from the mineral resource management in terms of knowing and predicting what you are going to mine and blast, all the way to analyzing the losses and everything in the processing system. (African Director)

Although these are certainly process innovations, measurement of them and quantifying their impact on the business is a challenge for researchers and mine managers. Although new product development can be isolated from business activities, process innovations tend to be done on site within the business and are harder to analytically separate from non-innovative activities. As highlighted in the previous quote, process innovation can have a huge impact but also tends to be an accumulation of smaller innovations over time. Consequently, this accumulation of small process innovations can be invisible to senior executives and sometimes an early target in a round of budget cutting.

The other signature of innovation in the mining industry is that mining companies are predominantly consumers of innovation produced by the METS sector. Measuring patents shows that the METS sector produces vastly more innovation processes than the mining companies and that this gap between them is growing (Figure 1.1).

The innovativeness of the METS sector comes with diversity as providers compete across different technologies and customers. In a study done for the Minerals Council of Australia, Scott-Kemmis (2013) mapped the interrelationships between different subsectors of the METS sector (Figure 1.2). This map represents a wide array of products and services with some services such as consulting and design and project management playing a connecting role between other specializations. This map shows a diverse and connected industrial ecosystem within the mining supply chain that accounts for much of its innovativeness relative to the mining companies.

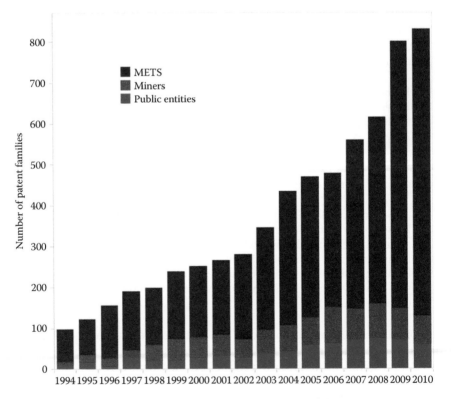

FIGURE 1.1
Ownership of patent families by mining companies versus METS sector and other public research providers. (From Francis, E., *The Australian Mining Industry: More Than Just Shovels and Being the Lucky Country*, IP Australia, Canberra, Australia, 2015.)

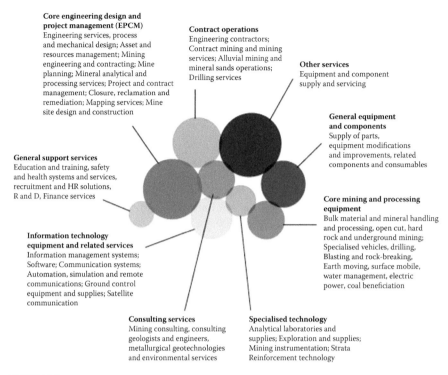

Core engineering design and project management (EPCM)
Engineering services, process and mechanical design; Asset and resources management; Mining engineering and contracting; Mine planning; Mineral analytical and processing services; Project and contract management; Closure, reclamation and remediation; Mapping services; Mine site design and construction

Contract operations
Engineering contractors; Contract mining and mining services; Alluvial mining and mineral sands operations; Drilling services

Other services
Equipment and component supply and servicing

General equipment and components
Supply of parts, equipment modifications and improvements, related components and consumables

General support services
Education and training, safety and health systems and services, recruitment and HR solutions, R and D, Finance services

Core mining and processing equipment
Bulk material and mineral handling and processing, open cut, hard rock and underground mining; Specialised vehicles, drilling, Blasting and rock-breaking, Earth moving, surface mobile, water management, electric power, coal beneficiation

Information technology equipment and related services
Information management systems; Software; Communication systems; Automation, simulation and remote communications; Ground control equipment and supplies; Satellite communication

Consulting services
Mining consulting, consulting geologists and engineers, metallurgical geotechnologies and environmental services

Specialised technology
Analytical laboratories and supplies; Exploration and supplies; Mining instrumentation; Strata Reinforcement technology

FIGURE 1.2
Relationships between subsectors of the METS sector (From Scott-Kemis, D., *How About those METS? Leveraging Australia's Mining Equipment, Technology and Services Sector*, Minerals Council of Australia, Canberra, Australia, 2013.)

In this diagram (Figure 1.2) from 2013, the category "Information technology and related services" is one of the smaller categories of subsectors on the periphery of the cluster. However, this is deceptive because we know that this sector has not only become bigger but is also embedded in the technology and capabilities of all the other service sectors (Francis 2015). With the blurring of technological boundaries, it becomes harder to define the technological boundaries of METS sector because a range of new information technology (IT) service companies such as SAP, IBM, and GE Digital have become part of the mining supply chain.

If the METS sector is providing innovation for mining, then how can mining companies make more of this innovative capacity? At least two considerations suggest how mining companies can create a more innovative industrial ecosystem. The first of these considerations is the principle of absorptive capacity (Cohen and Levinthal 1990) which states that organizations are unable to absorb innovation unless they are capable of producing innovations and implementing them within the business. The second of these considerations is collaborative supply chain relationships with innovation incentives and risk sharing/risk bearing agreements (Caldwell 2009; Steen et al. 2017).

Traditionally, mining companies have transactional relationships with the supply chain where procurement managers endeavor to find technology and capabilities at the lowest price and structure the contract to transfer delivery risk to the supplier. In other industries, such as construction, this approach has been shown to stifle innovation because the supplier will do just enough to fulfil the requirements of the contract. With no incentive for finding a better solution, the introduction of a new process by the contractor simply becomes risky and financially unviable for them (Davies et al. 2014).

Typically, digital technology is considered as an outcome of innovation but there are many examples of digital technology being a catalyst for innovation, especially in the areas of improving absorptive capacity and facilitating collaboration between businesses. Dodgson et al. (2005) coined the term "innovation technologies" which span a range of forms and applications including data mining, 3D printing, computer-aided design (CAD), computer simulation, and virtual collaboration (Salter et al. 2005). Innovation technologies improve innovation performance in settings such as car manufacturing, architecture, and services such as hospitals and also may be especially important in large, complex, capital intensive businesses like mining.

Prototyping and experimentation are important for innovation which might be easy for small consumer products but how can we experiment in a mine without risking safety and production? Simulation and visualization can enable performance of experiments in virtual mines without the risk of performing real experiments. Mine simulation technology already exists to assist with optimizing scheduling of mine development and load and haul decisions, so Dodgson et al. (2007) have shown how simulation can gather important information about the cascading effects of new technology in complex technological systems and can foster collaboration within the organization and with external stakeholders. Visualization allows these stakeholders to see the results of the new technology and offer their input into the design process (Gann and Dodgson 2008).

The internet has become the basis for many forms of innovation technology, especially for opening up a business' unsolved problems and intellectual property base. Proctor and Gamble developed an internet portal to find external partners to commercialize their unused intellectual property (Dodgson et al. 2006) and Goldcorp famously developed their Red Lake mine in 2000 by posting geological data on the internet and offering a prize for the best model of the resource (Saefong 2016).

Innovation technologies also can change the relationship between businesses and their supply chains by encouraging more collaboration and risk sharing. Forsythe et al. (2015) studied the effects of construction simulation technology (building information modeling [BIM]) on the relationship between builders and their suppliers. They found that BIM enabled the sharing of information so that contracts were more easily managed, and a higher level of collaboration could exist between project owners and contractors.

However, innovation technologies alone will not make mining companies more innovative. The way that mines and mining companies are organized also will need to change to become better integrated.

1.4 Complexity and Silos as Barriers to Innovation in Mining

One barrier to increased process integration in mining may be explained by the hierarchical structure that is characteristic of most major mining companies. Mine sites are typically structured following the Mintzberg (1981) Machine Bureaucracy (i.e., a large operating core conducting low-skilled and specialized work supported by a large middle management and a small upper management). This structure is effective for optimizing individual system components (e.g., mining, processing, environment, and community relations). However, this structure creates challenges for optimization across the value chain. Many mines suffer from communication "silos" across departmental functions (Kemp and Owen 2013; Kunz et al. 2017) such that individuals are so specialized that they can no longer see the big picture.

> I made a comment to a manager who was running a coal plant for me when I was running the mine there and I said, "Your job is to optimize the coal plant..." and I was responsible for the coal plant, the tech services, the mine and the likes...and my job is not to let you. My job is to optimize across all of them but with the tools you've got you're supposed to try and optimize what you've got there." (UK GM)

Engineering efforts to overcome this integration challenge are widespread. For example, concepts such as "mine-to-mill" and the more recent "cave-to-mill" have highlighted the importance of optimizing net present value (NPV) from ore extraction through to the processing circuit (Nadolski et al. 2015). Recently, scholars have also encouraged mining companies to adopt an integrated approach to designing mining tailings in a way that anticipates and prevents environmental legacies that have historically arisen at mine closure (Edraki et al. 2014).

The integration challenge will become ever-more critical for mining companies in the future. Technical complexity is set to increase due to declining ore grades, more complex mineralogy, and deeper deposits which will broaden the environmental footprint of mines. Mining companies will need new technological innovations and mass-mining methods to extract ore from more remote locations. The increased social conflicts associated across the mining industry sector also will exert growing pressure on mining companies to minimize their environmental impacts (Hodge 2014). This effort will create a challenge because complex socio-technical systems are inherently challenging to optimize and control:

> ...if you want to rely on human heuristics to do this integrated value optimisation, then you can only do that within relatively limited size of teams. Once you get above a thousand people in a team then it becomes very difficult to get that connectivity.

The "Machine Bureaucracy" may have been an effective structure for the mining industry of the past when it operated in traditional command-and-control paradigms; however, its rigidity impedes adaptability and resilience. A further challenge is disconnects in vertical integration from the CEO to operator (Kunz 2016). Systems optimization is therefore also reliant on middle management to connect tactics and strategy. Unfortunately, research reveals that many mining companies have a lack of strategic thinkers at these levels. As reported in Mitchell et al. (2014), several of the CEOs who we interviewed observed that most system thinkers in the mining industry sector are at such a senior level that they are side-lined from the day-to-day operations of the mine. Kunz (2013) coined this problem as the "missing middle" and found that it could have significant implications for managing water on mines, an example of an issue that requires a systems approach.

Addressing the mine integration problem therefore requires not only technical innovation but also innovations in the human systems responsible for management. Incentive structures including key performance indicators (KPIs) may need a rethink as they may be impeding innovation. These incentive structures must be linked to systemic goals, not only to task functions. As we were told by one R&D executive:

> I said before that there is not going to be real innovation in mining for as long as engineers are paid by the hour, because the whole model, the supply chain procurement model that pulls together the mining system is not able to embrace technology risk effectively, in a controlled manner. (North American R&D director)

1.5 Managing Mine Innovation in a Digital Environment: Lessons from Other Industries

The fundamental enabler of digital technology is the exponentially increasing processing power of computer chips which at the same time are becoming smaller and cheaper. What this means is that data can be generated, processed, and integrated in a way that was not previously possible. This is the age of the "Internet of things" where people and machines are connected through the constant collection and transmission of data.

The digitization of mining is not the normal incremental or radical type of innovation that mining firms are used to dealing with. Instead, digitization represents a systemic architectural innovation (Henderson and Clark 1990) that changes the way in which the components of a mining system are linked together. This type of innovation is recognized as being difficult to respond to and manage because, although it changes the way components are linked together, it leaves the identity and knowledge of individual components relatively untouched and can thus be hard to recognize until it is too late to adapt. In our interviews, several executives commented on the problem of getting value from data. Mines can generate a huge quantity of information but verifying the quality of the data and connecting it to processes and better decisions is a problem according to many mining executives. The following quote typical of many similar comments:

> We collect ten times more data than we even look at. So why collect it? Because we thought it was valuable to do so. But yet we do not look at it. So, let's start looking at the data we already collect. (Australian Operations Manager)

Digital technology is an enabler of a new integrated business architecture and the linking of different systems within a mine to dynamically optimize production but this transformation is not straightforward.

> Building the model sounds so simple, but I've seldom come across it ever actually, a fully integrated model that goes right across the mine value chain from one end to the other that doesn't have other feed ins from spreadsheets and all sorts of other things along the way - that fully integrated model that's actually picking up information independently from the processes and running real time is still something that seems to be very, very difficult to put together. (South African Director)

However, the technology alone will not produce this integration across the mine functions. Without changes to the organization in terms of decision responsibilities, roles, and information flow, the digital revolution in mining has so far been disappointing.

> Things like SAP are a business tool and, by themselves, add zero value. It is the processes you put around them that may or may not add value. In the way it was done in the organizations that I have worked in, they were a complete disaster, in my opinion. Value destructive, in my opinion. (Australian Director)

Digitization is fundamentally transforming the knowledge required to construct, maintain, and orchestrate linkages within big mining businesses and within the supply chain. Those that control this knowledge are likely to dominate the future of mining. One interviewee, a North American R&D

director, used the example of aircraft manufacturing to show how dominant businesses were active curators of an innovative industrial network.

> You can look at integration as an internal system problem. But then to reap the real benefits you need to look at external integration, and that means risk allocation across the equipment supply chain. Let's look at an aircraft manufacturer putting together a new model of aircraft, a new aircraft product. They do not generally go out with a prospectus to raise $5 billion to create this new aircraft, it is just routine business, and a lot of their suppliers pay to participate in consortia or in commercial teams that share the development risk. (North American R&D director)

To understand the potential trajectory of digital technology in mining and how it might influence innovation, we turn to two exemplar industries that share some similarities with the mining industry. From these cases we draw some conclusions on how digital technology might change the mining innovation ecosystem.

1.5.1 Lessons from Digital Technology in Agriculture: Supply Chain and Value Chain Coordination

The agricultural industry has many parallels to the mining industry: being exposed to cyclical commodity prices, the variability of the inputs to downstream processing, complex four-dimensional (space and time) production systems, production cost pressures with a drive to produce yield and quality, and historically slow rates of adoption of innovation. Some sectors of the agricultural industry, particularly broad acre crops and viticulture have been undergoing a digital transformation since the 1980s, commencing with precision agriculture and evolving in recent times to big data. Productivity gains of between 5% and 35% have been reported for various crops with half coming from input efficiencies and half from increases in rates of production output (Watcharaanantapong et al. 2014; Keogh and Henry 2016; Castle et al. 2016). These gains are similar to those suggested to be possible by the mining industry, with digitization enabling improved integration through architectural innovation (McKee 2016; Mitchell et al. 2017).

Precision agriculture can be defined as the application of information technologies to improve the management of inputs, the quantity and quality of outputs, and profitability by enabling the right management strategy at the right place at the right time (Pierce and Nowak 1999; Sonka 2016). This precision agriculture is achieved by measuring key characteristics of the soil, crop quality, and quantity at high spatial-temporal resolutions across the lifecycle of production including pre-sowing, sowing, in-crop, and harvesting. The analogous approach in the mining industry would be measuring key characteristics of the mineral deposit such as mineralogy, texture, hardness, potential acid generation, acid consumption, and size distributions in addition to the standard measures of grades (elemental content) at improved resolutions, coupled with integrated planning across planning, production, and disposal. Such an approach is currently rare due to the lack of sensors that can measure these inputs and the inability to analyse processes in real time with true integrated planning systems.

Studies of new technology and innovation within the agricultural sector indicate that usefulness and ease of use are important for adoption but with the caveat that there is not a significant increase in production cost, irrespective of improvements in revenue (Castle et al. 2016; Pierpaoli et al. 2013). An additional factor is the perception of the risk in adopting the new technologies (Marra et al. 2003). The perception is based on several factors including a producer's existing practices and the uncertainty surrounding the cost savings that will be achieved. A study of corn producers in the United States found that the advanced precision agricultural systems resulted in variable cost savings relative to the intermediate systems, although in some cases, the advanced systems can lead to increased outputs and profits (Schimmelpfenning and Ebel 2016). This variability in outcomes is typical for the application of new technology in complex production systems. Exacerbating the situation is that for both the agricultural and mining industry, significant difficulties exist in estimating the financial benefits associated with the adoption of digital innovation due to the predominance of static and conservative financial modelling tools such as discounted cash flow analysis which does not easily accommodate estimations of the value of innovation investments (Hayward et al. 2017).

Although it can be argued that big data has been used in precision agriculture for many years, it is enhancing the ability to enabling learning, prediction, and optimization of farm production not only at individual farm scale but also at district and regional level such as the American Midwest (Sonka 2016; Wolfert et al. 2017). Open data platforms have been embraced in the United States which has resulted in ease of data transfer between businesses and software platforms. This result also led to new applications and services in data storage and management. However, big data can only reliably deliver long-term business advantage when fully integrated with traditional data management and governance processes (Wolfert et al. 2017). The focus to date within the mining industry is primarily at the mine site level but a gradual move to open data standards has gathered momentum over the past few years through bodies such as the Global Mining and Standards Guidelines Group (GMSG 2017).

The consensus in agriculture is that it will take 5–7 years from the acquisition of relevant data (e.g., soil data, yield, weather, and other inputs) to demonstrate clear observable improvements in outcomes for the industry. This consensus reflects the time required to acquire data from sufficient seasons and develop useful and robust models, algorithms, and analytics. Thus, the payback from the initial investment in digital technology is relatively long term and without adequate capital can be a barrier to adoption. A similar time frame is likely for the mining industry across the integrated value chain but with some improvements likely in the short term for specific functions such as asset management and fleet maintenance.

The industrial networks in agribusiness exhibit a high degree of dynamism with new players entering the industry and the incumbents assuming different roles. New relationships have been formed among competing and collaborating firms, suppliers and customers, and stakeholders (Keogh and Henry 2016; Long et al. 2016; Wolfert et al. 2017). Two different scenarios appear to play out within agriculture value chains: (1) closed, proprietary systems in which the producer is part of a highly integrated food supply chain and (2) open, collaborative flexible systems (Wolfert et al. 2017). These scenarios are influenced by the architecture and infrastructure of big data solutions and the control of data. If these scenarios are valid in mining, then mining companies will need to change the way they work with their supply chain and technology providers. Regardless of whether they build deep, long-term networks that are closed to competition or flexible, open, collaborative platforms, new business capabilities will be required that will challenge the current transactional lowest cost, lowest risk procurement models.

1.5.2 Lessons from Digital Technology in Aerospace: New Management Capabilities and Network Governance

Complex product systems (CoPS) are customized, one off or small batch capital goods which have large physical size, high investment costs, long life cycles, and engineering complexity with interconnected subsystems (Hobday 1998) such as power stations, airports, hospitals, and flight simulators (Acha et al. 2004). Mining operations involve many forms of CoPS such as processing plants, railways, and ports. Arguably a mine is also a complex product system being a one-off design with high complexity and interconnected subsystems, so comparisons with other CoPS industries are useful to understand the future of mining. Increasingly software and embedded intelligence are being integrated into CoPS industries, which allows comparisons to be made with the digitization of mining.

The aerospace industry can be classified as a CoPS with large multi-organizational, multinational projects to create and build aerospace vehicles. In this section two contrasting cases of innovation in the aerospace industry and parallels with the mining industry will be considered. Boeing's Dreamliner was an innovative project, particularly in terms of aviation technology with new avionics and computing systems that had never been used on large commercial aircraft. It also was innovative in the coordination of its design and production by globally outsourcing a significant proportion of design, engineering, manufacturing, and production along with new risk-revenue arrangements with these suppliers. However, the project suffered significant delays and an overrun in development cost of 100%. This is not uncommon for larger CoPS projects, including mining projects (Merrow 2011; Flyvbjerg, 2015).

In an analysis of the project, Shenhar (2016) concluded that Boeing underestimated the level of complexity of the interdependencies between the technological innovations and the supplier network. Boeing treated the project as having a low level of dynamics and socio-political complexity, as if things are quite stable and the international cultural environment is mostly homogeneous. Given the globally distributed nature of the project, this was far from the case; however, there is no generally accepted best practice for managing such projects. Traditional project and program management tools rarely deal with change and managing innovation within a project (Davies et al. 2015; Steen et al. 2017). As Boeing found, managing a transition to a new way of organizing production requiring different technologies and supply chains with a high level of international sub-system integration requires new management capabilities.

Airbus has been a successful player in the aerospace industry with a track record of innovation in a CoPS environment. Key to Airbus' success was the development of a supply-chain organization model to maintain technical innovation (Kechidi 2013). Airbus evolved from an aircraft manufacturer to a technology system integrator. The stability of key partners in terms of the organizational model, while evolving and changing, resulted in a strong relationship where the suppliers evolved with Airbus. Although widespread, these partners were predominately located within Europe.

Airbus introduced technological innovation with each new aircraft which also shaped Airbus' organizational model. The modularization of technology at Airbus was based on subsystem components governed by subsystem integrator firms. This allowed Airbus to focus on managing the system and interacting with a smaller number of external firms. Generally, the mining industry has a very transactional and cost driven approach to its suppliers. As with Airbus, a mining company's governing role in the innovation ecosystem can shape the evolution of new technologies with suppliers. Developing strong stable relationships through a coordinated network with suppliers can accelerate and reduce risk in the innovation process in a CoPS environment.

Over the past 2 years, Airbus has undergone a more aggressive transformation as it sees enhanced competition and a risk of being disrupted and so is adopting American style management and business practices. This transformation includes trying different approaches to innovation in technology and product development, a focus on sensors, digital design, and digital manufacturing. The jury is out on whether Airbus will make a successful transition to this new hyper-digital environment (Gelaine 2017).

Looking at these two cases from industries that are more advanced in the adoption of digital technology yet similar to mining, we can see the important implications for the future of established mining companies:

1. Digital technology can generate significant productivity gains but only if the technology enhances the coordination of production across the value chain and supply chain. Applying digital technology to specific points in the value chain will have limited impact.
2. Digitization and coordination of an integrated supply network requires new business and management capabilities.
3. The accumulation of data and experience will take time. Digitizing a mining operation and creating a new industrial ecosystem around digital technology is a long-term project.
4. Miners have an important role to play in shaping the new digital ecosystem. Rather than being passive buyers of technology and services, they need to become system integrators to capture the value of innovation across the industrial network and consolidate their competitive position by leveraging the intellectual power of this network.

However, these changes are also potentially disruptive for the incumbent dominant businesses because they have the potential to change the nature of competitive advantage and tip bargaining power in favor of businesses that supply technology to the miners. In the next section, we consider how technological changes within the industry can transform the dynamics of competition.

1.6 The Digitization of Architectural Knowledge and the Impending Competition for Its Control

An outcome of digitization is the growing competition for control of architectural knowledge within the mining innovation ecosystem. The architectural knowledge we have in mind is that which enables the skillful coordination of different components of the mining system (e.g., development, production, processing, and distribution) in an attempt to maintain an optimal system state. Traditionally, this knowledge was embedded in the people, systems, and tools controlled by mining companies or, more recently, IT service companies. However, as the mining process becomes increasingly digitized and automated, traditional boundaries and roles are coming under pressure.

Digitization is happening at the level of the components and the systems linking these components together. At the component level, the functionality of mobile, heavy mining equipment has become critically dependent on software. Autonomous trucks, such as those developed by Komatsu and Caterpillar, provide insight into how knowledge boundaries are shifting at the component level. Traditionally, these firms would sell a mining company an asset. The mining company would then have its engineers and operators work to customize, tinker, hack, and otherwise innovate their way to improving the asset's performance in use. In terms of engineering, this mainly required mechanical and electrical systems. However, as digital control systems become central to asset performance, knowledge of software is obviously required to continue improving performance.

What is less well understood is that the access to and jurisdiction over this knowledge is problematic for mining companies because the software in question is usually provided, maintained, and protected by the original equipment manufacturers (OEMs). For example, OEMs can use obfuscation techniques to prevent miners from digitally upgrading or customizing the control system's source code, and thus restrict their ability to improve or adapt the asset's performance. Even if mining companies could access this code, OEMs claiming proprietary information can place licensing conditions on purchasers that limit who can access and edit this code. We are not aware of controversy around this yet in the mining industry, but the "Right to Repair" debate in the United States provides an illustration of how restrictive such licensing agreements can be. For example, John Deere locks farmers into license agreements that forbid them from attempting to repair their own tractors by requiring that they channel repairs through manufacturers and authorized dealers (Solon 2017). Digitization makes it possible to quarantine islands of knowledge at the component-level within the mining system and shift control of these islands to OEMs.

At the same time as digital islands are being walled off from mining companies, an influx of new entrants into the mining industry are seeking to build, codify, and control architectural knowledge at the level of the mining system. The abundance of digital data being produced within the mining system creates opportunities for firms such as IBM. These firms can draw on proprietary techniques to integrate, analyze, simulate, and predict mine production at the system level. It was with precisely this goal in mind that GE and South32 (2017) formed a 3-year strategic partnership in April 2017 in an attempt to leverage these techniques to drive performance improvements through better mine integration. These initiatives are likely to focus on both the codification of existing architectural knowledge and its augmentation through new processes. In doing so, firms such as GE and IBM are taking an important step towards gaining access to the architectural knowledge required to run a mine site. It remains to be seen how much of this knowledge they will end up controlling. However, similar to what we see playing out at the component level, a knowledge-based shift in the boundaries of the industry is currently underway.

These changes leave mining companies vulnerably placed with the eventual ownership of mining's architectural knowledge, and the margins that goes with it, in flux. It also asks interesting question regarding the nature of intellectual property and how it is secured. In the digital era, when mining systems and their components are so dependent on software programs for their competitive advantage, who will own the proprietary systems for running them? And in an era awash with industrial espionage, how will these highly codified digital programs be protected from competitors? The mining company of the future may look very different from what it is now, and the term company may be replaced with alliance network where one integrated digital platform competes with others in a similar way that Android competes with iOS. One interviewee, a North American R&D director, was already considering the disruption of the large integrated mining company:

> If we imagine a world in which a lot of the functionality of a mining system is delivered as a service instead of a product, then you'd be in a very different investment game with mining. It wouldn't be a major sort of CAPEX driven operation, it would be an operational excellence driven operation, and that would change a lot of things. First of all it would probably make it unnecessary to have large mining companies, which may be one of the difficulties in persuading them to do it. (North American R&D director)

1.7 Conclusion

The mining industry currently has an idiosyncratic innovation ecosystem. It largely relies on importing innovation from suppliers and much of this innovation is embedded in improved processes so the innovation is frequently overlooked. However, mining will need to become more innovative to meet the inexorable challenges of rising costs, lower grade reserves in more difficult locations, and increased environmental and social scrutiny. Business as usual will not be an option.

Although the current short-term focus on the prospects for digital technology in mining revolves around cost reduction, we see a much broader impact of digital technologies as enablers of innovation such as simulation, modeling, real-time decision-making, and collaborative problem solving. Advanced automation will promote systemic solutions and improved efficiency across mine-to-mill, including critical areas such as water and energy use, which have traditionally been impeded by cross-department integration challenges.

Digital integration also will bring in suppliers as closer partners in mine performance as they use proprietary data collection and analysis methods to reveal ways to improve mining operations. As the challenges from energy costs, water, remote locations, and social license to operate escalate, mining companies will need to reinvent themselves to remain viable. This will mean a fundamental transformation away from large integrated mining companies to more agile networks of companies based on technology platforms and data for competitive advantage.

However, this increasing reliance upon supply chain companies to manage data will shift the balance of competitive power towards powerful tech companies like GE, IBM, and SAP. When Anglo American CEO, Mark Cutifani, foreshadowed at the 2013 World Mining Congress that innovative companies like GE might take over the mining industry, his words may prove to be prophetic rather than provocative.

References

Acha, V., Davies, A., Hobday, M., and Salter, A. 2004. Exploring the capital goods economy: Complex product systems in the UK. *Industrial and Corporate Change* 14(3):505–529.

Caldwell, N., Roehrich, J., and Davies, A. 2009. *Procuring complex performance in construction: London Heathrow Terminal 5 and a Private Finance Initiative hospital. Journal of Purchasing and Supply Chain Management* 15(3):178–186.

Castle, M.H., Lubben, B.D., Bradley, D., and Luck, J.D. 2016. Factors influencing the adoption of precision agriculture technologies by Nebraska producers, Presentations, Working Papers, and Gray Literature. *Agricultural Economics* Paper 49. http://digitalcommons.unl.edu/ageconworkpap/49.

Cohen, W., and Levinthal, D. 1990. Absorptive capacity: A new perspective on learning and innovation. *Administrative Science Quarterly* 35:128–152.

Davies, A., Macaulay, S., DeBarro, T., and Thurston, M. 2015. Making innovation happen in a megaproject: London's crossrail suburban railway system. *Project Management Journal* 45(6):25–37.

Dodgson, M., Gann, D., and Salter, A. 2006. The role of technology in the shift towards open innovation: The case of Proctor & Gamble. *R&D Management* 36(3):333–346.

Dodgson, M., Gann, D., and Salter, A. 2005. *Think, Play, Do: Technology, Innovation, and Organization.* Oxford: Oxford University Press.

Dodgson, M., Gann, D., and Salter, A. 2007. In case of fire, please use the elevator: Simulation technology and organization in fire engineering. *Organization Science* 18(5):849–864.

Dodgson, M., and Steen, J. 2008. New innovation models and Australia's old economy. In *Creating Wealth from Knowledge: Meeting the Innovation Challenge,* J. Besant and T. Venables (Eds.), 105–124. Cheltenham, UK: Edward Elgar.

Edraki, M., Baumgartl, T., Manlapig, E., Bradshaw, D., Franks, D.M., and Moran, C.J. 2014. Designing mine tailings for better environmental, social and economic outcomes: A review of alternative approaches. *Journal of Cleaner Production* 84:411–420.

Eslake, S., and Walsh, M. 2011. *Australia's Productivity Challenge*. Melbourne, Australia: Grattan Institute.

Flyvbjerg, B. 2015. What you should know about megaprojects and why. An overview. *Project Management Journal* 45(2):6–19.

Forsythe, P., Sankaran, S., and Biesenthal, C. 2015. How far can BIM reduce information asymmetry in the Australian construction context? *Project Management Journal* 46(3):75–87.

Francis, E. 2015. *The Australian Mining Industry: More Than Just Shovels and Being the Lucky Country*. Canberra, Australia: IP Australia.

Gann, D., and Dodgson, M. 2008. Innovate with vision. *Ingenia* 36:45–59.

GE and South32. 2017. South32 and GE enter digital transformation strategic partnership. https://www.south32.net/docs/default-source/media-releases/south32-and-ge-enter-digital-transformation-strategic-partnership.pdf?sfvrsn=2bd6a0af_7.

Gelaine A. 2017. A critical look at Airbus's push for disruption: In its bid for self-disruption, is Airbus going too far? *Aviation Week & Space Technology*, December 6, 2017, 1–3.

Hanson, D., Steen, J., and Liesch, P. 1997. Reluctance to innovate: A case study of the titanium dioxide industry. *Prometheus* 15(3):345–356.

Hayward, M., Caldwell, A., Steen, J. Liesch, P., and Gow, D. 2017. Entrepreneurs capital budgeting orientations and innovation outputs: Evidence from Australian biotechnology firms. *Long Range Planning* 50(2):121–133.

Henderson, R., and Clark, K. 1990. Architectural innovation: The reconfiguration of existing product technologies and the failure of established firms. *Administrative Science Quarterly* 35:9–30.

Hobday, M. 1998. Product complexity, innovation and industrial organization. *Research Policy* 26:689–710.

Hodge, R.A. 2014. Mining company performance and community conflict: Moving beyond a seeming paradox. *Journal of Cleaner Production* 84:27–33.

Kastelle, T., and Steen, J. 2011. Ideas are not innovations. *Prometheus* 29(1):39–50.

Kechidi, M. 2013. From "aircraft manufacturer" to "architect-integrator": Airbus's industrial organisational model. *International Journal of Technology and Globalisation* 7(2/3):8–22.

Kemp, D., and Owen, J.R., 2013. Community relations and mining: Core to business but not "core business." *Resources Policy* 38:523–531.

Keogh, M., and Henry, M. 2016. The implications of digital agriculture and big data for Australian agriculture, research report. Sydney, Australia: Australian Farm Institute.

Kunz, N.C. 2013. Sustainable water management by coupling human and engineered systems. PhD thesis. Sustainable Minerals Institute, The University of Queensland, St Lucia, Brisbane, Australia.

Kunz, N.C. 2016. Catchment-based water management in the mining industry: Challenges and solutions. *The Extractive Industries and Society* 3(4):972–977.

Kunz, N.C., Kastelle, T., and Moran, C.J. 2017. Social network analysis reveals that communication gaps may prevent effective water management in the mining sector. *Journal of Cleaner Production* 14:915–922.

Long, T. B., Blok, V., and Coninx, I. 2016. Barriers to the adoption and diffusion of technological innovations for climate-smart agriculture in Europe: Evidence from the Netherlands, France, Switzerland and Italy. *Journal of Cleaner Production* 112:9–21.

Marra M., Pannell, D. J., and Ghadim, A.A. 2003. The economics of risk, uncertainty and learning in the adoption of new agricultural technologies: where are we on the learning curve? *Agricultural Systems* 75:215–234.

McKee, D., 2016. Understanding Mine to Mill. Canberra: The Cooperative Research Centre for Optimising Resource Extraction (CRC ORE). Brisbane, Australia.

Merrow, E.W. 2011. *Industrial Megaprojects: Concepts, Strategies and Practices for Success*. New York: John Wiley and Sons.

Mintzberg, H. 1981. Organization design: Fashion or fit? *Harvard Business Review* 59:103–116.

Mitchell, P., Bradbrook, M., Higgins, L., Steen, J., Henderson, C., Kastelle, T., Moran, C.J., Macaulay, S., and Kunz, N.C. 2014. *Productivity in Mining: Now comes the hard part, a global survey*. Sydney, Australia: Ernst and Young.

Mitchell, P., and Steen, J. 2014. Productivity in mining: A case for broad transformation. Sydney, Australia: Ernst and Young.

Mitchell. P., Steen, J., Sartorio, A., Bolton, W., MacAaulay, S., Higgins, L., Kunz, N.C., Yameogo, T., Hoogedeure, W., and Jackson, J. 2017. How do you prepare for tomorrow's mine today? Sydney, Australia: Ernst and Young.

Nadolski, S., Klein, B., Elmo, D., and Scoble, M. 2015. Cave-to-mill: A mine-to-mill approach for block cave mines. *Mining Technology* 124:47–55.

Pierce, F.J., and Nowak, P. 1999. Aspects of precision agriculture. *Advances in Agronomy* 67:1–86.

Pierpaoli, E., Carli, G., Pignatti, E., and Canavan, M. 2013. Drivers of precision agriculture technologies adoption: A literature review. *Procedia Technology* 8:61–69.

Saefong, M. 2016. How a goldminer turned a $1 million investment into $3 billion. *Market Watch*.

Salter, A., Gann, G., and Dodgson, M. 2005. *Think, Play, Do*. Oxford, UK: Oxford University Press.

Schimmelpfenning D., and Ebel R. 2016. Sequential adoption and cost savings from precision agriculture. *Journal of Agricultural and Resource Economics* 41:97–115.

Schumpeter, J. 1912. *The Theory of Economic Development*. New Brunswick, NJ: Transaction Publishers. (Tenth printing, 2004.)

Scott-Kemis, D. 2013. *How About those METS? Leveraging Australia's Mining Equipment, Technology and Services Sector*. Canberra: Minerals Council of Australia.

Shenhar, A. J., Holzmann, V., Melamed, B., and Zhao, Y. 2016. The challenge of innovation in highly complex projects: What can we learn from Boeing's Dreamliner experience? *Project Management Journal* 47(2):62–78.

Shook, A. 2015. Innovation in mining—Are we different? *AusIMM Bulletin*, April. https://www.ausimmbulletin.com/feature/innovation-in-mining/.

Solon, O. 2017. *A right to repair: Why Nebraska farmers are taking on John Deere and Apple*. London: The Guardian. https://www.theguardian.com/environment/2017/mar/06/nebraska-farmers-right-to-repair-john-deere-apple.

Sonka, S.T. 2016. Big data: Fueling the next evolution of agricultural innovation. *Journal of Innovation Management* 4(1):114–136.

Steen, J., Ford, J., and Verreynne, M. 2017. Symbols, sublimes, solutions and problems: A garbage can model of megaprojects. *Project Management Journal* 48(6):117–131.

Syed, A., Graftan, Q., and Kalirajan, K. 2013. *Productivity in the Australian Mining Sector*. Canberra, Australia: BREE.

Tilton, J., and Landsberg, H. 1997. Innovation, productivity growth and the survival of the US copper industry. Washington, DC: Resources for the Future Discussion paper, pp. 97–41.

Wolfert, A., Lan, G., Verdouw, C., and Bogaardt, M. 2017. Big data in smart farming: A review. *Agricultural Systems* 153:69–80.

Watcharaanantapong, P., Roberts, R. K., Lambert, D. M., Larson, J. A., Velandia, M., English B. C., Rejesus, R. M., and Wang, C. 2014. Timing of precision agriculture technology adoption in US cotton production. *Precision Agriculture* 15:427–446.

2

Mining Innovation: Barriers and Imperatives

Kane Usher and Ian Dover

CONTENTS

2.1 Introduction

Over the course of an 18-month period leading up to April 2016, METS Ignited Australia, the government-funded, industry-led, national growth centre for the Mining Equipment, Technology and Services (METS) sector, was engaged with the METS and Mining sectors, to understand and unlock the mining innovation system for the benefit of Australian METS sector providers. Such a journey was at times confusing and frustrating: the mining innovation ecosystem is complex, containing many actors with differing objectives. Furthermore, views vary as to how to make it more effective, where technology is headed, and how to address the challenges within the system. MI was tasked with the challenge of synthesizing masses of data, opinions, and information into a coherent strategy for Australian METS sector industry growth.

2.2 Mining Innovation Ecosystem

The mining innovation ecosystem can broadly be segmented into "research and development providers," "METS sector providers," and mining companies, as illustrated in Figure 2.1. Within each segment, further categorization is possible. Highlighted in Figure 2.1 are some key issues in the system as identified by MI [1]. Not shown in Figure 2.1 are the numerous industry bodies or the state and federal government acting as advocates and funding providers. In our discussion, we focus on the barriers to innovation from a mining company's perspective.

Others have written with eloquence on the need for innovation to sustain the mining industry. Key elements of the innovation imperatives in mining are illustrated in Figure 2.2. One of the key drivers for the mining industry is that most of the economic ore bodies have already been discovered and are being exploited. Ore bodies are becoming deeper, lower grade, and have increasingly complex mineralogy and ground conditions. By way of example, global gold grades are now 2 g/tn compared to 12 g/tn in the 1960s and 1970s [2], while average copper grades have declined 25% in the last 10 years [3]. Compounding the issues of declining grade, mining productivity and return on capital delivers a less-than-favorable comparison with other similar industries. For example, overall equipment effectiveness (OEE) in mining lags behind most other similar industries with OEEs in mining of 27%, 39%, and 69% for underground mining, open-pit mining, and crushing and grinding, respectively, *versus* OEEs in comparable industries of 88% for upstream oil and gas, 90% for steel, and 92% for oil refining [4]. In other words, we are mining material that has lower valuable mineral content per unit at a higher per unit cost.

Social license needs also are increasing with governments and communities demanding more transparency in operations, increased returns, and

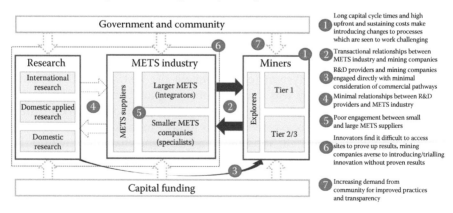

FIGURE 2.1
METS/mining innovation ecosystem and some of the challenges within it.

More challenging or ore bodies
• Ore bodies are deeper, lower grade and have increasingly complex mineralogy and ground conditions

Declining productivity and return on capital invested
• Productivity and return on capital in mining is not competitive with other industries

Increasing social license expectations
• Increasing social considerations covering communities, energy, emissions, land, waste treatment, and water

Changing demographics and socio-political landscape
• Aging workforces, urbanisation
• Transition of emerging economies from industrial to technical

Changing business models
• Open enterprises, supply chain participation, the rise of contract mining?
• Knowledge and digital economies

Digitalization and automation
• Connectedness, open platforms, sensing technoligies, analytics, and decision support
• Increasing availability of process/equipment automation and rabotics

FIGURE 2.2
Drivers and trends influencing mining.

minimized negative environmental, safety, and community impacts. Before even considering innovative mining approaches, these demands impact on the ability to attract investment, as return on capital invested and free cash flow performance of mining companies is generally poorer than in other sectors. Further pressures include sociopolitical changes with the emergence of developing economies, changing workforce demographics and urbanization, the potential for business model disruption, and the rise of digitalization and automation.

Over the last few years, depressed commodity prices have highlighted these drivers and trends, some of which can be masked in good times. Mining is cyclic and history shows generally three waves of response to commodity price challenges in mining, as shown in Figure 2.3. The first wave response usually involves scaling back of exploration and use of contractors, and a refocus on the basics of safety, production, and costs. The second wave response, often driven through large consulting firms, is to "sweat the assets"—maximizing the value from installed capital, further squeezing suppliers, and reducing personnel numbers. It is surprising to many during this period that even with less people and equipment, production normally goes up. The third-wave response is where "innovation" and process re-engineering come into focus. Arguably, we have been in the third wave response for several years. However, the cyclic nature of mining means that the time spent in the "third wave" response, in which momentum is building toward real innovation, is limited.

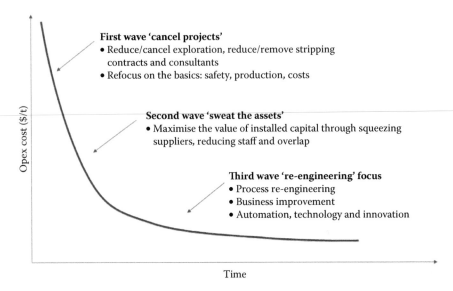

FIGURE 2.3
The three waves of cost cutting. (Figure by Kirby Johnson, in: The critical role of the mining tech-nology professional, *AusIMM Bulletin* (June 2016): Figure 2.1. Courtesy of *AusIMM Bulletin*. https://www.ausimmbulletin.com/feature/the-critical-role-of-the-mining-technology-professional/.)

It is clear that innovation is required in mining if it is to continue to attract investment. However, investors are typically seeking short-term benefits which can be at odds with the longer terms generally required to see innovation benefits play out in mining. Even within mining com-panies, competition for capital (growth and sustaining) is fierce, with innovation projects competing alongside "keeping the lights on" projects. Despite all the rhetoric and urging from informed commentators on the "innovation imperative," the mining industry is unlikely to respond until it is absolutely necessary.

Lower commodity prices have sparked early discussion on innovation. However, commodity price turns have historically been just around the corner, and the experience is that innovation momentum is quickly lost to the drive towards increased production. Ultimately, then, mining compa-nies will deal with the innovation imperatives when they have to, whether this is through necessity, substitution, disruption, or a mining industry catastrophe. In considering mining industry views on innovation, the fol-lowing quote from Sandeep Biswas, CEO of Newcrest Mining, in a recent address to the Melbourne Mining Club [2] is enlightening:

> "As I said earlier, we are driven by necessity. We will act when under pressure. When we have a difficult deposit that cannot be economically developed without it. And we are typically very risk averse. And inno-vation involves risks. We like innovations that have been trialled suc-cessfully by others. We all want to be fast followers. It is easier to mount

a business case when you can point to Exhibit A. I know this because I have been guilty of asking to see Exhibit A myself. And innovation by definition means disrupting our established ways of doing things. And that will often test our determination to get a result. So, it is not easy. But we are going to have to get used to it. Because whether we like it or not we are going to have to innovate faster as an industry to address the challenges ahead."

2.3 Barriers to Mining Innovation

With all the good reasons to innovate, it is informative to investigate why innovation is so difficult in the mining industry. The mining industry has long been castigated when it comes to innovation. However, this castigation is often from commentators who may not understand the underlying drivers or the long history of innovation within mining. Obvious analogies to manufacturing and other industries are frequently made. Despite the similarities and things that the mining industry needs to learn from other industries, there are some key differences. Hollitt [6] proposes that no other sector has the following combination of characteristics:

- Very long product lifecycles (with some mineral and metal commodities having been traded for millennia)
- High capital intensity (with capital, finance, and energy-related costs typically dominating net cash outflows for production)
- A finite resource base underlying each investment, with high asset redundancy at depletion
- Very long capital cycles (typically from 8 to 20+ years) relative to business cycles
- Relatively short business cycles
- Fixed costs that are high relative to movements in revenue

These characteristics give innovation in the mining industry a particular flavor. Hollitt [6] also posits that, as opposed to the "creative destructionist" approach of the manufacturing, electronics, and software sectors, the mining sector is better supported by "creative preservation." That is, short cycles to capital replacement are not rewarded in mining; but rather, smarter ways to extend asset life, reduce costs, and increase production with minimal incremental capital are what is rewarded by investors.

Shook [7] provides another lens by comparing mining innovation with other industries including manufacturing, military operations, and oil and gas by questioning why "military operations, manufacturing, and oil and gas look vastly different to what they did 50 years ago, and yet mining remains startlingly similar (aside from larger machinery)." The author cites three key reasons for the mining industry appearing to lag behind other industries: first, the high cost of failure which breeds known, conventional approaches; second, the success of the conventional processes makes any subsequent changes difficult to justify; and, third, a lack of information, both for in-ground resources, and monitoring and characterizing solid materials while flowing through the system. In other words [7]: "if fear of failure causes us to start with proven (old) technology, and success causes us to stay with it as long as possible, then I believe that one other factor—information (or rather, lack of it)—is the overriding reason why we have stayed this way as long as we have."

The framing of the challenges by Shook (2015) is a good representation and a highly recommended read. However, there are several other factors including integration barriers, and culture and organization, which are also critical. Partially adapting the work of Shook (2015), we build on the discussion in the sections which follow.[1]

These discussions do paint a somewhat bleak picture. This is intentional. Innovation in mining has a long history but it is usually driven by a particular property of an ore body. The discussion is also generalized for which there are always exceptions. There are also nuances depending on the size of the mining organization in question.

2.3.1 Technical Risk

The cost of failure in mining is high. For example, establishing the case to invest in new mines, or indeed expansions, relies on patchy information about mineralization which can be hundreds to thousands of meters beneath the surface. Drilling and otherwise probing for further information is expensive. Similarly, profitability relies on future, often volatile, commodity prices, as we have experienced acutely in recent times. Mines are usually mid- to long-term ventures (decades), so commodity price risks are a big exposure.

Adding to ore body uncertainty and product price risks are the high capital costs of establishing operations, which are usually in remote areas, and which also must appropriately deal with people, process, and environmental risks. The costs of ongoing operations and sustaining/expanding capital are similarly significant. Decisions on investments are predominately based on net present value (NPV), meaning that delays in selling product at the required volume and quality quickly erode expected returns. Compounding these known risks with unknown technological risks is usually avoided unless it is absolutely necessary (e.g., due to a particular property of the ore body which might require a new approach or technology).

---

[1] With thanks to Tony Filmer and Melinda Hodkiewicz for helping to shape this taxonomy.

Equally, mining can be a victim of its own success. Once operating with tried and true methods, there is rarely a "burning platform" to make a change. Capital cycles are long, as plant and equipment lifecycles are usually of the order of 10+ years. Also, process and equipment change-outs rarely occur all at once due to a need to stagger capital investments and minimize production impacts. Once capital infrastructure is installed, these decisions lead to "embedded constraints" that can be long lived [6]. Further, when it comes to capital investment (growth or sustaining), individual mines are usually competing within a portfolio, so the case for change needs to be strong. Appetite for innovation also may be dependent on the expected mine life—mines with shorter expected life generally will opt for conventional, proven technologies, or require an extraordinary rate of return on the new technology. Finally, there is also a paradox in that when times are good, any loss or decrease in production to introduce new technology outweighs the possible value which might be generated by that technology; when times are bad, capital is scarce and there is a laser focus on costs and efficiency, again making it challenging to justify capital for new technologies.

The result of this "aversion to failure" and "sticking to successes" is that mines usually start with conventional technologies, running these through good and bad times until catastrophic failure or obsolescence. Unless there is a serious crisis or a major development, minor technology tweaks are usually all that occur until mines close. Faced with uncertain information on ore bodies, volatility in commodity prices, and the high costs of investment, mining companies usually elect not to compound this with further technical risk. For either green-field or existing mines, if returns are acceptable using incumbent or conventional technology, then justifying capital expenditure for new or novel technologies is difficult.

2.3.2 Integration Risk

Mining is complicated and has perhaps evolved to be more complicated than it needs to be. It requires the orchestration of a multitude of people, assets, processes, and systems to access, extract, and beneficiate ore bodies about which there is patchy information, selling into markets that are historically volatile. The processes within mining operate over different parts of the mining lifecycle (from exploration, development, production, and closure), and within different parts of the mining value chain (exploration, planning, operations, processing, maintenance, support functions, and logistics). Processes can operate in series and parallel, over multiple time horizons, with cost and value drivers that also may vary over time.

Innovators often approach mining companies with a solution which addresses a small part of the whole. A catch-phrase from many mining companies is that they often operate "under a hail of silver bullets" from suppliers. As well as integration risks, innovators (and mining companies for that matter) do not always appreciate the potential up- and down-stream process benefits and risks of technology introduction. Further, large and incumbent suppliers and original equipment manufacturers (OEMs) have historically applied proprietary interfacing technologies, whether these are equipment, hardware, or software. Changes or inserting nonproprietary products and services can require significant engineering and incumbent vendor relationship effort.

Anecdotally, for every $1 a mining company spends on an innovative product or service, there are a further $8–9 to be spent on integrating into the existing processes and workflows. Some of the integration challenges may not be visible upfront. Integration challenges apply to operating mines as well as new or expanded mines considering novel technologies. Integration challenges and costs are, therefore, often poorly understood. In many cases, gain is not sufficient from a new or different technology to justify the mining company rejigging the wider system to fit the new technology into a production environment.

2.3.3 Information-Challenged

Mining generates vast volumes of data; however, much of this data is not translated into information. Further, much of the data currently generated may not be the *right* data. For example, does a mining company really need to sample turbo-charger temperature on a haul truck at 1 Hz. This might be good for the truck manufacturer but is it all that useful for the mining company? Figure 2.4 illustrates some of the challenges with data use in mining.

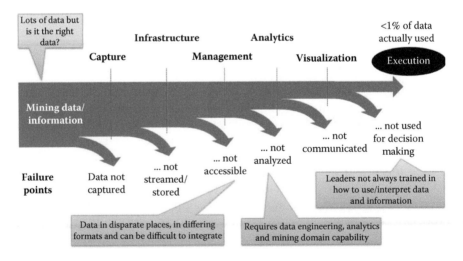

FIGURE 2.4

Data to information in mining. (From Durrant-Whyte, H. et al., How digital innovation can improve mining productivity, (Online). Available: http://www.mckinsey.com/industries/metals-and-mining/our-insights/how-digital-innovation-can-improve-mining-productivity.)

In terms of the right data, we have already discussed the uncertainty in ore body information perhaps leading to a preference for conventional processes over new technologies. There is also the extreme challenge of measuring what actually has been mined, which is usually in the form of solid materials made up of particles of varying size and composition, further complicated by the presence of clays, sludge, and water. It can be sticky, dusty, and abrasive, and its properties can vary over time. Measuring flows and characterizing solids is notoriously difficult (size, composition, mass flow, chemical properties, and so on). This contrasts with oil and gas, which deal with liquids and gases which are relatively straightforward to measure and meter, which in turn more readily enable sophisticated control approaches and "loop closure." The poor knowledge about the material being dealt with in mining has led to a number of "coping strategies" across the planning and operational domains:

- Averages and blending used over different spatial and time horizons to remove noise, which unfortunately also removes information
- Moving slowly, in low resolution, and in large chunks which are unlikely to be wrong
- Mining companies have gotten extraordinarily good at responding to surprises, including relying on flexible equipment and processes and a propensity to not over-plan (or at least not follow plans)

What is more, and as discussed earlier, many of the "opportunities" within mining are hidden or invisible. We simply do not know if, let alone where or how, valuable material is being lost as it traverses the value chain. Even if we do find out, it is often too late to act.

In some cases, the data may be there but it is not acted upon. As an example, the experience from several large mining operations indicates that up to 10% of haul truck loads which have been correctly dispatched end up going to the wrong place—that is, ore to waste (lost product) and waste to ore (unnecessary processing). Further to this, across major users of Fleet Management Systems (which are intended to "optimize" haul truck fleet operations) the optimization function is turned off for upwards of 70% of the time, inferring that dispatchers believe that they do a better job than the automated systems.

In summary, mining is a data-rich but information-poor environment, particularly when it comes to understanding the ore body and tracking it as it is extracted. The absence of the right information about the right things means that the technical risks and benefits around adopting new technologies can be poorly understood. This acts to reinforce the use of proven, conventional technologies.

2.3.4 Culture and Organization

Mining organizations have been set up to deal with what are truly daunting business challenges: finding, mining, processing, and transporting of resources about which there is little information or certainty, often in very remote locations. Teams of people covering different exploration, development, planning, operational, technical, support, and management disciplines are required, often operating over different time horizons. Despite best efforts at alignment, each team can have differing objectives and key performance indicators (KPIs), which in turn can drive conflicting outcomes. There is also often a corporate overlay to site operations. Again, KPIs for these teams may not be in alignment with each other or the sites. KPI achievement is usually tied to personal bonuses and despite best efforts at alignment between teams, there are often unintended outcomes.

Of course, it is the procurement area where METS companies most acutely feel the pain point of misalignment, usually as a result of a "cost" versus "value" focus. This misalignment also is felt internally within mining. The effort required for an already busy operations person to on-board a new vendor or technology often is overwhelming, and as a result innovation can fall by the wayside.

Another lens on the cultural challenge is that of self-interest. Improvement and innovation usually mean more efficiency, which can have the effect of reducing personnel numbers. Self-interest can stymie the success of innovation projects. There are numerous anecdotes of people waiting out their latest manager's tenure so that things can "go back to normal." Protection of one's own patch (and annual bonus) can make the introduction of innovation challenging. An additional factor is a culture which values flexibility and an ability to respond to evolving circumstances (firefighting) rather than rigid plans. This factor in turn drives preferences for flexible approaches to mining and processing. Flexibility and agility are prized in other industries such as manufacturing and software services, so what is the correct balance in mining of flexibility, systemization, and overall complexity?

Further, the benefits from innovation ideas can sometimes be difficult to communicate. Organizationally, mining companies can be complicated. Innovation ideas may span different teams and functions, all with competing demands and priorities. There also may not be sufficient knowledge or capability to appreciate the benefits of a particular innovation. As a result, marketing an idea and getting consensus among teams can be challenging. Then there is the task of reaching a decision maker. There are usually plenty of people who can say "no" and very few who can say "yes."

2.4 What Are the Common Characteristics of Successful Innovation in Mining?

Keeping in mind the barriers to innovation, we briefly examine the characteristics of successful innovation in mining. The key driver for successful innovation in mining is sustained championship from senior levels in mining organizations. Usually, this sustained championship is due to an economic need and, in many cases, a result of the properties of a particular ore body (or set of ore bodies) that require innovation in order to make the innovation economic or push down the cost curve. New mines having new challenges (e.g., lower grades and more overburden) provide the imperative for new approaches [6].

A common framework for innovation is the "three horizons of growth" (Baghai, Coley, and White 2000) as shown in Figure 2.5. Horizon 1 activity relates to execution improvements on existing processes. This activity might include changes to maintenance strategies, scaling of equipment (although this is generally longer term for the manufacturer), and value chain optimization type work. Success at this scale relies on having the ongoing support of a range of operational and corporate personnel within a mining house, provided there is sufficient certainty in technical risk and a clear path for dealing with integration risk. This activity requires a very clear benefits case, articulated in language which helps mining companies communicate the case internally.

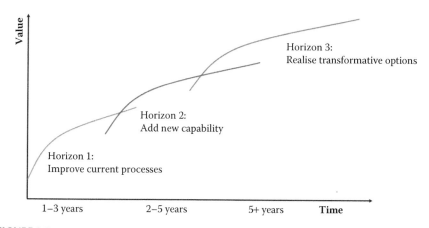

FIGURE 2.5
The three horizons of growth framework. (From Baghai, M. et al., *The Alchemy of Growth: Practical Insights for Building the Enduring Enterprise*, Perseus Books, Cambridge, MA, 2000.)

Horizon 2 activities relate to a higher level of return built over a longer time period. This activity could include, for example, the introduction of automated fleet and surrounding process integration. Success here relies on a very clear value case and mitigations for technical and integration risk. Sustained support across a wider audience generally is needed at a much higher level in the mining organization. Again, a clear benefits case is required which is aligned with the mining company's strategy.

For Horizon 3 or "new platform" scale innovation, Hollitt [6] proposes that the following characteristics are necessary (but not always sufficient):

- Attachment, at good value, to at least one new resource project or significant expansion (there is no need for "creative destruction" of existing capital).
- The new approach is critical to the fortunes of at least one such project or expansion (i.e., there is no other possibility of a satisfactory or sufficient project).
- The new approach is uniquely suited to avoiding loss of previous gains, including continued growth from resource acquisition or market options, which gains are otherwise under clear threat (the strategic imperative).
- It is expected that these necessary conditions will still be present in future business cycles considering other industry or regulatory developments.
- Sufficient finance or operating cash flows are available (equity rather than debt backed investment) to provide for development across several business cycles.

Importantly, the time for penetration for "new platforms" in mining and minerals processing is typically on the order of decades. Sustaining innovation momentum for such periods requires very high-level strategic support which outlives the typically shorter tenures of chief executive officers (CEOs) and senior leadership.

2.5 Discussion, Implications, and Conclusions

This discussion paints a gloomy picture for innovation in mining. This is intentional as the aim of this chapter is to inform the reader that there are indeed characteristics in mining that genuinely make innovation challenging. In truth, it is not all bad; mining innovation has a rich history (see, e.g., [9,6] and others for an overview) to which we cannot do justice within the scope of this discussion. But, innovation in mining is not as straightforward as it is in other industries,

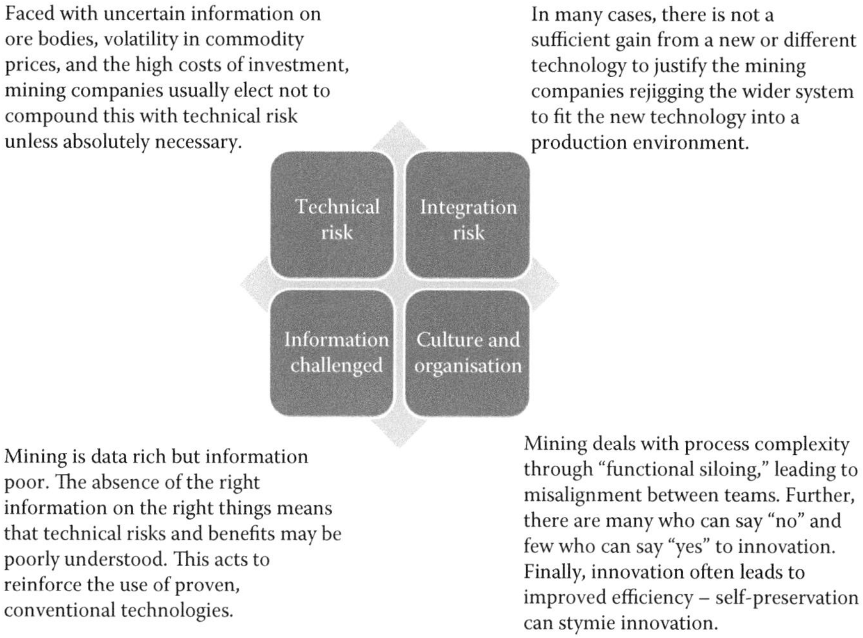

Faced with uncertain information on ore bodies, volatility in commodity prices, and the high costs of investment, mining companies usually elect not to compound this with technical risk unless absolutely necessary.

In many cases, there is not a sufficient gain from a new or different technology to justify the mining companies rejigging the wider system to fit the new technology into a production environment.

Technical risk

Integration risk

Information challenged

Culture and organisation

Mining is data rich but information poor. The absence of the right information on the right things means that technical risks and benefits may be poorly understood. This acts to reinforce the use of proven, conventional technologies.

Mining deals with process complexity through "functional siloing," leading to misalignment between teams. Further, there are many who can say "no" and few who can say "yes" to innovation. Finally, innovation often leads to improved efficiency – self-preservation can stymie innovation.

FIGURE 2.6
Taxonomy of mining innovation barriers.

nor as straightforward as it could be. This reality is a result of a mix of unique foundational and structural issues, as well as the distinctive cultures and organizations which have evolved to execute and support the mining process.

There are numerous ways of describing the innovation challenges. Here we have used the taxonomy detailed in Figure 2.6. Understanding the barriers and challenges to mining innovation is an important step towards addressing them. In fact, this understanding may provide insights to some important innovation opportunities, which may in turn further unlock innovation in mining.

For example, a more granular and certain understanding of the ore body may reduce the technical risk to novel extraction and processing technologies such that the perceived gain will outweigh the risk. Another example would be understanding that integration risks may lead to the identification of "market failures" and potentially different business models to orchestrate innovation.

One of the keys to success in mining innovation is ensuring that the idea is genuinely addressing a business need. Communication will require very clear value cases and business drivers, articulated in language that makes it easy for mining companies to communicate the benefits internally. Also, a sound understanding of the culture and organization within mining companies will be required to navigate to those who can say "yes."

There will need to be consideration of the technical risk of introducing innovation, recognizing the already significant risks associated with mining (ore body uncertainty, commodity price volatility, expected mine life, health, safety, environment, community, and so on). Further, providers and suppliers will need to consider solutions, solution offerings, and collaborations which act to reduce integration risks. This consideration may include investing in pilot scale activities with the recognition that risk sharing will likely be required.

Culture and organization is an issue for the entire mining industry to grapple with and it will not be easy. Most mining CEOs and senior leadership are attuned to the need to innovate. However, mining culture, systems, and organizational structure need to be shaken up if we are to see significant performance improvement through innovation.

References

1. G. Stanway, METS strategy development, Final report (Detailed), GoVCI, 2016.
2. S. Biswas, *The Age of Innovators*, Address to the Melbourne Mining Club, 2017 http://www.melbournemining club.com/wp.content/uploads/2016/11/Presentation_Sandeep_Biswas_Newest_9-Feb-2017.pdf.
3. G. Calvo, Decreasing ore grades in global metallic mining: A theoretical issue or a global reality, *Resources*, 5(4): 36, 2016.
4. H. Durrant-Whyte, R. Geraghty, F. Pujol and R. Sells, How digital innovation can improve mining productivity, (Online). Available: http://www.mckinsey.com/industries/metals-and-mining/our-insights/how-digital-innovation-can-improve-mining-productivity. Accessed Feb 2017.
5. DPI Mining, The Development Partner Institute For Mining—A Call To Action, 2016. Online. Available: http://www.dpimining.org/wp-content/uploads/2017/02/20170212-DPI-Overview-Booklet.pdf. Accessed Feb 2017.
6. M. Hollitt, Innovation and growth—Keeping pace in a virtuous cycle, in *International Mineral Processing Congress*, New Delhi, India, 2012.
7. A. Shook, Innovation in mining—Are we different?, *AusIMM Bulletin Magazine*, April 2015.
8. M. Baghai, S. Coley and D. White, *The Alchemy of Growth: Practical Insights for Building the Enduring Enterprise*, Perseus Books, Cambridge, MA, 2000.
9. R. Batterham, The mine of the future—Even more sustainable, *Minerals Engineering*, 107: 2–7, 2016.

3

How Innovation and Research & Development Happen in the Upstream Oil and Gas Industry: Insights from a Global Survey*

Robert K. Perrons

CONTENTS

3.1 Introduction

Oil and gas have been mainstays of the world's energy mix for decades (British Petroleum, 2012) and this trend will probably endure for many years to come (Longwell, 2002; Cook, 2007; Fischer, 2007; World Economic Forum, 2008; Yergin, 2009; Bullis, 2009). While the global demand for these energy sources continues,

---

* Reprinted by permission of the publisher, Elsevier, from Perrons, R. K. (2014). How Innovation and R&D happen in the upstream oil & gas industry: Insights from a global survey. *Journal of Petroleum Science and Engineering* 124:301–312.

however, the industry that provides them is changing in two fundamental ways. First, with much of the world's "easy oil" already consumed (Urstadt, 2006; Weijermars, 2009), upstream oil and gas companies will have to use increasingly sophisticated technologies to find and produce tomorrow's hydrocarbons (Tillerson, 2006; Lord, 2007; Paul, 2007). Future oil and gas resources—especially in non-OPEC (Organization of the Oil Producing Countries) countries—will tend to be deeper, harder to find, and in environments that are significantly more difficult to access than they used to be (Managi et al., 2004, 2005b; Hinton, 2010). Second, high-profile disasters like the *Piper Alpha* incident in 1988 (Paté-Cornell, 1993), the *Exxon Valdez* oil spill in 1989 (Plater, 2011; Coll, 2012), Shell's *Brent Spar* incident in 1995 (Frynas, 2003; Sluyterman, 2007, 2010), and the recent *Deepwater Horizon* accident (Flournoy, 2011; Perrons, 2013) have brought about a marked change in the expectations placed upon oil and gas companies with regards to environmental stewardship, safety, and human welfare (Mirvis, 2000; Managi et al., 2005a; Hofmeister, 2010). In the face of these kinds of challenges, technology will clearly play a pivotal role in the success or failure of tomorrow's oil and gas firms (Longwell, 2002; Mitchell et al., 2012).

Despite the strong case for technology, however, the industry has a reputation for being slow to develop and adopt innovations. The shared equity structure of many upstream oil and gas assets frequently makes it difficult for companies to keep new innovations proprietary (Acha, 2002; Sharma, 2005; Perrons and Watts, 2008), thereby creating a problem of "free ridership" within the sector that frequently erodes the competitive advantage that technology might otherwise deliver to an innovating firm. Also, the extreme risks (Daneshy, 2003a, b; Rao and Rodriguez, 2005) and high cost of failure associated with being a first user of new technologies are such that companies frequently prefer to be "fast followers" (Daneshy and Donnelly, 2004, 28) and the industry's innovations consequently take an average of 16 years to progress from the concept phase to widespread commercial adoption (National Petroleum Council [NPC], 2007).[1] The sector has accordingly been characterized in the literature as "slow clockspeed" (Fine, 1998, 239), "low- and medium-tech" (von Tunzelmann and Acha, 2006, 408), and "technologically timid" (Lashinsky, 2010, 88). Oil and gas producers also have been categorized as "low research and development (R&D) intensity" because they have historically invested less than 1% of their net revenue in R&D (von Tunzelmann and Acha, 2006; Moncada-Paternò-Castello et al., 2010).[2]

---

[1] While I do consider this to be a fair generalization of the industry, I also recognize that some companies within the sector are considerably more aggressive than others in developing and deploying innovations (Anderson 2000; Bohi 1998) and that companies within the industry often have different motivations for pursuing new technologies (Acha and Finch 2003; Daneshy and Shook 2004).

[2] It is important to note, however, that this is not true for most oilfield service companies. Whereas oil and gas producers in the United States have a R&D intensity of 0.21%, the country's oil equipment and service providers have a R&D intensity of 2.24% (Moncada-Paternò-Castello et al., 2010). Also, there are many people—like ExxonMobil CEO Rex Tillerson (2006), for example— who contend that the industry is more high-tech than is reflected in these kinds of statistical

The industry seems to be changing, however. Several international oil companies (IOCs) have pointed to technology as an increasingly important strategic priority (e.g., Kulkarni, 2011; Parshall, 2011; Chazan 2013), and spending on innovation and R&D by IOCs and national oil companies (NOCs) has risen dramatically over the past few years (Thuriaux-Alemán et al., 2010). These efforts to increase the amount and pace of innovation within the upstream oil and gas sector give rise to a few important questions. How does technology happen in the industry? Specifically, what ideas and inputs flow from which parts of the industry's value network, and where do these inputs go? And how do firms and organizations from different countries contribute differently to this process? The literature offers no shortage of anecdotal evidence (Daneshy and Donnelly, 2004), perspectives (Donnelly, 2006), and stories about individual technology programs (e.g., Artigas et al., 2012; Rassenfoss, 2013), but fails to give a comprehensive and holistic snapshot of how the industry's innovation system works overall.[3] What is more, the sheer size of the industry—seven out of the ten largest publicly listed firms in the world by revenues in 2011 were oil and gas companies with significant upstream operations (Fortune, 2012)—makes this sector an important part of the global economy. The specific mechanics of how new oil and gas technologies are created may therefore also be of interest outside of the industry because of the larger strategic and geopolitical role that the sector often assumes.

As a step towards improving how the upstream oil and gas sector develops and deploys new technologies in the future, I set out to deepen the understanding of how R&D happens within the industry at present. I will first review the existing literature connected to R&D and technology management within the industry. Then I will describe a survey that was put together with the Society of Petroleum Engineers (SPE) to shed light on several different aspects of how the industry conducts R&D, and I will present the results. Finally, I will explain how these data constructively add to the existing body of research in this field, underline the practical implications of this evidence, and recommend potentially fruitful directions for future investigations.

---

indicators. Chazan (2013) offers just one of many compelling pieces of evidence to support this alternative point of view: "BP's supercomputer complex in Houston was the world's first commercial research center to achieve a petaflop of processing speed."

[3] Helfat (1994a, b, 1997) makes several valuable contributions in this area by offering a very rigorous quantitative analysis of R&D in the oil and gas industry. But these investigations were based on datasets focusing on the period from 1974 to 1981 and, as explained in this paper, the R&D landscape of the industry has changed dramatically since that time. Also, much like the contributions of Grant (2003), Cibin and Grant (1996), Grant and Cibin (1996), Bastian and Tucci (2010), and Ollinger's (1994) in this area, the work of Helfat (1994a, b, 1997) looks more or less exclusively at large oil-producing companies and pays little attention to the service companies and other members of the upstream oil and gas ecosystem that play such an important role in the sector's technology development and deployment efforts today. As Martin (1996) points out, the upstream oil and gas sector consists of a highly interconnected system of organizations and is, therefore, most appropriately considered in a system-wide way. Finally, I should point out that Enos (1958, 1962, and 2002) offers some extremely detailed accounts of R&D and innovation in the downstream and refining parts of the oil and gas industry (which are frequently quite detached from what happens in the upstream part).

3.2 Literature Review and Research Questions

Prior to the IOCs' reduction in their in-house technology and innovation programs in the 1980s and 1990s, more than 80% of the industry's overall R&D investment was borne by just 11 oil and gas producers (Economides and Oligney, 2000). Technology historically had been an important strategic priority for several of the IOCs before to this period (Wilkins, 1975; Howarth and Jonker, 2007; Priest, 2007) and most of them had previously supported fairly comprehensive in-house R&D programs (Sharma, 2005).

But things have changed significantly since that time. The costs associated with modern-day R&D projects in any industry are an increasingly daunting proposition (Manders and Brenner, 1995; Kumpe and Bolwijn, 1988), and technology "has become so sophisticated, broad, and expensive that even the largest companies can't afford to do it all themselves" (Leonard-Barton, 1995, 135). Whereas major breakthroughs in many industries frequently used to come about through in-house R&D teams working within a single company, today's researchers often reach out to outside organizations to broaden the radius of new ideas to which they can gain access (Quinn and Hilmer, 1994; Rigby and Zook, 2002). To these ends, many companies within the upstream oil and gas industry have embraced the concept of "open innovation" (Chesbrough, 2003a, c) and more collaborative models of R&D that welcome ideas from other industries and technical domains (e.g., Verloop, 2006; Ramírez et al., 2011; Dennis et al., 2012). Oilfield service companies and a broad range of vendors, government agencies, and universities now potentially play important roles in the sector's R&D activities (Acha, 2002; Acha and Cusmano, 2005). Some firms in the upstream oil and gas industry have begun to experiment with various forms of venture capital to support potentially promising concepts outside of their in-house R&D activities (Hansen and Birkinshaw, 2007; Shah et al., 2008), and several companies in the sector have even forged R&D alliances with direct competitors (Crump, 1997).

While the industry's innovation processes are far more collaborative than they used to be, the specific details of these collaborations are less obvious. Different parts of the upstream oil and gas "ecosystem" have different resources and skill sets, and may therefore turn to different sources of information and knowledge throughout their innovation-related activities.

Research Question 1: What sources of information and knowledge do different types of organization use for innovation-related activities within the upstream oil and gas industry?

And with such a high degree of heterogeneity among the large number of organizations playing a role in the sector's innovation-related activities, each of these constituent groups may contribute differently to R&D outputs such as patents and deployed innovations.

Research Question 2a: What is the relative contribution of technology-related patents from each type of organization within the upstream oil and gas industry?

Research Question 2b: What is the relative contribution of deployed technologies from each type of organization within the upstream oil and gas industry?

Not all new technologies are the same, however. One frequently recurring basis for analysis among technology management researchers is the degree of change brought about by an innovation. Some technologies are characterized in the literature as "radical" because (1) they require the innovating companies to acquire fundamentally new skill sets (Afuah, 1998), (2) they add entirely new performance features, dramatically improve existing performance features, or significantly reduce costs (Leifer et al., 2000), or (3) they dramatically and obviously change the world around them by creating entirely new lines of business (Bozdogan et al., 1998; McDermott, 1999; Gilbert, 2003). "Incremental" innovations, by stark contrast, usually offer comparatively modest cost or feature improvements and move things ahead in a way that more or less preserves the status quo (Leifer et al., 2000). Prior discussions about the upstream oil and gas industry explain that the inherent riskiness of the sector has resulted in a pronounced emphasis on incremental innovation over the years (Daneshy and Donnelly, 2004), but more radical breakthroughs such as 3D seismic mapping and horizontal drilling have appeared from time to time (Martin, 1996; Managi et al., 2005b; Yergin, 2011). However, much of the literature in this area is highly anecdotal, and relatively little has been said about the origins of these new technologies on an industry-wide basis.

Research Question 3: What is the relative contribution of radical innovations from each type of organization within the upstream oil and gas industry?

Technology management literature applies an important distinction between product innovations and process innovations (Afuah, 1998; Tidd et al., 2001; Burgelman et al., 2004). As Schilling (2010) explains, "product innovations are embodied in the outputs of organizations—its goods or services… [while] process innovations are innovations in the way an organization conducts its business, such as in the techniques of producing or marketing goods or services" (p. 50). But here, too, the literature sheds relatively little light on the specifics of the upstream oil and gas sector on a worldwide basis.

Research Question 4: What are the relative contributions of product- and process-based innovations from each type of organization within the upstream oil and gas industry?

A considerable amount of research in the technology management domain also examines the geographic aspects of innovation (e.g., Stuart and Sorenson, 2003; Feldman, 2010; Fifarek and Veloso, 2010). The uncommonly global nature of the upstream oil gas industry (Yergin, 1991; Hatakenaka et al., 2006; Goldstein, 2009) makes this sector a particularly interesting backdrop for investigations concerning the spatial dimensions of R&D.[4] Although the sector's R&D efforts occur in many places around the world, however, these activities are by no means evenly distributed. Barlow (2000) notes that the upstream oil and gas industry has seen "a high degree of geographical clustering" (p. 980), and much of the R&D-related research that specifically examines the sector has consequently focused on this cluster. There have, for example, been a broad range of investigations into the myriad technology hubs and clusters that have emerged in different geographic locations around the world, including Texas (Elliott, 2011; Hinton, 2012), Australia (Steen et al., 2013), the United Kingdom (Bower and Young, 1995; Crabtree et al., 2000; Cumbers, 2000; Cumbers and Martin, 2001; Cumbers et al., 2003; Chapman et al., 2004; MacKinnon et al., 2004), Norway (Fagerberg et al., 2009; Hatakenaka et al., 2006, 2011), Brazil (Dantas and Bell, 2009, 2011; Silvestre and Dalcol, 2009, 2010), France (Furtado, 1997), the Middle East (Henni, 2013), and Nigeria (Vaaland et al., 2012). Far less is known, however, about how the industry's innovation processes happen on a global level.

Research Question 5: What is the relative contribution of upstream oil and gas innovations from different countries?

3.3 Method

3.3.1 The Survey

An online survey was carried out in collaboration with the SPE to answer the research questions presented. With more than 110,000 members in 141 countries, the SPE is the largest individual-member organization within the upstream oil and gas industry around the world. A "data firewall" was established so that I did not have access to any of the specific details of the survey participants. I helped to set up the survey and assisted with processing the results, but SPE did not divulge the name and company behind each completed survey.

---

[4] Although the majority of research concerning the geographic aspects of innovation within the upstream R&D industry has tended to focus on spatial phenomena, Bastian (2009) argues that the sector's vast geographic reach necessarily carries with it a broad array of political risks that also impact the technology strategies of the industry's firms.

Although the upstream oil and gas industry includes several large multinational firms, these companies often have a noticeably different approach to managing innovation and new technologies from one part of the world to the next throughout their global operations.[5] To capture these region-by-region differences, this survey asked questions about how technology-related and innovation-related activities are managed at the business unit level. Smaller companies and organizations that develop and deploy upstream oil and gas technologies in a consistent way throughout all their operations around the world were instructed to consider their entire organization as a "business unit" for the purposes of this survey.

Consultancies, universities, and governments also play a potentially valuable role in the innovation and R&D processes within the upstream oil and gas industry. This survey therefore included them, too. Throughout the survey, their "business unit" was the part of their organization that interacts with upstream oil and gas companies in their region.

The 23-question survey asked respondents about several aspects of their business unit's R&D and innovation-related activities. The survey also asked for several self-reported measures of R&D output from their business unit. Respondents were informed before completing the survey, however, that their results would be made anonymous and aggregated with data from other respondents, thereby removing any incentive to distort their responses or provide untrue data.

The survey and corresponding delivery strategy were put together according to the principles outlined in Dillman's (2000) "Tailored Design Method."[6] One practical concession had to be made that was a clear departure from the prescribed formula, whereas Dillman (2000) recommends a four-contact model for maximizing survey return rates, the SPE was uncomfortable with contacting its members that many times. Instead, the SPE allowed three contacts in February 2012: an official e-mail from the SPE inviting people to answer questions about the explanatory variables, a reminder 1 week later, and then a final e-mail 2 weeks after the start of the survey to ask questions about the dependent variables and close out the survey. Questions asking about explanatory and dependent variables were separated in time to minimize the impact of common method bias (Podsakoff et al., 2003).

Prior to its release, the survey was tested by six people—three from the oil and gas industry and three from academia—who were familiar with questionnaires and survey-based research. The survey's questions were

---

[5] For example, Shell's Smart Fields digital oilfield program has noticeable differences in deployment strategy from one region to the next, and BP's use of the WITSML drilling data exchange protocol in the North Sea is markedly different from what the company does in the Gulf of Mexico.

[6] This protocol is essentially an updated, Internet-savvy version of Dillman's. (1978) "Total Design Method," which has been a workhorse of survey-based research for decades.

iteratively refined and improved based on this feedback, thereby reducing the potential for measurement error in the survey instrument (Lindner et al., 2001). Respondents were asked at the end of the survey if they would object to being asked a few clarifying questions about their responses. Several said yes, and five follow-up discussions were carried out later to deepen our understanding of the survey results.

3.3.2 Sample

Potential respondents were initially identified from SPE membership records. These individuals had indicated in their SPE profiles that their positions were somehow related to R&D or technology. From this subset of the SPE population, 469 individuals were invited to participate in the survey. Invited participants were typically high-ranking managers who played a significant role with regards to R&D and/or technology deployment in their business unit. Only one potential participant was chosen from each business unit, but several large organizations had respondents from multiple business units in different parts of the world. Candidates were invited to participate via an e-mail sent from the SPE. On clicking on a link in the e-mail, respondents were directed to a web-based survey.

Of the 469 people invited to participate, a total of 199 people completed both the explanatory and dependent variables within the survey, yielding an overall usable response rate of 42.4%. The "extrapolation method" (Armstrong and Overton, 1977) was used to test for nonresponse bias. Respondents were grouped as early (first 20%) or late (last 20%) in the timing of their reply, and responses from the two groups were compared using t-tests (Lindner et al., 2001). No significant differences were found between the two groups' responses, so the results can be reasonably generalized to the target population (Miller and Smith, 1983). Figure 3.1 outlines the breakdown of respondents according to employment by organization type. Table 3.1 shows the location of the worldwide headquarters for the respondents' employing organizations and Table 3.2 shows the geographic location of respondents' business units. Figure 3.2 outlines the breakdown of respondents according to the number of people who were directly employed by their employing organization around the world.

It should be noted that this pool of respondents clearly does not provide a comprehensive picture of the entire industry's R&D activities, and the statistics captured herein do not reflect the totality of the industry's output with regards to innovation and new technologies. Nonetheless, the survey does provide a potentially valuable snapshot of the industry's R&D-related activities around the world. Absolute figures gleaned from this survey—like, for example, the total number of innovations or patents reported by respondents—are of questionable value in and of themselves. But the *relative* measures and comparisons presented here do point to some interesting trends.

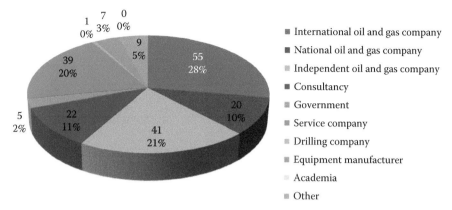

FIGURE 3.1
Breakdown of respondents by type of organization.

TABLE 3.1

Breakdown of Respondents by Country Where Employing Organization's Global Headquarters Is Located

Country[a]	Number of Respondents	Percentage %
Australia	4	2.0
Austria	3	1.5
Canada	23	11.6
China	2	1.0
Denmark	6	3.0
India	6	3.0
Italy	3	1.5
Malaysia	2	1.0
The Netherlands	23	11.6
Nigeria	4	2.0
Norway	8	4.0
Oman	4	2.0
Pakistan	3	1.5
Switzerland	3	1.5
United Arab Emirates	4	2.0
United Kingdom	18	9.0
United States	71	35.7
Other	12	6.0
Total	199	100.0

[a] Countries included in "Other" category only had one respondent in them.

TABLE 3.2

Breakdown of Respondents by Location of Their Business Unit

Country[a]	Number of Respondents	Percentage %
Australia	7	3.5
Austria	2	1.0
Brunei	3	1.5
Canada	26	13.1
Denmark	3	1.5
France	2	1.0
India	7	3.5
Indonesia	2	1.0
Malaysia	8	4.0
The Netherlands	10	5.0
Nigeria	4	2.0
Norway	6	3.0
Oman	7	3.5
Pakistan	3	1.5
Qatar	2	1.0
United Arab Emirates	3	1.5
United Kingdom	18	9.0
United States	74	37.2
Other	12	6.0
Total	199	100.0

[a] Countries included in "Other" category only had one respondent in them.

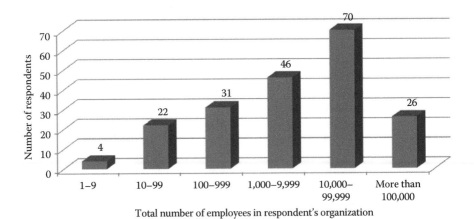

FIGURE 3.2
Size of respondents employing worldwide organizations.

3.3.3 Measures

3.3.3.1 Explanatory Variables

Much of the survey focused on the sources of knowledge that organizations rely upon throughout their R&D-related processes. What sources of information, data, and knowledge are most important as they develop new technologies? This part of the survey was modeled after the Eurostat Community Innovation Survey (CIS) of innovation, which has been used in over 60 academic articles for measuring the knowledge inputs that go into innovation-related activities (Laursen and Salter, 2005). As shown in Table 3.3, the framework consists of 16 potential sources of knowledge. Respondents were asked to identify the degree to which they had used each of the sources throughout the past 3 years, ranging from "not used" to "high use." In addition to the knowledge sources contained within the Eurostat framework, five more independent variables were added:

1. Country in which the world headquarters for the respondent's company or organization resides.
2. Country in which the local headquarters for the respondent's business unit resides.
3. Number of employees in respondent's worldwide organization (this is conceptually similar to the Laursen and Salter (2005) "LOGEMP" variable).
4. Number of employees in respondent's business unit (this is also similar to the Laursen and Salter (2005) "LOGEMP" variable).
5. Type of organization for which the respondent works.

   The type of organization" was gleaned from the SPE. Throughout their own surveys and data-gathering exercises over the years, the SPE has found that the following list contains organizational labels that are exhaustive, mutually exclusive, and that are understood by most people in the industry:

   a. International oil (or gas) company
   b. National oil (or gas) company
   c. Independent oil and/or gas company
   d. Consultancy
   e. Government
   f. Service company
   g. Drilling company
   h. Equipment manufacturer
   i. Academia
   j. Other

3.3.3.2 Dependent Variables

I used five proxies aimed at reflecting various types of innovative performance by business unit:

1. Following Chesbrough (2003b), the number of patents awarded in the last 3 years for which respondent's business unit played a leading role.

2. The number of new technologies deployed within the last 3 years for which respondent's business unit played a leading role. Focusing on the rate of introduction of new products (Hagedoorn and Cloodt, 2003), this variable was used as an alternative measure of R&D output because patent statistics are frequently derided in the literature as being unreliable (Archibugi, 1992). A "deployed technology" was defined in the survey as an innovation that successfully went through field trials and is ready for use in revenue-generating activities.

3. The number of radical innovations deployed within the past 3 years for which the respondent's business unit played a leading role. Using the definition put forward by Leifer et al. (2000), a "radical innovation" was defined in the survey as a new technology that fulfilled at least one of these three criteria:

    a. Delivered an entirely new set of performance features to the marketplace that simply were not available before.

    b. Brought about an improvement in existing performance features of five times or greater.

    c. Delivered a significant (30% or greater) reduction in cost.

4. The number of innovations deployed within the past 3 years that were "new to the world" and for which the respondent's business unit played a leading role (this is conceptually similar to the Laursen and Salter (2005) "INNWORLD" variable).

5. Characterization of nature of innovations created by respondent's business unit throughout the past 3 years. Options included:

    a. Majority were product or component innovations

    b. Majority were process innovations

    c. An almost even mix of product and process innovations

    d. Not applicable—didn't create any innovations

3.3.3.3 Control Variables

I added six control variables to the survey so that I could assess the role of other potential factors and environmental influences that might affect the results.

1. Did the global organization come into existence after 2008 (and should therefore be considered a start up)? (Yes/No). This control variable is conceptually similar to the Laursen and Salter (2005) "STARTUP" variable.

2. Did the business unit come into existence after 2008? (Yes/No). This control variable is also similar to the Laursen and Salter (2005) "STARTUP" variable.

3. Following the Laursen and Salter (2005) "GEOMARKET" variable, what is the characterization of the largest market for respondent's business unit. Answers were limited to:

   a. Local

   b. Regional

   c. National

   d. International

4. Much like for the Laursen and Salter (2005) "COLLAB" variable, has the respondent's worldwide organization been involved in any kind of collaboration arrangements pertaining to innovation-related activities—like, for example, a consortium, industry discussion group, or formal R&D partnership—within the past 3 years? (Yes/No)

5. Also as in the Laursen and Salter (2005) "COLLAB," has the respondent's business unit been involved in any kind of collaboration arrangements pertaining to innovation-related activities—like, for example, a consortium, industry discussion group, or formal R&D partnership—within the past 3 years? (Yes/No)

6. In how many countries does the respondent's organization have offices and employees other than where the world headquarters are located?

   a. 0

   b. 1–5

   c. 5–10

   d. 11–50

   e. 51–100

   f. More than 100.

3.4 Results and Discussion

Based on the 16-item framework from the Eurostat CIS survey, Table 3.3 shows which sources of information, data, and knowledge were most important throughout the respondents' R&D and innovation-related activities.

TABLE 3.3

Sources of Information and Knowledge for Innovation-Related Activities across All Respondents

Type	Knowledge Source	Number of Responses	Not Used	Low Use	Medium Use	High Use
			Percentages			
Market	Suppliers of equipment, materials, components, or software	143	6.2	23.1	33.6	37.1
	Clients or customers	141	19.2	16.3	33.3	31.2
	Competitors	142	19.7	45.8	28.2	6.3
	Consultants	144	18.8	45.8	23.6	11.8
	Commercial laboratories/R & D enterprises	141	24.8	37.6	22.7	14.9
Institutional	Universities or other higher education institutes	145	17.2	39.3	26.9	16.6
	Government research organizations	143	37.1	38.5	18.2	6.3
	Other public sector, for example, business links, government offices	142	40.8	40.1	13.4	5.6
	Private research institutes	141	41.8	36.9	16.3	5.0
Other	Professional conferences, meetings	142	3.5	23.2	43.7	29.6
	Trade associations	142	31.1	38.7	22.5	7.7
	Technical/trade press, computer databases	141	15.6	31.9	36.9	15.6
	Fairs, exhibitions	141	15.6	36.2	38.3	9.9
Specialized	Technical standards	142	13.4	31.0	36.6	19.0
	Health and safety standards and regulations	142	16.1	31.0	26.1	26.8
	Environmental standards and regulations	141	14.8	29.8	28.4	27.0

As noted earlier, respondents were asked to identify the degree to which they used each of the sources throughout the past 3 years, ranging from "not used" to "high use." The data show that the largest sources of the industry's knowledge and inputs for innovation-related activities are suppliers and clients. Professional conferences, health and safety standards, and environmental standards are also considered to be very important. By contrast, the industry places very little emphasis on government research organizations, universities, or public sector organizations where R&D inputs and knowledge are concerned.

Of the 16 potential knowledge sources, five were selected for more in-depth analysis because more than 25% of the total respondents indicated that they relied on these particular knowledge sources as "high use" inputs. Table 3.4 answers Research Question 1 by showing how different types of organization rely differently on these top five knowledge sources.

Of particular interest is the fact that IOCs relied on their suppliers more than any other knowledge source, but service companies relied very little on their suppliers. The tables are turned on "clients or customers," however. IOCs did not consider this to be a particularly valuable knowledge source, but service companies did. This result suggests that IOCs' innovation-related activities are more guided by suppliers' activities than by feedback from their customers. Table 3.4 also shows that the industry's consultants seem to rely quite heavily on professional conferences and meetings as knowledge sources. NOCs and government respondents put noticeably more emphasis on health and safety standards and environmental standards as knowledge sources than did other types of organizations.

In answering Research Question 2a, I examined the numbers of patents reported by the respondents. As shown in Figure 3.3, service companies generated about 80% of the patents reported in the survey. This statistic is even more impressive when you consider that slightly less than 20% of the respondents worked for service companies. By contrast, relatively few patents were reported by independent oil and gas companies, effectively signaling that the developing of proprietary technologies was not a strategic priority for these firms. Also, significantly fewer patents were reported per responding business unit from NOCs than from IOCs.

To answer Research Question 2b, respondents were also asked about the number of technologies deployed within the past 3 years for which their business unit played a leading role. Perhaps unsurprisingly, Figure 3.4 shows that the types of organizations deploying the most innovations also tended to file the most patents.

But an interesting trend emerges when the underlying numbers behind Figures 3.3 and 3.4 are examined together: the relative number of patents filed per deployed technology varies quite significantly from one type of organization to the next. Table 3.5 shows that service companies filed an average of 8.0 patents per deployed technology; IOCs, by stark contrast, filed only 4.4; and most other types of organization produced even less. While the data point to a clear trend, however, they fail to explain why service companies put so much more emphasis on patenting their innovations than other parts of the industry. Upon reviewing the survey data, an executive from a service company offered this explanation:

> "IP (that is, intellectual property) is typically used to defend a space in the marketplace for future business or defend products and service evolution in the businesses we are in. If you start to plot out "competitive

TABLE 3.4

Percentage of Respondents Indicating "High Use" for Various Knowledge Sources Throughout Innovation-Related Activities in Past 3 Years

Type of Organization	Total Responses to This Question	Suppliers of Equipment, Materials, Components, or Software %	Clients or Customers %	Professional Conferences or Meetings %	Health and Safety Standards and Regulations %	Environmental Standards and Regulations %
International Oil and Gas Company	31	51.6	25.8	25.8	29.0	29.0
National Oil and Gas Company	14	42.9	21.4	28.6	42.9	42.9
Independent Oil and Gas Company	32	43.8	9.4	15.6	21.9	21.9
Consultancy	15	40.0	40.0	53.3	20.0	13.3
Government	5	20.0	60.0	40.0	60.0	60.0
Service Company	30	20.0	43.3	30.0	30.0	33.3
Equipment Manufacturer	7	28.6	42.9	14.3	0.0	0.0

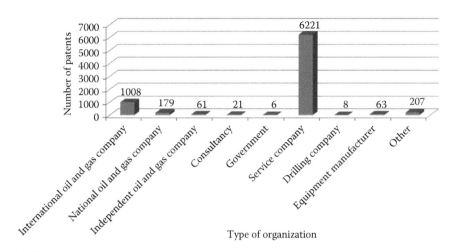

FIGURE 3.3
Number of patents awarded in the 3 years for which respondent's business unit played a leading role.

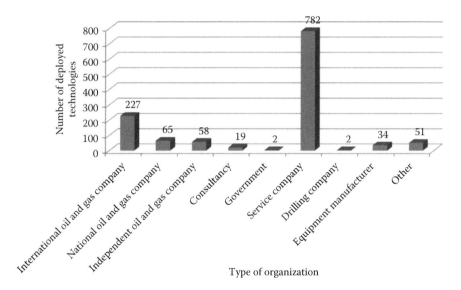

FIGURE 3.4
Number of deployed technologies in the past 3 years for which respondent's business unit played a leading role.

threat vectors," the service side—especially the integrated service side—have the most degrees of competition (from direct competition, niche service players, tech start-ups, academia and customers)… As such, we typically have IP strategies that build protective layers around core ideas to make it more difficult for competitors to "design around."

TABLE 3.5
"Patent Intensity" of Innovation: Average Number of Patents Received per Deployed Technology by Organization Type

Type of Organization	Number of Deployed Technologies in Past 3 Years for Which Respondent's Business Unit Played a Leading Role	Number of Patents Awarded in Past 3 Years for Which Respondent's Business Unit Played a Leading Role	Patents/ Deployed Technologies
International Oil and Gas Company	227	1008	4.4
National Oil and Gas Company	65	179	2.8
Independent Oil and Gas Company	58	61	1.1
Consultancy	19	21	1.1
Government	2	6	3.0
Service Company	782	6221	8.0
Drilling Company	2	8	4.0
Equipment Manufacturer	34	63	1.9
Other	51	207	4.1
Total	1240	7774	6.3

This particular finding is important insofar as it highlights the distortions that can arise when investigations about the industry's R&D rely principally on patent statistics as a direct proxy for innovative output. Because service companies typically file many more patents per innovation than other types of organizations in the sector, any analysis on R&D activity based only on patent figures is quite likely to overstate the relative contribution of service companies.

Another important aspect of Research Questions 2a and 2b concerns whether or not the size of the respondent's organization impacted its R&D output. Are large organizations better positioned to create new innovations in the upstream oil and gas industry than small ones?

Table 3.6 shows that most deployed technologies (74.8%) and patents (79.2%) from the past 3 years that were reported by respondents came from organizations with between 10,000 and 99,999 employees. This is not a surprising result, however, when one considers that many of the larger service companies fit comfortably within this profile.

To answer Research Question 3, respondents were asked to report the number of radical innovations that their organizations deployed throughout the past 3 years. As shown in Table 3.7, smaller firms with fewer employees contributed relatively more to the creation of the industry's radical innovations than larger firms did. Nearly 15% of the reported radical technologies came from companies with less than 1,000 employees, but these firms were

TABLE 3.6

Innovative Output of Firms versus Size of Respondents' Worldwide Organizations

Number of Employees within Organization	Number of Respondents	Percentage (%) of Respondents	Number of Deployed Technologies in Past 3 Years for Which Respondent's Business Unit Played a Leading Role	Percentage (%) of Total Deployed Technologies in Past 3 Years	Number of Patents Awarded in Past 3 Years for Which Respondent's Business Unit Played a Leading Role	Percentage of Total Reported Patents in Past 3 Years
1–9	4	2.0	2	0.2	7	0.1
10–99	22	11.1	49	4.0	38	0.5
100–999	31	15.6	40	3.2	650	8.4
1,000–9,999	46	23.1	103	8.3	374	4.8
10,000–99,999	70	35.2	928	74.8	6156	79.2
More than 100,000	26	13.1	118	9.5	549	7.1
Total	199	100.0	1240	100.0	7774	100.0

TABLE 3.7

Output of Radical Innovations versus Size of Respondents' Worldwide Organizations

Number of Employees within Organization	Number of Radical Innovations Deployed in Past 3 Years for Which Respondent's Business Unit Played a Leading Role	Percentage (%) of Total Radical Technologies in Past 3 Years
1–9	1	0.3
10–99	29	7.9
100–999	25	6.8
1,000–9,999	39	10.6
10,000–99,999	241	65.7
More than 100,000	32	8.7
Total	367	100.0

responsible for less than 8% of the total number of deployed technologies during that same period.

But the data also suggest that the same large firms—that is, those with 10,000–99,999 employees—that create most of the upstream oil and gas sector's new technologies also seem to be responsible for nearly two-thirds of the radical innovations. In other words, they may indeed have contributed fewer radical innovations on a proportional basis, but their sheer size and the overwhelming volume of new technologies that they provide in the industry mean that large companies contributed most of the industry's radical innovations in absolute terms.

Table 3.8 shows the different emphasis on radical innovations by different types of organization within the industry. There is no shortage of qualitative evidence in the literature (e.g., Daneshy and Donnelly, 2004) suggesting that service companies tend to steer their R&D portfolios towards more incremental technologies that are essentially iterative improvements on existing product lines. The data presented here do not contradict this widely held belief. NOCs and independent oil and gas companies, on the other hand, behave very differently in that they do not create and deploy large numbers of technologies overall, but they consider what they do create to be fairly radical in nature.

In answering Research Question 4 concerning the relative focus on process—versus product-based innovation in the industry, Figure 3.5 shows that many respondents reported that their business units were almost evenly focused on product- and component-based innovations and process-based innovations. Service companies had the highest fraction of respondents (40%) who believed that their business units were more focused on product/component types of innovation.

As explained earlier, the uncommonly global nature of the upstream oil and gas industry makes this sector a particularly interesting backdrop for investigations concerning the spatial dimensions of R&D.

TABLE 3.8

Radicalness of Innovations by Organization Type

Type of Organization	Number of Deployed Technologies in Past 3 Years for Which Respondent's Business Unit Played a Leading Role	Number of Radical Innovations in Past 3 Years for Which Respondent's Business Unit Played a Leading Role	Percentage (%) of Deployed Technologies Considered to Be Radical
International Oil and Gas Company	227	83	36.6
National Oil and Gas Company	65	54	83.1
Independent Oil and Gas Company	58	30	51.7
Consultancy	19	12	63.2
Government	2	1	50.0
Service Company	782	155	19.8
Drilling Company	2	2	100.0
Equipment Manufacturer	34	11	32.4
Other	51	19	37.3
Total	1,240	367	29.6

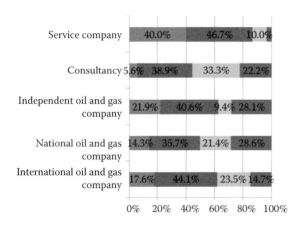

FIGURE 3.5
Relative fraction of respondents focusing on process- and product-based innovations.

Table 3.9 addresses Research Question 5 by showing the geographic origins of the deployed technologies according to the country where the creating organization's headquarters were located. Although only 35.7% of the respondents were from the United States, more than 60% of

TABLE 3.9

Geographic Origin of Headquarters for Innovating Organizations

Country	Number of Respondents Whose Organization's World Headquarters Is in That Country	Percentage (%) of Total Respondents from This Country	Number of Deployed Technologies in Past 3 Years from Organizations Whose Worldwide Headquarters in That Country	Percentage (%) of Deployed Technologies That Came from Organizations Whose Headquarters Are in This Country
Australia	4	2.0	6	0.5
Austria	3	1.5	7	0.6
Canada	23	11.6	37	3.0
China	2	1.0	0	0.0
Denmark	6	3.0	8	0.6
India	6	3.0	4	0.3
Italy	3	1.5	45	3.6
Malaysia	2	1.0	25	2.0
The Netherlands	23	11.6	125	10.1
Nigeria	4	2.0	1	0.1
Norway	8	4.0	25	2.0
Oman	4	2.0	3	0.2
Pakistan	3	1.5	5	0.4
Switzerland	3	1.5	80	6.5
United Arab Emirates	4	2.0	8	0.6
United Kingdom	18	9.0	48	3.9
United States[a]	71	35.7	748	60.3
Other	12	6.0	65	5.2
Total	199	100.0	1240	100.0

[a] Although Schlumberger has principal offices in Houston, Paris, and The Hague, all the respondents from that company pointed to the United States as their world headquarters.

the reported deployed technologies came from companies based in the United States. This leaves little doubt that the United States is still largely the epicenter of innovation and new technologies in the upstream oil and gas sector. The United States dominance is supported further by the data in Table 3.10, which shows a breakdown of the deployed technologies according to the geographic location of each respondent's business unit. Almost certainly owing to Shell's presence, the Netherlands plays an important role in the industry's R&D activities as well.

TABLE 3.10

Geographic Origin of Business Units Where Innovations Were Deployed

Country	Number of Respondents	Percentage (%) of Total Respondents from This Country	Number of Deployed Technologies in Past 3 Years from Respondents Whose Business Units Are in That Country	Percentage (%) of Deployed Technologies That Came from Respondents Whose Business Units Are in This Country
Australia	7	3.5	8	0.6
Austria	2	1.0	5	0.4
Brunei	3	1.5	0	0.0
Canada	26	13.1	39	3.1
Denmark	3	1.5	2	0.2
France	2	1.0	50	4.0
India	7	3.5	4	0.3
Indonesia	2	1.0	1	0.1
Malaysia	8	4.0	33	2.7
The Netherlands	10	5.0	93	7.5
Nigeria	4	2.0	5	0.4
Norway	6	3.0	25	2.0
Oman	7	3.5	14	1.1
Pakistan	3	1.5	5	0.4
Qatar	2	1.0	6	0.5
United Arab Emirates	3	1.5	5	0.4
United Kingdom	18	9.0	60	4.8
United States	74	37.2	527	42.5
Other	12	6.0	358	28.9
Total	199	100.0	1240	100.0

The relative contribution of new technologies from the United Kingdom was also surprising because of the meager number of innovations reported from that country. Despite the significant footprint of the United Kingdom in the upstream oil and gas sector and the considerable body of research about oil and gas technology hubs in that country (Bower and Young, 1995; Crabtree et al., 2000; Cumbers, 2000; Cumbers and Martin, 2001; Cumbers et al., 2003; MacKinnon et al., 2004; Chapman et al., 2004), respondents from organizations based in the United Kingdom reported relatively few deployed innovations. A total of 9% of the survey's respondents were employed by companies domiciled in the United Kingdom, but only 3.9% of the deployed innovations captured in the survey came from those firms. And although Switzerland has comparatively little in the way of domestic oil and gas resources, respondents from organizations based in Switzerland reported a surprisingly large number of deployed technologies related to the upstream oil and gas sector.

3.5 Conclusions and Recommendations

3.5.1 Implications for Theory

By providing a detailed snapshot of how innovation happens in the upstream oil and gas sector, this paper provides a valuable foundation for future investigations and discussions aimed at improving how R&D and technology deployment are managed within the industry. Of the many statistics and trends discussed, one stands out as being particularly helpful from a theoretical point of view: over 63% of the deployed innovations reported in the survey originated in service companies. As noted earlier, technology will clearly play a pivotal role in the success or failure of tomorrow's oil and gas firms, and the shift in the industry's technological center of gravity away from the IOCs towards the service companies may therefore go some way towards explaining the enormous transfer of market power that has occurred within the industry. As recently as 1972, seven IOCs—specifically, Exxon, Texaco, Socal, Gulf, Mobil, BP, and Shell—directly controlled 70% of the world's total oil production (Sampson, 1975, 241), but western IOCs now manage less than 10% (Jaffe and Soligo, 2007). By contrast, the largest of the service companies, Schlumberger, has increased in value fourfold throughout the past decade (Economist, 2012) and, with a market capitalization of U.S.$92 billion, is now bigger than all but the largest of its customers (PFC Energy, 2013). One would expect that the economic rewards in an industry that is increasingly technology driven would increasingly go to the firms that create most of the innovations—and, indeed, that seems to be what is happening. In this way, the evidence presented in this paper usefully adds to theoretical discussions in the literature about this dramatic transfer of market power within one of the world's largest industries.

3.5.2 Implications for Industry and Policy

One other interesting finding from the survey was that neither universities nor government-led research organizations were valuable sources of new information and knowledge in the industry's R&D initiatives. Towards unlocking their local oil gas reserves, several countries' government agencies and publicly funded universities—including high-profile institutions like the U.S. Department of Energy, the United Kingdom's Natural Environment Research Council, and the Norwegian Petroleum Directorate—currently spend many millions of dollars every year on R&D programs focusing on a wide variety of topics like offshore drilling and enhanced oil recovery. The evidence presented here draws into question the effectiveness of these types of investment strategies within this domain, however. These publicly funded bodies would therefore do well to find out why their R&D investments have been met with such a lukewarm reception by the industry before investing more money in these areas.

And the sheer size and uncommonly high profile of the industry cause some of the results presented here to carry with them larger strategic and geopolitical consequences, too. The dominant role of the United States in the industry's overall R&D and technology deployment activities is an important example of this. As discussed earlier, over 60% of the reported technologies deployed within the 3-year timeframe of the survey originated in organizations whose headquarters were in the United States. This finding underlines the efficacy of American-led sanctions of oil-rich countries like Iran (e.g., Amuzegar, 1997; Jacobson, 2008) since it shows how difficult it would be for these countries to operate without any kind of American technology or expertise. Sanctions led by the United States tend to cause economic hardship and distress for practically any nation on the receiving end of this tactic (Selden, 1999), but the evidence presented here strongly hints that these policies are considerably more potent against nations whose economies heavily depend on oil and gas.

Also, beyond merely improving the future performance of technology development and deployment in the upstream oil and gas industry, the data presented here are also relevant to the topic of CO_2 mitigation and climate change. Carbon capture and sequestration (CCS) has been explored at length in the literature as an extremely promising strategy for reducing the amount of CO_2 emitted into the atmosphere (Anderson and Newell, 2004; Holloway, 2005; Gibbins and Chalmers, 2008). The oil and gas sector has considerable experience and know-how in many of the technical disciplines that will likely contribute to technical improvements in CCS, such as reservoir engineering and modeling, downhole measurement, and well engineering. It therefore follows that any gains that can be achieved with regards to how the upstream oil and gas sector manages innovation could also potentially translate to advances in the CCS domain. In this way, an improved understanding of how new technologies come about in the upstream oil and gas industry could be parlayed into better strategies for reducing the amount of CO_2 emitted into the atmosphere.

3.5.3 Recommendations for Future Research

Despite the potentially valuable insights that the data provide, however, this survey was clearly not without limitations. One unavoidable consequence of identifying prospective survey participants from SPE membership records is that it created a significant coverage bias within the sample. Coverage biases arise when the list or frame from which the sample was drawn fails to contain all subjects within the target population (Lindner et al., 2001). Using dues-paying members of the SPE to sample the population of R&D managers and executives within the upstream oil and gas industry clearly does introduce this kind of experimental error. Thus, even though I have no doubt that the number and quality of the respondents who participated in this survey were profoundly improved because of the SPE's participation in the project, I must concede that this strategy essentially created a different kind of methodological weakness. Moreover, despite the dozens of petroleum engineering departments and thousands of academics working in this domain at universities around the world, the pool of survey respondents did not include any academics working in this research space. Future investigations in this area should attempt to overcome these methodological shortcomings.

Finally, as diagnosed earlier, the survey data reveal that neither universities nor government-led research organizations were valuable sources of new information and knowledge in the industry's R&D initiatives, but this investigation fails to explain why this is the case. Yin (1994) points out that qualitative research methods are "the preferred strategy when "how" or "why" questions are being posed" (p. 1). Thus, one potentially fruitful line of questioning for the future would be to approach this phenomenon with a more qualitative methodology to explain why this widespread perception exists within the industry.

Acknowledgments

A practitioner-focused version of this paper, entitled "How Does Innovation Happen in the Upstream Oil & Gas Industry? Insights from a Global Survey" (paper no. SPE-166084-MS), was presented at the SPE' Annual Technical Conference & Exhibition in New Orleans, USA, from September 30 to October 2, 2013. The highlights of this survey were also discussed in an article entitled "Who Drives E&P Innovation?" in the December 2012 issue of the SPE's *Journal of Petroleum Technology*, an oil and gas trade magazine. Special thanks to the SPE's John Donnelly and Jason Davis for their assistance throughout this project. They are the "invisible co-authors" of this document. I also wish to thank Robin Mills and two anonymous reviewers from this journal for comments made on earlier versions of this paper.

References

Acha, V. (2002). Framing the Past and Future: The Development and Deployment of Technological Capabilities by the Oil Majors in the Upstream Petroleum Industry. PhD dissertation, Science and Technology Policy Research Unit (SPRU). Brighton, UK: University of Sussex.

Acha, V., and Cusmano, L. (2005). Governance and coordination of distributed innovation processes: Patterns of R&D cooperation in the upstream petroleum industry. *Economics of Innovation and New Technology* 14(1–2):1–21.

Acha, V., and Finch, J. (2003). Paths to deepwater in the international petroleum industry. Paper presented at the DRUID Summer Conference, June 12–14, 2003. Copenhagen, Denmark: Theme B--Knowledge Transfer Within and Between Firms.

Afuah, A. (1998). *Innovation Management: Strategies, Implementation, and Profits.* New York: Oxford University Press.

Amuzegar, J. (1997). Iran's economy and the U.S. sanctions. *The Middle East Journal* 51(2):185–199.

Anderson, R. (2000). Technical innovation: An E&P business perspective. *The Leading Edge* 19(6):632–635.

Anderson, S., and Newell, R. (2004). Prospects for carbon capture and storage technologies. *Annual Review of Environment and Resources* 29:109–142.

Archibugi, D. (1992). Patenting as an indicator of technological innovation: A review. *Science and Public Policy* 19(6):357–368.

Armstrong, J. S., and Overton, T. S. (1977). Estimating nonresponse bias in mail surveys. *Journal of Marketing Research* 14:396–402.

Artigas, G., Ramadan, Z. J., El-Fouty, I., Bahgat, A., and Degrange, J.-M. (2012). Sourceless formation evaluation reduces HSE risks. *E&P* 85(12):84–86.

Barlow, J. (2000). Innovation and learning in complex offshore construction projects. *Research Policy* 29(7):973–989.

Bastian, B. L. (2009). Technological innovation strategy in natural resource based industries. PhD dissertation submitted to the, École Polytechnique Fédérale de Lausanne, Lausanne, Switzerland.

Bastian, B. L., and Tucci, C. (2010). An empirical investigation on the effects of political risk on technology strategies of firms. *Paper presented at the DRUID Summer Conference.* London, UK: Imperial College Business School, June 16–18, 2010.

Bohi, D. R. (1998). Changing productivity in U.S. petroleum exploration and development. Discussion Paper no. 98–38. Washington, DC: Resources for the Future.

Bower, D. J., and Young, A. (1995). Influences on technology strategy in the UK oil- and gas-related industry network. *Technology Analysis & Strategic Management* 7(4):407–415.

Bozdogan, K., Deyst, J., Hoult, D., and Lucas, M. (1998). Architectural innovation in product development through early supplier integration. *R&D Management* 28(3):163–173.

British Petroleum. (2012). *BP Statistical Review of World Energy,* June 2012. Retrieved from bp.com/statisticalreview.

Bullis, K. 2009. Petroleum's long good-bye. *Technology Review,* 112(6):60–61.

Burgelman, R. A., Christensen, C. M., and Wheelwright, S. C. 2004. *Strategic Management of Technology and Innovation—International Edition.* 4th ed. Singapore: McGraw-Hill.

Chapman, K., MacKinnon, D., and Cumbers, A. (2004). Adjustment or renewal in regional clusters? A study of diversification amongst SMEs in the aberdeen oil complex. *Transactions of the Institute of British Geographers* 29(3):382–396.

Chazan, G. 2013. Cutting-edge technology plays key role for repsol in hunt for oil. *Financial Times* (May 5). Retrieved on May 6, 2013 from http://www.ft.com/intl/cms/s/0/a20c0066-b420-11e2-b5a5-00144feabdc0.html#axzz2SSDP6umg (accessed May 6, 2013).

Chesbrough, H. W. (2003a). The era of open innovation. *MIT Sloan Management Review* 44:35–41.

Chesbrough, H. W. 2003b. The governance and performance of xerox's technology spin-off companies. *Research Policy* 32(3):403–421.

Chesbrough, H. W. 2003c. *Open Innovation: The New Imperative for Creating and Profiting from Technology.* Cambridge, MA: Harvard Business School Publishing.

Cibin, R., and Grant, R. M. 1996. Restructuring among the world's leading oil companies, 1980–1992. *British Journal of Management* 7(4):283–307.

Coll, S. 2012. *Private Empire: ExxonMobil and American Power.* New York: Penguin Press.

Cook, L. 2007. Six unvarnished truths: China and the global energy challenge. *World Energy* 10(3):74–76.

Crabtree, E. A., Bower, D. J., and Keogh, W. 2000. Manufacturing strategies of small technology-based firms in the UK oil industry. *International Journal of Manufacturing Technology and Management* 1(4/5):455–463.

Crump, J. G. 1997. Strategic alliances fit pattern of industry innovation. *Oil & Gas Journal* 95(13):59–63.

Cumbers, A. 2000. Globalization, local economic development and the branch plant region: The case of the aberdeen oil complex. *Regional Studies* 34(4):371–382.

Cumbers, A., and Martin, S. 2001. Changing relationships between multinational companies and their host regions? A case study of aberdeen and the international oil industry. *Scottish Geographic Journal* 117(1):31–48.

Cumbers, A., MacKinnon, D., and Chapman, K. 2003. Innovation, collaboration, and learning in regional clusters: A study of SMEs in the aberdeen oil complex. *Environment and Planning A* 35(9):1689–1706.

Daneshy, A. 2003a. Dynamics of innovation in the upstream oil and gas industry (Guest Editorial). *Journal of Petroleum Technology* 55(November):16–18.

Daneshy, A. 2003b. Evolution of technology in the upstream oil and gas industry. *Journal of Petroleum Technology* 55(May):14–16.

Daneshy, A., and Donnelly, J. 2004. A JPT roundtable: The funding and uptake of new upstream technology. *Journal of Petroleum Technology* 56(6): 28–30.

Daneshy, A., and Shook, M. 2004. Issues stifle technology adoption. *E&P* (February): 11–12.

Dantas, E., and Bell, M. 2009. Latecomer firms and the emergence and development of knowledge networks: The case of petrobras in Brazil. *Research Policy* 38(5):829–844.

Dantas, E., and Bell, M. 2011. The co-evolution of firm-centered knowledge networks and capabilities in late industrializing countries: The case of petrobras in the offshore oil innovation system in Brazil. *World Development* 39(9):1570–1591.

Dennis, R., Jones, T., and Roodhart, L. 2012. Technology foresight: The evolution of the shell game changer technology futures program. In *Sustaining Innovation: Collaboration Models for a Complex World* (Eds.) S. P. MacGregor and T. Carleton, pp. 153–165). New York: Springer.

Dillman, D. A. 1978. *Mail and Telephone Surveys: The Total Design Method*. New York: John Wiley & Sons.

Dillman, D. A. 2000. *Mail and Internet Surveys: The Tailored Design Method*. (nd ed. New York: John Wiley & Sons.

Donnelly, J. 2006. Q&A: Rick fontova, senior vice president, enventure global technology. *Journal of Petroleum Technology* 58(11):20–23.

Economides, M., and Oligney, R. 2000. *The Color of Oil: The History, the Money, and the Politics of the World's Biggest Business*. Katy, TX: Round Oak Publishing Company.

The Economist. 2012. Oilfield services: The unsung masters of the oil industry. *The Economist* 404:51–52.

Elliott, A. R. 2011. *Cracking Energy's Puzzle Dig Deep: George Mitchell's Fracking Lifted the Industry*. Los Angeles, CA: Investor's Business Daily, November 7, p. A05.

Enos, J. L. 1958. A measure of the rate of technological progress in the petroleum refining industry. *The Journal of Industrial Economics* 6(3):180–197.

Enos, J. L. 1962. Invention and innovation in the petroleum refining industry. In *The Rate and Direction of Inventive Activity: Economic and Social Factors*, (Ed.) National Bureau of Economic Research, pp. 299–322. Princeton, NJPrinceton University Press.

Enos, J. L. 2002. *Technical Progress and Profits: Process Improvements in Petroleum Refining*. New York: Oxford University Press.

Fagerberg, J., Mowery, D. C., and Verspagen, B. (2009). The evolution of norway's national innovation system. *Science and Public Policy* 36(6):431–444.

Feldman, M. P. 2010. *The Geography of Innovation*. Dordrecht, the Netherlands: Kluwer Academic Publishers.

Fifarek, B. J., and Veloso, F. M. 2010. Offshoring and the global geography of innovation. *Journal of Economic Geography* 10(4):559–578.

Fine, C. H. 1998. *Clockspeed: Winning Industry Control in the Age of Temporary Advantage*. Reading, MA: Perseus Books.

Fischer, P. A. 2007. Editorial comment: History lessons. *World Oil* 228(4):11.

Flournoy, A. C. 2011. Three meta-lessons government and industry should learn from the BP deepwater horizon disaster and why they will not. *Boston College Environmental Affairs Law Review* 38:281–303.

Fortune. 2012. Global 500: The world's largest corporations. *Fortune* (Asia-Pacific edition) 166(2):F1–F11.

Frynas, J. G. 2003. Royal Dutch/Shell. *New Political Economy* 8(2):275–285.

Furtado, A. 1997. The french system of innovation in the oil industry some lessons about the role of public policies and sectoral patterns of technological change in innovation networking. *Research Policy* 25(8):1243–1259.

Gibbins, J., and Chalmers, H. 2008. Carbon capture and storage. *Energy Policy* 36(12):4317–4322.

Gilbert, C. 2003. The disruption opportunity. *MIT Sloan Management Review* 44:27–32.

Goldstein, A. 2009. New multinationals from emerging Asia: The case of national oil companies. *Asian Development Review* 26(2):26–56.

Grant, R. M. 2003. Strategic planning in a turbulent environment: Evidence from the oil majors. *Strategic Management Journal* 24(6):491–517.

Grant, R. M., and Cibin, R. 1996. Strategy, structure and market turbulence: The international oil majors, 1970–1991. *Scandinavian Journal of Management* 12(2):165–188.

Hagedoorn, J., and Cloodt, M. 2003. Measuring innovative performance: Is there an advantage in using multiple indicators? *Research Policy* 32(8):1365–1379.

Hansen, M. T., and Birkinshaw, J. 2007. The innovation value chain. *Harvard Business Review* 85(6):121–130.

Hatakenaka, S., Westnes, P., Gjelsvik, M., and Lester, R. K. 2006. From "Black Gold" to "Human Gold": A comparative case study of the transition from a resource-based to a knowledge economy in Stavanger and Aberdeen. Report no. MIT-IPC-06-004. Cambridge, MA: Massachusetts Institute of Technology Industrial Performance Center.

Hatakenaka, S., Westnes, P., Gjelsvik, M., and Lester, R. 2011. The regional dynamics of innovation: A comparative study of oil and gas industry development in stavanger and aberdeen. *International Journal of Innovation and Regional Development* 3(3/4): 305–323.

Helfat, C. E. 1994a. Evolutionary trajectories in petroleum firm R&D. *Management Science* 40(12):1720–1747.

Helfat, C. E. 1994b. Firm-specificity in corporate applied R&D. *Organization Science* 5(2):173–184.

Helfat, C. E. 1997. Know-how and asset complementarity and dynamic capability accumulation: The case of R&D. *Strategic Management Journal* 18(5):339–360.

Henni, A. 2013. The middle east's flourishing technology hubs. *Journal of Petroleum Technology* 65(6):60–63.

Hinton, D. D. 2010. Introduction. *Business History Review* 84(2):195–201.

Hinton, D. D. 2012. The seventeen-year overnight wonder: George mitchell and unlocking the Barnett Shale. *Journal of American History* 99(1):229–235.

Hofmeister, J. 2010. *Why We Hate the Oil Companies*. New York: Palgrave MacMillan.

Holloway, S. 2005. Underground sequestration of carbon dioxide—A viable greenhouse gas mitigation option. *Energy* 30(11/12):2318–2333.

Howarth, S., and Jonker, J. 2007. *Powering the Hydrocarbon Revolution, 1939–1973: A History of Royal Dutch Shell*. Vol. 2. Oxford, UK: Oxford University Press.

Jacobson, M. 2008. Sanctions against iran: A promising struggle. *The Washington Quarterly* 31(3):69–88.

Jaffe, A. M., and Soligo, R. 2007. *The International Oil Companies*. Houston, TX: The James A. Baker III Institute for Public Policy, Rice University.

Kulkarni, P. 2011. Organizing for innovation. *World Oil* 232(3):69–71.

Kumpe, T., and Bolwijn, P. T. 1988. Manufacturing: The new case for vertical integration. *Harvard Business Review* 66 (March–April): 75–81.

Lashinsky, A. 2010. There will be oil. *Fortune* 161(5):86–94.

Laursen, K., and Salter, A. 2005. Open for innovation: The role of openness in explaining innovation performance among U.K. manufacturing firms. *Strategic Management Journal* 27(2):131–150.

Leifer, R., McDermott, C. M., Colarelli O'Connor, G., Peters, L. S., Rice, M., and Veryzer, R. W. 2000. *Radical Innovation: How Mature Companies Can Outsmart Upstarts*. Boston, MA: Harvard Business School Press.

Leonard-Barton, D. 1995. *Wellsprings of Knowledge: Building and Sustaining the Sources of Innovation*. Boston, MA: Harvard Business School Press.

Lindner, J. R., Murphy, T. H., and Briers, G. E. 2001. Handling nonresponse in social science research. *Journal of Agricultural Education* 42(4):43–53.

Longwell, H. J. 2002. The future of the oil and gas industry: Past approaches, new challenges. *World Energy* 5(3):100–104.

Lord, R. 2007. Technological breakthroughs advanced upstream E&P's evolution. *Journal of Petroleum Technology* 59(10):111–116.

MacKinnon, D., Chapman, K., and Cumbers, A. 2004. Networking, trust and embeddedness amongst SMEs in the aberdeen oil complex. *Entrepreneurship & Regional Development* 16(2):87–106.

Managi, S., Opaluch, J. J., Jin, D., and Grigalunas, T. A. 2004. Technological change and depletion in offshore oil & gas. *Journal of Environmental Economics and Management* 47(2):388–409.

Managi, S., Opaluch, J. J., Jin, D., and Grigalunas, T. A. 2005a. Environmental regulations and technological change in the offshore oil and gas industry. *Land Economics* 81(2):303–319.

Managi, S., Opaluch, J. J., Jin, D., and Grigalunas, T. A. 2005b. Technological change and petroleum exploration in the Gulf of Mexico. *Energy Policy* 33(5):619–632.

Manders, A. J. C., and Brenner, Y. S. 1995. "Make or Buy": The potential subversion of corporate strategy—The case of philips. *International Journal of Social Economics* 22(4):4–11.

Martin, J.-M. 1996. Energy technologies: Systemic aspects, technological trajectories, and institutional frameworks. *Technological Forecasting and Social Change* 53(1):81–95.

McDermott, C. M. 1999. Managing radical product development in large manufacturing firms: A longitudinal study. *Journal of Operations Management* 17(6):631–644.

Miller, L. E., and Smith, K. L. 1983. Handling nonresponse issues. *Journal of Extension* 21(5):45–50.

Mirvis, P. H. 2000. Transformation at shell: Commerce and citizenship. *Business and Society Review* 105(1):63–84.

Mitchell, J., Marcel, V., and Mitchell, B. 2012. *What Next for the Oil and Gas Industry?* London, UK: Chatham House.

Moncada-Paternò-Castello, P., Ciupagea, C., Smith, K., Tübke, A., and Tubbs, M. 2010. Does Europe perform too little corporate R&D? A comparison of EU and Non-EU corporate R&D performance. *Research Policy* 39(4):523–536.

National Petroleum Council. 2007. Facing the hard truths about energy: A comprehensive view to 2030 of global oil and natural gas. Washington, DC: National Petroleum Council.

Ollinger, M. 1994. The limits of growth of the multidivisional firm: A case study of the U.S. oil industry from 1930–1990. *Strategic Management Journal* 15(7):503–520.

Parshall, J. 2011. Shell: Leadership built on innovation and technology. *Journal of Petroleum Technology* 63(1):32–38.

Paté-Cornell, M. E. 1993. Learning from the piper alpha accident: A postmortem analysis of technical and organizational factors. *Risk Analysis* 13(2):215–232.

Paul, D. L. 2007. Technology to meet the challenge of future energy supplies. *Journal of Petroleum Technology* 59(10):153–155.

Perrons, R. K. 2013. Assessing the damage caused by *Deepwater Horizon*: Not just another *Exxon Valdez*. *Marine Pollution Bulletin* 71(1/2):20–22.

Perrons, R. K., and Watts, L. 2008. The selfish technology gene (Guest Editorial). *Journal of Petroleum Technology* 60(4):20–22.

PFC Energy. 2013. *PFC Energy 50*. Washington, DC: PFC Energy.

Plater, Z. J. B. 2011. The *Exxon Valdez* resurfaces in the Gulf of Mexico... and the hazards of "Megasystem centripetal di-polarity". *Boston College Environmental Affairs Law Review* 38:391–416.

Podsakoff, P. M., MacKenzie, S. B., Lee, J.-Y., and Podsakoff, N. P. 2003. Common method biases in behavioral research: A critical review of the literature and recommended remedies. *Journal of Applied Psychology* 88(5):879–903.

Priest, T. 2007. *The Offshore Imperative: Shell Oil's Search for Petroleum in Post-War America*. College Station, TX: Texas A&M University Press.

Quinn, J. B., and Hilmer, F. G. 1994. Strategic outsourcing. *Sloan Management Review* 35(Summer):43–55.

Ramírez, R., Roodhart, L., and Manders, W. 2011. How shell's domains link innovation and strategy. *Long Range Planning* 44(4):250–270.

Rao, V., and Rodriguez, R. 2005. Accelerating technology acceptance: Hypotheses and remedies for risk-averse behavior in technology acceptance. *Paper presented at the Society of Petroleum Engineers (SPE) Annual Technical Conference and Exhibition*, Dallas, TX.

Rassenfoss, S. 2013. Seismic unwired: Going wireless in difficult terrain. *Journal of Petroleum Technology* 65(3):57–59.

Rigby, D., and Zook, C. 2002. Open-market innovation. *Harvard Business Review* 80 (October):5–12.

Sampson, A. 1975. *The Seven Sisters: The Great Oil Companies and the World They Shaped*. New York: Bantam Books.

Schilling, M. A. 2010. *Strategic Management of Technological Innovation*. 3rd ed. New York: McGraw-Hill.

Selden, Z. 1999. *Economic Sanctions as Instruments of American Foreign Policy*. Westport, CT: Praeger Publishers.

Shah, C. M., Zegveld, M. A., and Roodhart, L. 2008. Designing ventures that work. *Research-Technology Management* 51(2):17–25.

Sharma, A. K. 2005. *Technology Collaboration Between Competitive Firms in the Upstream Oil Industry: The Effect of Operational Control on Collaborative Behavior*. Cleveland, OH: Executive Doctor of Management Program, Case Western Reserve University.

Silvestre, B. D. S., and Dalcol, P. R. T. 2009. Geographical proximity and innovation: Evidences from the campos basin oil & gas industrial agglomeration--Brazil. *Technovation* 29(8):546–561.

Silvestre, B. D. S., and Dalcol, P. R. T. 2010. Innovation in natural resource-based industrial clusters: A study of the brazilian oil and gas sector. *International Journal of Management* 27(3):713–727.

Sluyterman, K. 2007. *Keeping Competitive in Turbulent Markets, 1973–2007: A History of Royal Dutch Shell* (Vol. 3). Oxford, UK: Oxford University Press.

Sluyterman, K. 2010. Royal dutch shell: Company strategies for dealing with environmental issues. *Business History Review* 84(2):203–226.

Steen, J., Ford, J., Curtin, R., Farrell, B., and Naicker, S. 2013. *Delivering a Step Change in Organisational Productivity: Findings from the Australian Oil & Gas Productivity and Innovation Survey*. Brisbane, Australia: University of Queensland Business School, the University of Queensland Centre for Coal Seam Gas, and Ernst & Young for the Australian Petroleum Production and Exploration Association.

Stuart, T., and Sorenson, O. 2003. The geography of opportunity: Spatial heterogeneity in founding rates and the performance of biotechnology firms. *Research Policy* 32(2):229–253.

Thuriaux-Alemán, B., Salisbury, S., and Dutto, P. R. 2010. R&D investment trends and the rise of NOCs. *Journal of Petroleum Technology* 62(10):30–32.

Tidd, J., Bessant, J., and Pavitt, K. 2001. *Managing Innovation: Integrating Technological, Market, and Organizational Change*. 2nd ed. Chichester, West Sussex: John Wiley & Sons.

Tillerson, R. 2006. The high-tech reality of oil. *Newsweek* ("Issues 2007" Special Edition):54–55.

Urstadt, B. 2006. The oil frontier. *Technology Review* 109(3):44–51.

Vaaland, T. I., Soneye, A. S. O., and Owusu, R. A. 2012. Local content and struggling suppliers: A network analysis of nigerian oil and gas industry. *African Journal of Business Management* 6(15):5399–5413.

Verloop, J. 2006. The shell way to innovate. *International Journal of Technology Management* 34(3/4):243–259.

von Tunzelmann, N., and Acha, V. 2006. Innovation in "Low-Tech" industries. In *The Oxford Handbook of Innovation*, J. Fagerberg, D. C. Mowery, and R. R. Nelson (Eds.), 407–432). Oxford, UK: Oxford University Press.

Weijermars, R. 2009. Accelerating the three dimensions of E&P clockspeed—A novel strategy for optimizing utility in the oil & gas industry. *Applied Energy* 86(10):2222–2243.

Wilkins, M. 1975. The oil companies in perspective. *Daedalus* 104(4):159–178.

World Economic Forum. 2008. *Energy Vision Update 2008*. Geneva, Switzerland: World Economic Forum.

Yergin, D. 1991. *The Prize: The Epic Quest for Oil, Money, and Power*. New York: Touchstone.

Yergin, D. 2009. It's still the one. *Foreign Policy* 19(September–October): 89–95.

Yergin, D. 2011. *The Quest: Energy, Security, and the Remaking of the Modern World*. New York: Penguin Press.

Yin, R. K. 1994. *Case Study Research: Design and Methods*. 2nd ed. Thousand Oaks, CA: Sage Publications.

4

*Vignette: Technological Innovation
in Mexico's Hydrocarbon Sector*

Victor Gerardo Ortiz Gallardo

In your experience, has there been an inertia toward technological innovation in the/Mexico's hydrocarbons sector?

Mexico is developing technology innovations for the hydrocarbons sector to a limited extent. The Mexican Petroleum Institute (IMP) is a very active organization in this matter. Many of its technologies are, in fact, already being used in the industry, particularly by PEMEX (Mexico's state-owned oil company).

However, it worth pointing out that the rate of diffusion of innovations to the industry is limited by several situations. First, the dominance of PEMEX dictates that most technologies have been developed specifically to solve their specific operational issues. PEMEX, for example, has been IMP's sole customer for years, but this has also largely applied to other service providers such as Halliburton or Schlumberger. Second, IMP is not active at licensing technologies to third parties, with these innovations usually being embedded into technical services that offer to the industry. Third, exploration and production activities in Mexico have been stymied by current depressed economic conditions in the country. Finally, this is the status quo until around 2019–2020, when the effects of Mexico's energy reform are likely to start impacting. However, as a consequence of these reforms, the landscape is gradually changing, with new companies ramping up to starting operations, and organizations like IMP looking for different ways to exploit and tie into innovations associated with this context.

Have other barriers to entry to developing and incorporating innovation in the sector also been significant?

Yes. The broader, aforementioned, lack of competition and business culture focused on exploiting new technologies does not allow for the creation of new service firms specialized on specific issues or solutions to the industry with an international perspective.

What have been the most significant recent achievements in terms of innovation in the hydrocarbons industry?

Speaking solely on the behalf of IMP, it has developed high value-added solutions to the industry based on proprietary technologies and know-how. Some of the products and services launched in the last 5 years include:

- Multifunctional foaming agents for additional oil recovery
- Antiscales chemicals
- Multifunctional foaming agents for gas wells having liquid-loading problems
- Flow Pattern Enhancer System (MPFV®) in condensate production control and reservoir optimization
- Surface electromagnetic inspection (TIEMS®): a nondestructive pipeline-integrity technology
- System for leak detection and illegal connections
- Chemicals inhibitors for corrosion control
- Antifouling chemicals
- Gel systems to control water invasion
- Real time hydraulic fracturing monitoring system

What further shifts in international, domestic or industry-led policies would have the most beneficial impacts in encouraging adoption of innovation in the petrochemical sector?

As indicated by other contributors to this volume, policies enforcing environmental sustainability are arguably the most influential drivers to adopt new technologies in the petrochemical sector. Such policies may ask industries to produce and use degradable materials, environmentally friendly materials and substances, recycled materials, as well as to consume less hydrocarbons as "feedstock."

One of the main justifications given for legislation opening up Mexico's oil fields to greater competition in 2015 was to benefit from "technology transfers" that might boost the country's lagging record on sector innovation: are there are early indications that this decision has been vindicated?

So far, unfortunately, there is insufficient evidence that can be used to answer this question. However, there are some signs that the energy reform is contributing toward closing technological gaps that exist between Mexico and the global average, particularly in exploiting deep-water resources. Experienced private companies have won licenses to explore and exploit the country's substantial deep water oil reserves. Under the previous legislation, PEMEX may had taken several years to exploit these resources on its own. The bid process took place only very recently, so successful companies are still finalizing the contractual process. Exploration and production activities are likely to begin shortly after. From this point onwards, we will see whether the speculated technology transfer process is an effective one in the case of deep-water oil production.

5

Vignette: Innovation in Australia's Extractive Industries—Current Context and Future Requirements

Miranda Taylor

What are the particular circumstances relating to the Australian extractive industries that make a focus on and clear strategy for innovative thinking and practice so pertinent?

Australia, after 25 years of sustained economic growth, strongly underpinned by the energy resources sector, is experiencing a serious decline in productivity while adapting to accelerating technological change and disruption. The energy resources sector is concluding a massive investment phase, while simultaneously facing its own perfect storm of disruptive challenges, including low commodity prices, increasing community opposition (with moratoriums and bans on unconventional onshore gas developments leading to a temporary shortage of gas supply into the East Coast market), global commitments to sources of decarbonised energy, and digital disruption.

How does National Energy Resources Australia (NERA)'s work and the organization's Sector Competitiveness Plan (SCP) tie into facilitating innovative practice in Australia's energy resources sector?

NERA's role is to maximize value to the Australian economy by having a globally competitive energy resources sector that is sustainable, innovative, and diverse. Our Sector Competitiveness Plan provides a 10-year strategic road map for what is needed to achieve this, identifying the trends and priorities for knowledge acquisition, collaborative action, and innovation. It also lays out the leadership and culture and the business and operating models required by the energy resources sector to both facilitate the commercialization and rapid adoption of "disruptive technologies" and to secure a globally competitive and sustainable future. These models are underpinned by an emphasis on collaboration and innovation, combined with a shift to a customer-centric rather than commodity-centric focus.

The background material to the SCP notes that Australia has been relatively slow in adopting innovation: what factors have been behind this trend?

While the energy resources sector has a long and proud history of pioneering, discovery, research, problem-solving and invention, particularly in exploration, over the past 20 years the increasing capital intensity and high-risk nature of complex projects has seen the rise of large "majos" and a lower ability to easily uptake risky innovation.

These complex large projects often do, in fact, involve proportionately significant risks and innovation at the onset. However, the standard practice during the much longer operational phase has largely relied on incremental change, with the occasional adoption of a major technological innovation. While this approach provides a relatively low-risk, predictable environment, it is only capable of maintaining the status quo since these changes are also adopted by peers and competitors alike.

Yet faced with multiple challenges and the significant disruption and opportunity offered by digital technology, the industry is transforming. The question now, particularly in Australia but also elsewhere, is how fast can it transform in an age of abundant energy resources and increasingly intense competition (notwithstanding the temporary gas supply shortages experienced by the East Coast of Australia and high-energy prices).

What do you see as being the most prominent future requirements for innovation within Australia's natural resource industries in order to fulfil NERA's 10-year road map?

In a world of low commodity prices and increasing energy alternatives, the energy resources industry today needs to optimize efficiency and the value it can realize across the value chain for bringing every unit or molecule of energy to market. They need to be agile and responsive to digitally integrated and rapidly changing markets. Instead of operating as relatively rigid pipeline businesses where energy product is explored, developed, and produced to market, the industry needs to remodel itself to be more akin to technology companies based on open-sourced digital platforms, with integrated and networked "eco-systems" and high levels of customer focus. This will enable the industry to capitalize on the massive efficiency and value creation opportunities offered by explosive growth of technologies such as advanced manufacturing and automation, 3D printing, data collection by drones, the industrial internet of things, along with cloud, mobile, and embedded computing.

All this means the supply chain will be disrupted. The very technology that is being used to disrupt markets (think Uber and Airbnb) as well as the growth of "digital twins", "living labs," and simulators that can substantially mitigate the risks of adopting innovations, are also creating the opportunity for the industry to harvest a far wider range of external knowledge (e.g, beyond internal solutions and a few traditional supply chain innovators and research partners) and for entrepreneurs and innovators to connect with the global industry.

Are there any promising innovations currently underway within the country's Extractive Industries (EI) sector that might pave the way for the "transformational changes" outlined by NERA as being a necessity in ensuring the sector's competitiveness in years to come?

Yes. Encouragingly, there are, in fact, far too many to do justice to the current levels of innovation within the industry. Many case studies examples can, in fact, be found on NERA's website: https://www.nera.org.au/DataFilter?DataFilter_id=76&Action=View.

To give some prominent illustrations from here:

1. *Shell*: Through their "smart fields" approach, multinational energy major Shell is using advanced analytics to improve the production and productivity of their operations. By combining state of the art sensor technologies with advanced analytics techniques, they have been able to better monitor and optimize the operations of their production facilities, increasing total oil production from existing wells by up to 10% and total gas production by 5%. Since they began using this approach, which interconnects over 5,000 machines around the world, Shell estimates it has saved more than 3.5 million barrels of lost production.

2. *Woodside*: As part of a wider reorientation in company focus toward innovation and technology, Australian multinational Woodside has committed significant resources toward a "Future Lab" tasked with modeling a "Plant of the Future." This includes research clusters like plant designs produced with intelligent, automated design software to minimize the total piping required; reducing cost, faster lead-times, and better quality construction with robotic fabrication; producing methods for improved production and simplified plant designs using self-learning controllers; lower inventory costs, better availability and lighter, smaller plants by printing spares on demand using 3D printing technology; and using wireless advanced sensors to inform predictive analytics for more throughput.

3. *Subcon*: Subcon is a marine asset stabilization company. Established in 2011, the company has built an impressive, global resume of

projects by providing rapid installation of innovative products in the offshore and nearshore environments. Combining existing methods with technological innovation has been a key element to their success. For example, by designing and perfecting the necessary logistics behind providing 80 kg sandbags that could be easily handled by Remotely Operated Vehicles for underwater sandbagging, replacing the use of 20 kg bags placed by divers (a method used for 40 years), installation time was reduced by 75%, offering significant savings to clients paying U.S.$150,000 per day for a boat.

Preparing for the inevitable downturn in oil and gas made us think about diversifying, Subcon also analyzed their core competencies (hydrodynamics, marine concreting, and scour mitigation) and extended their subsea oil and gas capability to reef construction, coastal stabilization, and port infrastructure. They won a tender to construct an artificial reef by first learning what problems their potential clients were looking to solve, and then developing their own market leading design. Then they were asked to implement a fiber design to stabilize a pipeline in New Caledonia, on the basis it had been proven in the fishing reefs, and from there, oil and gas companies started to adopt it. As a result of their innovative and proactive practices, Subcon has grown from a start-up sharing a warehouse and a forklift with two other businesses, into an operation employing 40 staff and 30 contractors with their own offices in Perth and warehouses in four countries.

6

Vignette: The Need for Innovation in Mining and Potential Areas for Adopting New Technologies

John McGagh

CONTENTS

6.1 Innovation in Mining: What Makes "Now the Time"

There is an industry-wide consensus that mining is getting harder. The sector, in higher income economies at least, is faced with maturing mines and associated grade decreases and haul distance increase. Despite mineral demand being at an all time historical high, and projected to continue growing into the future driven by developing nation urbanization, ore body replacement rates are falling, new mine development times are increasing, and fundamental cost drivers are rising. These factors combine to deliver industry-wide, adverse, multifactor productivity outcomes and poses severe challenges to the ongoing viability of operations.

  The preceding 20 years, however, have been a time of sustained production expansion across the wider mining industry. Intense capital build programs delivered increases in product output at a time of extraordinary demand growth for minerals, driven mainly by China and other "emerging" nations. But recently growth in demand has significantly slowed and the industry must now manage against structurally lower commodity prices for the immediate and foreseeable future. Therefore, if the industry is to respond to increasingly challenging operational and market conditions, there is a pressing need to refocus efforts towards enhancing productivity and working the installed capacity in a more efficient, effective, and innovative manner.

Traditional strategies are struggling to fundamentally reverse the productivity decline. It did not help that, even in the relative "boom" experienced in recent years, capital was invested predominantly in "known" technologies that could deliver immediate output with little evidence of new or innovative technologies making their way into front-line production. Yet how exactly can or should the mining community react in a way that might boost returns at a time of depressed commodity prices? More of the same is not likely to provide the answer. Arguably, innovation across all facets of the industry (people, processes, technologies, etc.) will be required.

The apparent inertia of the mining sector toward innovation and technology adoption can surely be partially explained by a sustained period of buoyant demand and profit margins shielding it against the need to engage in the type of "arms race" that has resulted in an era of unparalleled innovation across many sectors of the economy: for example, healthcare, consumer electronics, telecommunications, social networking, and computing. The mining sector's current context, however, suggests that the need for step-change innovative thinking has never been greater. The industry is larger than ever, product demand is high, and growth has slowed. Management focus has swung from capital expansion to internal productivity, and there is the adopt and adapt pre-existing and emerging solutions from other industries. Sector leadership and superior returns await the players that successfully exploit these opportunities.

6.2 Promising Areas for Innovation and Technological Improvements

6.2.1 Mining in the Age of Big Data

Research indicates that the mining industry is extraordinarily rich in operational, real-time, sensor-derived "big data"; however, we see little evidence that this opportunity is being leveraged systematically across the wider mining industry. Other industries (oil and gas, aerospace, logistics etc.) have demonstrated that new "profit pools" can be created by exploiting "big data". Liberating new value from the significant amounts of previously unused or underutilized data provides a new opportunity for the mining industry. As is highlighted and analyzed in more detail within this edited collection, value from this opportunity will come from many areas, including within some of the following examples,

Remote monitoring, modeling, and optimizing. It is possible for experts remote from operations to receive near-real time (~0.5 second delayed) operational data, allowing suitably informed and virtually instantaneous interventions and process improvements. The collected information can also be utilized to undertake sophisticated real-time process modeling that can be shared across multiple location-independent sites. Adopting this approach means that interventions to capture value can be more easily made, process efficiency will be improved, and variance and waste minimized. Increasingly scarce expert human and intellectual capital can also be leveraged and shared. Currently, a very small number of large industry players are early adopters of this opportunity. One potential innovation required to unlock this value will be in the form of new service offerings through either suppliers or industry consortia.

Future probabilities and smart algorithms. Smart algorithms predicting future events derived from "big data" are very common. For example, every time you fly on a modern airliner you are sitting in a complex machine watched over by such algorithms. Patterns and trends embedded within mining company data can be analyzed by contemporary machine learning and mathematical techniques and converted into smart algorithms that selectively sample real-time data and predict (with a known probability) future operational events. Libraries of these smart algorithms have the potential to become extremely valuable, not only operationally, through streamlining a variety of processes, but commercially if they can be utilized by other extractive companies or related industries. Effectively predicting future outcomes will be valuable in several areas:

1. *Pit optimization*: Probability based real-time in-pit optimization of machine movements will further maximize efficiency and minimize cost.

2. *Smart maintenance*: Equipment will be scheduled for maintenance on the basis that smart algorithms can effectively predict a specific component probability of failure rather than traditional system-time or system-condition. This will deliver significantly lower maintenance costs and a higher availability of the asset.

3. *Non-bulk supply chain*: Advanced understanding of ore characteristics derived from fusing ore body model information with blast hole intelligent drill data in combination with

online sampling of ore will provide orders of magnitude more information around the critical mine/mill interface. Leveraging this new information will lead to the elimination of variance (lost value) across the mine-to-mill supply chain through better, predictive rather than reactive, processing of changing ore types.

4. *Bulk supply chain*: A new understanding of product character-istics derived from fusing ore body model information with blast hole "intelligent" drill data combined with on line sam-pling will lead to improved product recoveries and allow for real-time product blend modeling. This delivers more precise supply chain scheduling, thus, minimizing handling and improving product quality along the mine-to-port supply chain.

6.2.2 Benefits of Autonomy for Mass Haulage and Wider Productivity

Increases in the size of haulage equipment have delivered a significant pro-ductivity benefit. For example, in the 1960s, 100 ton trucks were the norm and now 340/360 ton trucks are common. But, the size trend has slowed or stopped; the Ultra class trucks of +320 tons were first introduced in 1995 (Komatsu 930E) with other manufacturers following. In the longer term, the banking on productivity improvements driven by equipment size is, natu-rally, unsustainable.

Autonomous Haulage Systems, which saw the first industrial deployment from 2008 to 2012, potentially offer the next opportunity for step-change pro-ductivity improvement in mass material haulage. Autonomous Haulage is currently only in production on a handful of sites. These operators claim 15%–18% productivity improvements over manned alternates. Advances in computing power, network bandwidth, software development, machine sen-sors, and competition among the vendor community will serve to lower costs and improve performance, thus, over time opening access to the technology across the wider mining industry.

Autonomy will take hold outside of the haulage apparatus. Autonomy means that machines do not have to stop to cater for a human requirement—they run until they either require fuel or maintenance; they run within their design specification; they do not damage themselves; and they perform the task exactly as instructed. A range of autonomous solutions (loading, doz-ing, grading, people movement etc.) will find a value-accretive home in the mining industry. The price point for such technology will also come down and the capability will improve as vendor competition increases and as technology transfers from other sectors (civilian vehicles, military, aero-space, etc.).

6.3 New Process Developments

Another area of the sector which could benefit hugely from innovation is fundamental concentrator process design. Recent developments or improvements have been extremely limited. For example, froth-flotation, discovered and adopted in the early 1900s, was arguably the last significant breakthrough in the mineral concentration process. Component machinery has been dramatically scaled up, but the core concentrating processes remain the same. With concentrators demanding a large capital investment, mineral concentration remains a capital intensive and energy wasteful process. In fact, experts estimate 4%–6% total energy efficiency of prevailing concentration systems.

Clearly, then, improvements in any part of concentration process systems would represent a gain in productivity and there is significant scope for this to be attained. Industry participants are collaborating in this area, but the research will require significant investment and possibly new risk-sharing models. Selective dry preconcentration of the mineral in combination with novel comminution solutions are highly prospective and could produce substantial step-change improvements. However, the research programs needed to turn these types of ideas into commercial outcomes will be substantial and will require a cross industry, long-term, and collaborative approach. This is an innovation challenge for the mining industry.

Section II

Advances in Mining, Oil, and Gas Technologies

7

Digital Mining: Past, Present, and Future

Jonathon C. Ralston, Craig A.R. James, and David W. Hainsworth

CONTENTS

7.1 Introduction

The ongoing need to deliver improved safety, productivity, and environmental benefits in the mining industry presents an open challenge as well
as a powerful incentive to develop new and improved solutions. Critical to
the success of this agenda is the ability to identify new approaches that can
be applied to improve high-value processes in the mining value chain. This
chapter explores the evolution and impact of digital methodologies and technologies through the lens of a 50-year journey in mining automation innovation in underground longwall coal operations. The outcomes gained through
past and present developments provide critical insights and lessons to help
understand the value of emerging digital technologies for future mining.

7.1.1 Motivation and Context

The Commonwealth Scientific and Industrial Research Organisation (CSIRO) is
Australia's national science organization which focuses on the delivery of innovative solutions, domestically and globally, across a wide range of industries [1].
In the mining domain, CSIRO undertakes mission-directed research to promote
transformational change in the Australian mining and resource ecosystems.
The vision is to secure a clean energy future to sustain long-term benefits across
environmental, economic, and societal sectors [2]. The fundamental approach is
to collaborate with industry, research organizations, and government to create
cost-competitive, high-productivity, and low-emission energy technologies.

7.1.2 Industry-Led Drivers

The global coal mining industry is constantly driven by the need to
improve mining productivity, increase personnel safety, and achieve credible environmental stewardship [3]. This drive presents an open challenge
and a powerful incentive to develop innovation technology solutions. A
vital component of this strategy is the development and integration of key
underlying technologies to provide new levels of remote control and automated capability in the mining environment. Strong economic and social
drivers drive this technology development agenda. For example, coal
mining is a major economic contributor and accounts for around 24% of
employment and 27% of total revenue for the Australian mining sector [4].
It also remains the primary means of base-load energy generation as well as
providing an essential component in the manufacturing of steel and related
products. Longwall coal mining accounts for around 90% of Australia's
total underground coal production [5,6]. Improving the performance of
longwall operations is thus a major priority for research and development
(R&D). While key technical contributions have demonstrably influenced
the direction and capability of present-day automation solutions [7],

further innovation is clearly needed, and broader potential may be realized with the introduction of new and enabling digital technologies.

7.2 Digital Mining: Historical Development

This section provides an overview of major historical drivers and developments towards present-day longwall automation technology. While many initiatives did not manage to fully realize their overall aim or scope, collectively these efforts proved to positively influence both mining culture and industry confidence in the use of technology for longwall automation. This historical background provides an ideal context to describe the automation R&D undertaken by CSIRO to improve the safety and productivity of longwall mining.

Since the 1960s there were many attempts worldwide to develop longwall automation systems [3,8–10]. These attempts were largely unsuccessful because sensing methods had not yet been developed to accurately and reliably measure the positions in space of all the principal elements of the longwall. This information is fundamental to automation. Unless the current location of equipment is known it is impossible to command the equipment to travel on a desired path. In the late 1990s, the CSIRO research team realized that if they could measure the position of the longwall shearer continuously in three-dimensions, then the constrained path that the shearer takes in the mining process would then enable the calculation of the positions of all the individual powered supports and, moreover, the track of the mined roof and floor. This realization was a breakthrough in the evolution of mining automation.

7.2.1 The Remotely Operated Longwall Face

The drive to develop remote and automated capabilities for longwall operations can be traced back to the 1950s in the British underground coal industry. Interestingly, the initial driver for automation was not safety or efficiency but rather a strategy to support the rationalization of consistent wage structures for workers in the coal industry. At the time, mining was largely a pick-and-shovel era but mechanization was being progressively introduced. However, this "disruptive" mechanization technology was not systematically adopted across the mining industry which led to wide disparities in task equity and remuneration for mining personnel [10].

Ongoing negotiations involving the United Kingdom Government and the National Coal Board and Unions eventually lead to a National Agreement in 1955 based on the concept of a Remotely Operated Longwall Face (ROLF) [10]. They thought that if they could develop such a system, then task and wage equity would be readily achievable since it would be machines, not men, which dictated mining performance. By 1960, they

FIGURE 7.1
A ROLF console and operator at the maingate of the longwall, Bevercotes Colliery, Nottinghamshire, July 1963.

even discussed the notion of a "manless longwall face" together with the tantalizing prospect of coal production becoming a simple push-button operation. The National Coal Board was charged to undertake the research into the development of a ROLF system. The board conducted pioneering experiments in the early 1960s to develop the ROLF capability. Figure 7.1 shows an operator remotely operating a longwall shearer from the longwall maingate at Bevercotes Colliery in the United Kingdom in 1963. The operator console provided the first real demonstration of non-line-of-sight longwall shearer control, but a lack of any feedback to the operator meant that the control process was effectively open loop and so mining performance was difficult to maintain [11].

After many attempts at technology development, it was clear that ROLF as a system did not have the level of technology maturity or performance necessary for production conditions. The lack of suitable sensing, computation, and remote-control technologies significantly contributed to this outcome. ROLF was subsequently withdrawn from underground use due to limitations in achievable performance [10]. It should be noted that ROLF, as a technology concept, was well ahead of its time and so served to provide a compelling vision of future potential.

7.2.2 A Mine Operating System

After the ROLF initiative, the National Coal Board continued to conduct ongoing R&D to mechanize and automate coal mining processes. In the early 1970s, a new interest emerged to develop a framework that could support standardization, promote reusable software, and increase systems reliability [11] known as the mine operating system (MINOS). The intended scope of MINOS incorporated all mining operations including production, transport, environment, and ancillary equipment operation into a single modular configuration. MINOS aimed to reduce repair and maintenance costs, improve output quality, improve monitoring and information, eliminate human error, and increase safety levels for personnel [12].

The MINOS design incorporated a distributed systems architecture linking many mining subsystems that could be developed and managed as independent modules. It was the introduction of new digital microprocessor capabilities that directly supported the MINOS implementation. Figure 7.2 shows the MINOS system architecture consisting of set of core mining modules. Many experimental installations and pilots were conducted during the 1970s and early 1980s [13].

Like ROLF, MINOS lacked the critical sensing and digital processing capabilities that were essential to achieve effective closed-loop mining control. The concept was disbanded by the mid-1980s [3]. However, MINOS, like ROLF, was in many ways well ahead of its time because it was based on the concept of a distributed, collaborative, digital model, which is remarkably similar to the cloud-based services that are now commonplace today. Importantly, the introduction of MINOS promoted a step-change in the performance and cost-base of research in the industry.

7.2.3 Machine Information Display and Automation System

An important technology development to follow MINOS was the machine information display and automation system (MIDAS) [14]. Developed by the National Coal Board, MIDAS aimed to provide new automated vertical steering capability to fulfil a critical control function required for effective automation. The MIDAS architecture also offered new information-related functionality such as data capture, processing, and communications which

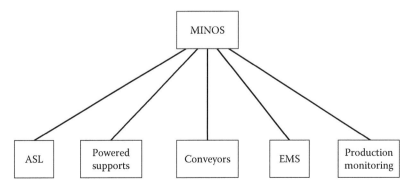

FIGURE 7.2
The MINOS concept consisting of a distributed and modular mining control architecture.

supported new remote monitoring and display options. Many MIDAS-enabled machines were deployed in collieries in the United Kingdom in the late 1970s and early 1980s [11,14]. Like MINOS, MIDAS was technologically advanced for the era, and served to direct attention towards the value of an information-driven future for mining.

7.2.4 The National Air and Space Administration Longwall Requirements Assessment

In the mid-1970s in the United States, the National Air and Space Administration (NASA) was approached to conduct a systematic review of system requirements for longwall automation. Utilizing capabilities and technologies developed through the space program, NASA identified that face alignment and horizon control were two priority candidates for R&D. NASA developed a technical report which introduced several new sensing solutions to the mining industry [15]. It is interesting to note that NASA concluded that it was not feasible to utilize inertial navigation for machine control, a capability later proved possible by CSIRO mining automation researchers [16,17] using high performance inertial technology originally developed for military applications. However, strong industry interest remained to develop new longwall automation capability and the goal was to explore improved sensing, processing, control, and visualisation technologies [11,18].

7.2.5 United States Bureau of Mines Mining Technology

Mining research undertaken by the United States Bureau of Mines (USBM) in the 1980s made many pioneering advances in the development of remote and automated mining capability. At the time, the USBM was without peer in terms of its scope of coal mining R&D. Among other research, USBM conducted many important technology trials to assess the performance of inertial navigation for mining guidance [19,20]. However, the researchers were unable to achieve the required accuracy and they did not develop any practical solutions. The closure of USBM in 1995 and the transfer of some of its activities to the National Institute of Occupational Safety and Health (NIOSH) resulted in drastic changes in the mining research focus, effectively terminating further development in mining guidance technology.

7.2.6 Commonwealth Scientific and Industrial Research Organisation Guidance System Development

In 1994, CSIRO reviewed global research into technologies to help automate and control longwall mining equipment. The systematic review showed that government research institutions in the United States, United Kingdom, Germany, France, and South Africa had mainly undertaken research. At that time there were few useful outcomes from this research. While it is possible that other individuals or organizations may have been attempting to develop technologies to support longwall automation, such outcomes, if any, were not apparent or accessible to industry. It is difficult to determine if or when such technologies might have emerged in the absence of CSIRO's activities [2].

During the 1990s, CSIRO convincingly demonstrated the use of inertial navigation techniques for the guidance of underground equipment in highwall mining [21]. Highwall mining is a remotely controlled mining method which extracts coal from coal seams at the base of an exposed highwall, typically through a series of parallel entries driven to depths greater than 400 m within the seam horizon [22]. The highwall mining scenario represented a very specific mining application where the motion of the mining equipment was largely constrained, meaning that navigation performance could be achieved using a conventional inertial navigation system (INS) and standard processing algorithms. The guidance technology was successfully applied to several different highwall mining scenarios using single auger-, dual auger-, and continuous miner-based highwall mining configurations [21]. This localization technology continues to be applied extensively for highwall mining guidance with major deployments in the United States [23]. Figure 7.3 shows the main display of the Mk4 Highwall Guidance system which provides the operator with an advanced touch screen graphical presentation of the key mining guidance parameters.

The confidence gained through the highwall mining guidance scenario led CSIRO to explore the technology for automating the longwall mining process.

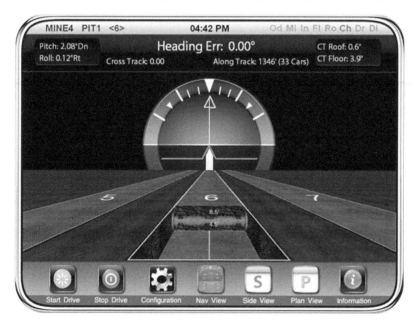

FIGURE 7.3
Main operator interface for the Mk4 highwall guidance system showing real-time information display for operators. (From Applied Mining Technologies, 2014.)

CSIRO's ground-breaking innovation in this area was to recognize and subsequently demonstrate that the position of all the relevant components in the longwall system could be inferred accurately from a 3D measurement of the position of the shearer component. In the late 1990s, CSIRO successfully patented an innovative application of INS for underground environments to accurately position and guide a longwall mining machine [7,24]. A successful short-term trial of the technology on a longwall face soon followed. This outcome proved to be a pivotal moment as it provided industry with the first real indication of the viability of the approach with relatively low technical risk [24,25].

7.2.7 Commonwealth Scientific and Industrial Research Organisation Longwall Automation

Despite the challenges and limitations associated with previous automation attempts, industry interest remained strong in supporting further innovation to realize a technological breakthrough. With the support of the Australian Coal Association Research Program (ACARP), CSIRO undertook a detailed analysis of what aspects of the longwall mining process they could realistically automate, as well as investigating what they could learn from previous attempts at technology development and deployment in this application. From this analysis they identified four key insights [24]:

1. Previous efforts had achieved limited success due to the reliance on single sensor technology and stand-alone systems that failed to sufficiently integrate with the existing longwall control systems and which were unsuitable for use in the harsh longwall mining environment.

2. The required system-level performance and reliability could only be achieved by combining the complimentary advantages of multiple and diverse sensor technologies with inertial navigation as the central enabling technology.

3. The resulting automation system needed to closely integrate with the proprietary control systems provided by each of the longwall equipment manufacturers.

4. The notion of digitally capturing information regarding the state of the mining process was a largely unexplored area which needed new architectures to meaningfully utilize data in a dynamic production context.

A series of highly aligned research projects were systematically undertaken from 2001 to 2007 to address critical gaps in technology capability, communications, Original Equipment Manufacturer (OEM) systems integration, and technology transfer.

The primary outcome of this sustained R&D effort was the development and commercialization of innovative technology that enabled a higher level in automated operation of underground longwall mining equipment. Following the introduction of the Landmark Automation Steering Committee (LASC) automation system, the mine site reported many immediate benefits. These benefits included [6,26]:

- Major increase in operational uptime gained through the removal of manual machine alignment activities.
- Improved safety by providing operators with options to operate away from immediately hazardous and dusty locations.
- Reduced maintenance by reducing machine mechanical stresses (due to over articulation and component misalignment).
- New digital information that systematically provided real-time machine location and state, enabling historical trending and analytics over the lifetime of the longwall panel.

As an indication of the consistency of achievable face straightening, Figure 7.4 shows a series of face profiles as the longwall retreated into the panel. The arrow indicates the level of face straightening achieved, with the most recent shear at the top of the display.

The adoption of the technology has so far been outstanding. Known to the industry as LASC Longwall automation, the technology has been adopted in approximately 70% of operating longwall coal mines in Australia. The generally accepted view of the industry is that a maximum of 80% of the longwall coal mines in the Australia are candidates to use mining machines which incorporate CSIRO longwall mining technology [2,6]. The main beneficiaries of the project include equipment manufacturers who benefit through the sale of the technology, mining companies who save on operating costs and achieve greater productivity, and employees of mining companies who experience improved safer working conditions.

7.2.8 Ongoing Longwall Mining Automation

Because of the positive outcomes associated with the deployment of the LASC automation technology, the industry is committed to continuing the process towards full-face automation [7]. This process will require further high-integrity sensing, processing, control, and visualization capability, and digital technologies are important enablers to meet this important goal.

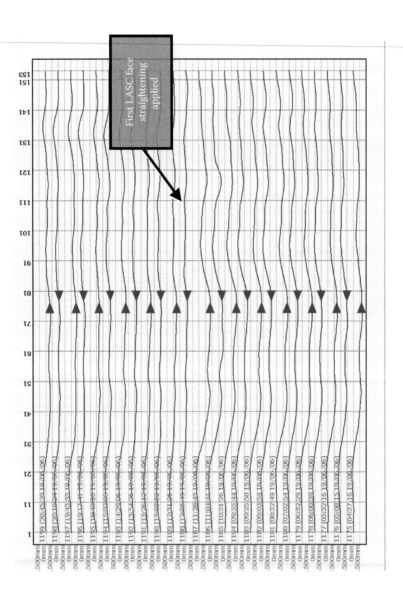

FIGURE 7.4
Historical display of the automated face alignment system over several longwall shear cycles with most recent mining activity at the top. Note the impact on face alignment performance when the LASC system is first enabled.

7.3 Industry Priorities and Technology Opportunities

7.3.1 Ongoing Mining Industry Needs

As previously stated, improving mining productivity, increasing person-nel safety, delivering effective environmental stewardship strategies, and maintaining competitiveness are key ongoing challenges in the mining industry.

For example, in the current Australian coal industry, the expected medium-term continuation of low coal prices will likely support a drive toward higher volumes from some market segments. Industry suggests that underground coal operations that are producing less than 3 month/year are closing, 5 month/year are marginal, and 8 month/year are productive [27]. As a result, most companies are focusing on production projects, or near-production projects, as these offer greater certainty for investors and shorter return-on-investment. The industry is presently largely risk-constrained in nature undertaking short-term (6–12 month) innovation projects to address more immediate efficiency issues, thus making the development and intro-duction of new innovations more difficult to deliver. There are many con-tributing drivers, factors, and perspectives across the mining sector around innovation, but three key target areas consistently emerge as priorities for fundamental improvement:

1. *Improving resource quality* through enhanced geoscience-based sens-ing capabilities to aid exploration, extraction, and processing opera-tions. This priority is an increasing challenge when operating in geologically complex environments and/or areas of uncertain struc-tural configuration.

2. *Increasing productivity* through the development of integrated plan-ning and operations capability, especially through integrated mine planning systems to intelligently inform automation ecosystems, as well as the integration of current information systems across opera-tions and business units.

3. *Social and environmental sustainability* strategy development, includ-ing delivery of low emission technologies, improved on-site energy efficiency utilization, and improved community engagement to ensure social license to operate.

To address these important challenges, the mining industry needs to develop new and enhanced technologies to provide new levels of mining situational awareness, decision support, and trusted autonomy.

7.3.2 Practical Challenges in Deploying Digital Technology in Mining

The rapid development of technologies in sensing, computation, remote operations, automation, and interoperability provide the prospect for similarly rapid deployment of innovative solutions [7]. However, the deployment of digital technologies into mining production zones is often very complex. There are several reasons for this:

1. The primary deliverable from these industries is not virtualized, information based, or service oriented but rather is a physical, tangible resource. This means that any digital intelligence must first be converted to actionable information, which can then be meaningfully executed by the operational plant.

2. The mining operational environment is effectively a "moving factory" with many large and interrelated components, which ultimately contribute to the overall quality and quantity of resource products. The dynamic nature of mining means that there always exists a measure of uncertainty, making digital process optimizations difficult to apply effectively in a generic way.

3. Statutory regulations and the unforgiving mining environment can make the introduction of new electronic devices into production systems challenging and costly, particularly in the case of underground coal mining, where safety and machine survivability and reliability are high priorities.

4. Ensuring end-to-end systems-interoperability across the enterprise is a major challenge because most mining processes consist of a mixture of legacy systems and processes. Increasingly, mine sites are outsourcing the provision of technology and this can further increase fragmentation of the mining information network.

7.3.3 Integrated Automation: A Way Forward

The next generation of improvements in mine productivity is likely to come from innovations in the digital economy, especially in the fields of sensing, automation, and seamless integration of the overall knowledge base across the mining value chain. In particular, interest is strong in physical-digital interaction to provide a fundamental framework from which new and highly efficient approaches can be derived to realize significant improvements across mining operations.

The pursuit of automation in mining often proves to be a catalyst, driving related developments that support new levels of overall digital capability. For example, the introduction of a new sensing device into a mining process requires applicable communications and interoperability functionality to support the utilization of the sensing data. New sensor information leads to a consideration of systems that can manage sensor data acquisition and storage, which in turn prompts the design of model-based systems. In time, the scale and complexity of these data models may evolve to require specific platform management approaches. To frame the role of emerging and enabling technologies in mining, the mining industry needs a generic system automation framework. Automation requires the integration of three key components: sensing, processing, and control [28]. These components interact in a continuous cycle as the underlying resource is dynamically sensed, modified, and transported through the mining process. This mining automation cycle in Figure 7.5 shows the three "sense, think, and act" components—sensing, processing, and control—dynamically operating in concert with the mining process. These three components respectively represent the underlying mechanisms that provide the situational awareness, decision support, and process management functions necessary to achieve a closed-loop mining automation cycle.

Given the importance of automation technology in terms of future mining innovation, the following automation levels have been identified [29] as a basic reference to clarify the maturity of a given automated system:

1. *Local manual control*: Where the operator is removed from the machine but has immediate line-of-sight vision and control of the machine and the mining process through a hard-wired, direct-connection machine control interface.

2. *Remote manual control*: Where the remote operator has immediate line-of-sight of the mining process and achieves control via a portable console interface and a wireless communications link to the equipment.

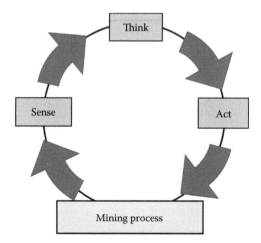

FIGURE 7.5
The mining automation cycle showing three key technology elements necessary to achieve remote situational awareness (sense), perception and decision support (think), and execution and control (act) for a longwall operation.

3. *Tele-operation*: Where the operator does not have line-of-sight to the machine and directly controls the mining process over a wireless link, achieving situational awareness through, for example, on-machine cameras and viewing monitors at the operator's remote location.

4. *Tele-supervision*: Where the operator does not have line-of-sight or immediate direct machine control but initiates and monitors the mining process at a supervisory level with immediate machine control facilitated by assistive automation technologies.

5. *Full automation*: Where the operator has no line-of-sight or immediate direct machine control and most (or all) components of the mining process are functioning autonomously.

In general, the automation capability of a system is strongly influenced by the lowest automation level of any given component in the overall automation cycle. For example, if a technology limitation exists in the sensing component, then this becomes the limit to the overall automation system capability.

7.3.4 Enabling Technologies and New Opportunities

The emergence of digital technologies and associated platforms is being strongly advocated across all sectors. For mining, the opportunity in "digital" lies in the opportunity to provide new and vastly more efficient solutions to help overcome current mining challenges. The primary value offered by digital technologies is an improved ability to connect information at process, system, and ecosystem levels through the introduction of new and enhanced mining capabilities and services.

Several roadmaps relating to the mining domain have been recently developed in response to the opportunities associated with emerging digital technologies [30,31]. The major recurring themes articulated by these roadmaps are:

1. *Data-driven mining decisions*: Next-generation analytics and decision support based on data, in addition to current process, culture, and experience.

2. *Mining automation and robotics*: High-integrity mining equipment and processes operating with trusted autonomy.

3. *Digitally enabled mining workforces*: Distributed, collaborative, remote operations.

4. *Integrated enterprise*: Augmenting platforms and ecosystem.

5. *Smart networked sensing*: Providing mining situational awareness.

These themes are important and so are explored and developed further in terms of technology evolution and casting a vision for the future digital mining ecosystem.

7.3.5 Emerging Technologies for Future Mining

Considerable interest and expectations currently exist around digital technologies such as Internet of Things (IoT), Machine Learning, Analytics, Automation, and Virtual Reality. However, many of these emerging innovations are relatively generic in nature or overstated in terms of potential impact (e.g., see [32]). As a result, many technologies have not yet met the required level of maturity, application specificity, or industry-wide acceptance for deployment wholesale into production mining contexts. Realizing the potential of these emerging digital technologies to benefit mining will require domain expertise to identify and develop solutions around high-value mining use-cases.

Overcoming the lack of comprehensive information regarding mining activities will require the development of well-integrated technologies and equipment to exploit resources efficiently with reduced environmental impact. This information is critical towards delivering safer, more , and more sustainable mining. For automation, four major areas are open for immediate R&D:

1. *Enhanced sensing* solutions across the mining value-chain to provide comprehensive resource characterization, accurate location and state of operating equipment, personnel location, vital signs, and process monitoring infrastructure management. As an example, Figure 7.6 shows a field-based sensing platform developed by the CSIRO that

FIGURE 7.6
An autonomous sensing platform and supervisor monitoring a real-time resource-sensing platform for machine guidance.

generates real-time information regarding the structure of mining resources as a new input to guide optimal mining extraction [33]. This type of sensing solution is important for providing new digital information streams to aid improved mining awareness and enable advance remote process control.

2. *Digital platforms* that process, interpret, and determine optimal mining sequences, incorporating safety, performance, economic, and sustainability measures. As an example of a relevant emerging industry development in digital platform architecture, the Industrial Internet Reference Architecture (IIRA) Committee developed a generic reference framework [34]. Figure 7.7 shows this architecture, key roles, and attributes from a digital information-centric viewpoint.

3. *Autonomous mining Systems* that can intelligently coordinate and execute tasks across multiple independent machines with a high-level of trusted autonomy. A key related research challenge is to provide ways to ensure equipment safety integrity levels (SIL) when operating in mixed (human-automation) or exceptional circumstances during the innovation transition (Longwall Automation Steering Committee 2015).

4. *Human systems interaction* to provide enhanced user awareness and visibility through natural user interfaces. Current technical innovations integrate immersive visualization, wearable technologies, and

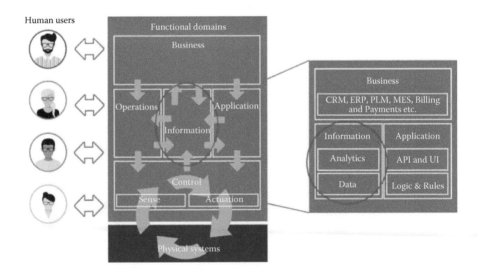

FIGURE 7.7
Information, analytics, and actions in the IIRA Functional Domain framework. Note the importance of integrated roles connecting people, process, data, and resources in the overall system architecture. (From Industrial Internet Consortium, 2017.)

FIGURE 7.8
Field operator utilizing a real-time remote-assistance system based on augmented-reality technology.

dynamic remote feedback mechanisms, using perception-based knowledge to increase worker efficiency and safety. An excellent example of this technology application is shown in Figure 7.8 in which a field technician can receive real-time assistive information from a remotely located expert through a collaborative augmented-reality system [35].

7.4 The Digital Mining Future

The ongoing evolution of digital mining will increasingly be information-driven with the ultimate view of achieving a closed-loop mining enterprise. The business needs for increased visibility and integration across the value chain drive this approach. Ongoing systematic development and introduction of digital technologies into mining are therefore needed to realize this vision [36].

Mining is a dynamic process consisting of a set of complex systems that operate under changing conditions which interact with each other in complex ways. An integrated mining system can be systematically developed by identifying and understanding each mining system and the larger set of process influences. For this to happen, a detailed understanding of the relationship between structure and behavior needs to be characterized to understand how these systems work. However, to provide credible foresight, it is first necessary to transcend the various technologies, practices, and processes used in the industry to deliver a broader mining vision for the future.

7.4.1 An Integrated Digital Mining Ecosystem

Our vision is to approach a fully integrated mining ecosystem by outlining the foundational principles and behaviors needed. This future mining vision comprises seven key themes:

1. Mission-directed Mining: This mining ecosystem is driven by digital information and operated as a mission. It is implemented using high-integrity systems which are enabled by decision-support services. Systems consistently monitor and maximize the utilization of assets to ensure continuous unit operations throughput through disciplined materials-flow. Intelligent process-monitoring and decision-support systems predict, react, and mitigate exceptions, providing automated recommendations and information delivered through natural interface technologies.

2. Integrated Enterprise Ecosystem: An integrated mine management framework guides mining from concept to completion, spanning prefeasibility, exploration, planning, opening, development, production, move, close out, and rehabilitation phases. This model supports operational consistency across business, process, culture, and enterprise drivers to deliver optimal production performance, personnel safety, and environmental governance. This ecosystem operates with the highest level of autonomy possible to meet business requirements.

3. Information Relevancy and Accessibility: A MINOS provides information that is accurate, contextual, and actionable. It is accessible from anywhere: machine, underground, surface, or remotely. This real-time mining information provides clarity on strategic directions and tactical decisions to support accurate, accessible, and actionable information at every level of the process from the chief executive officer to the operator.

4. Seamless Digital Systems Interoperability: The ecosystem utilizes an open mining system information architecture to ensure future-proof systems, seamless autonomy, and rapid ability to expand autonomy to more types of manual systems. Rather than insisting that one universal technology span the entire mining organization, the vision recognizes the intrinsic complexity and diversity of mining to meaningfully integrate a combination of different technology solutions that share a common information interface.

5. Zero-exposure Mining: Operations have zero operational personnel in underground environments. Complete visibility of all relevant system and mining processes will be afforded through natural and intuitive human-system interactions. High-fidelity, telepresence collaborative hubs will support new models of planning and operations.

6. High-integrity Autonomous Mining Systems: All mining processes and equipment operate with a high-level of trusted autonomy to provide continuous and consistent resource recovery. These systems incorporate self-monitoring capability that will automatically self-diagnose and repair, pre-empt failures, and provide graceful degradation on exceptions.

7. Environmentally Coherent: The digital mining ecosystem will have significantly reduced environmental impacts and emissions through optimal asset and resource utilization. Key resources such as water, energy, and related consumables will be efficiently stewarded through end-to-end monitoring. A comprehensive awareness of resource location and material characteristics will directly enhance optimal resource recovery strategies through enhanced selective mining capabilities, leading to vastly reduced generation of waste.

7.4.2 Ecosystem Rollout

The future digital mining ecosystem will require new levels of mining technology R&D. Based on current approaches, digital innovation will likely emerge in the industry through well-targeted projects in ways that are more evolutionary rather than revolutionary. The pace, however, of technology adoption and utilization will increase largely along the automation spectrum from manual control to fully autonomous operation.

The rapid adoption of longwall automation technology by Australian underground coal mines is an excellent indication that mine operators expect productivity gains will ensure return on investment. New and improved technologies are constantly in demand to meet the need for better remote and automated underground mining processes. Immediate short-term development therefore will focus on achieving greater levels of assistive automation to allow the mining process to be more effectively controlled from a remote location.

Support for much of the change will rely on the development of new information systems and supporting technologies that generate, analyze, and utilize this information. These systems are the architectures that intelligently integrate and manage the increased level of information associated with the mining process. In addition to digital technology developments, the adoption of consistent information reference architectures, such as the

IIRC framework, together with effective collaboration and partnerships, are essential to delivering practical benefits and impacts for the mining industry.

7.5 Summary

Digital innovation plays a significant role in the ongoing delivery of safer, more productive, and environmentally sustainable mining operations. A successful application in underground coal mining as a motivating use-case highlighted how sustained technology R&D can deliver significant operational benefits to industry. Highlights of current mining challenges and technology opportunities provided a sense of industry priority to inform ongoing development pathways. We outlined our vision for a future integrated mining ecosystem, together with the foundational behaviors that underpin the implementation of this vision. This future mining ecosystem is driven by digital information, operated as a mission, and has zero-exposure for both people and environment. We need to develop new digital systems to provide new sensing methods, new ways to dynamically assess risk, new decision-support functions, and machines with high levels of trusted-autonomy.

Mining, by its very nature, is a resource and asset-intensive business. The next generation improvements in mine productivity are likely to come from innovations in digital technologies, especially in the fields of automation and integration across the mining value chain. Digital productivity is core to this vision through its capacity to deal with large and complex data sets, extract knowledge from data, build intelligent and automated machines suitable for remote operation, and cohere the overall mining process. Immediate benefits from adopting a digital approach will facilitate a more effective use of human expertise, reduce process variance, and give rise to more informed and faster decision-making.

It will be exciting to see how ongoing developments in communications, sensing, analytics, automation, and augmented reality will accelerate delivery of digital mining to benefit the industry in the future. The call to deliver improved safety, productivity, and environmental outcomes for the mining industry remains as cogent as ever. The next decade will see an unprecedented drive towards a more agile, more dynamic, and more integrated coal mining business. Achieving this outcome will require a searching review of present mining assumptions and practice to identify what is essential and efficacious. An effective review will likely result in strategies that will challenge traditional mining cultures. It will also give rise to new operational models for vertical integration and service provision, convergence of mining operations and business operations, and opportunities for the deployment of innovative mining technologies.

Acknowledgments

The authors would like to thank the Australian Coal Association Research Program (ACARP) for its direct support to facilitate R&D in advancing longwall automation, and the Landmark Automation Steering Committee (LASC) for its invaluable industry direction and support.

References

1. Australia's Commonwealth Scientific and Industrial Research Organisation (CSIRO), https://www.csiro.au/en/About/We-are-CSIRO. Web. Accessed January 1, 2000.
2. CSIRO Renewables and Energy, https://www.csiro.au/en/Research/Energy. Web. Accessed July 21, 2014.
3. Brune, J. F., (Ed.). *Extracting the Science—A Century of Mining Research*, Society for Mining, Metallurgy and Exploration, Denver, CO, 2010, 544 p.
4. World Coal Association. Web. https://www.worldcoal.org. Accessed July 2016.
5. Mitchell, G. W. Longwall mining, Chapter 15, in *Australasian Coal Mining Practice*, 3rd ed., R. J. Kininmonth, E. Y. Baafi (Eds.), Carlton, Victoria, Australasian Institute of Mining and Metallurgy, Monograph series, no. 12, 2009, pp. 340–375.
6. ACIL Allen Consulting, CSIRO's *Impact and Value: An Independent Assessment*, ACIL Allen Consulting, 2014. Accessed online http://www.acilallen.com.au/cms_files/ACILAllen_CSIROAssessment_2014.pdf.
7. Ralston, J. C, Reid, D. C., Dunn, M. T. and Hainsworth, D. W. Longwall automation: Delivering enabling technology to achieve safer and more productive underground mining, *International Journal of Mining Science and Technology (IJMST)*, 25(6): 865–876, November 2015
8. Peng, S. S. *Longwall Mining*, 2nd ed., Society for Mining, Metallurgy, and Exploration, Inc. (SME), 2006, 621p.
9. Johnston, R., McIvor, A., *Miner's Lung: A History of Dust Disease in British Coal Mining*, Ashgate Publishing, Aldershot, UK, 2007, 374p.
10. Searle-Barnes, R. G. *Pay and Productivity Bargaining: A Study of the Effect of National Wage Agreements in the Nottinghamshire Coalfield*. Manchester University Press, Manchester, UK, 1969, 190p.
11. Cross, M. Coal has a future, *New Scientist*, pp. 12–16. Reed Business Information, London England, 1984.
12. Perkin, R. M. G. *MINOS: Systems Reliability and Reusable Software, Case Studies*, National Coal Board, London, UK, 1983.
13. Berzonsky, B. E., Breuil, F., and Yinglin, J. C. Evaluation of computer-based mine monitoring and control at barnes and tucker mine 20, *IEEE Transactions on Industry Applications*, IA–21(5): 1121–1126, September 1985.

14. Mackie, K. Design and development of longwall shearers for overseas application, in *Underground Mining Methods and Technology, Advances in Mining Science and Technology*, pp. 169–188, Vol 1. A.B. Szwilski, M.J. Richards (Eds.). Elsevier Science Publishers B. V., Amsterdam, the Netherlands, 1987.

15. NASA Technical Memorandum, Longwall Shearer Guidance and Control—Final Report, DOE/NASA TM-82466, January 1982.

16. Hainsworth, D. W., Reid, D. C. Mining Machine and Method, Australian Patent PQ7131 and US Patent, May 12, 2000.

17. LASC Longwall Automation Steering Committee. Lascautomation.org. Web. Accessed February 20, 2015.

18. Peng, S. S. Longwall mining in the US: Where do we go from here?, *Mining Engineering* 37:3 (March 1985): 232–234.

19. US Department of the Interior, *Bureau of Mines, Mechanized Longwall Mining. A Review Emphasising Foreign Technology*. Information Circular 8740, Washington, DC 1977.

20. US Department of the Interior, Bureau of Mines, Longwall automation. *A Ground Control Perspective*. Information Circular 9244, Washington, DC 1990.

21. Reid, D. C., Hainsworth, D. W., McPhee, R. J. Lateral guidance of highwall mining machinery using inertial navigation, *Proceedings of the 4th International Symposium on Mine Mechanisation and Automation*, 1997, pp. B6–1–B610 (Brisbane).

22. Shen, B. and Fama, M. Review of highwall mining experience in Australian and a case study, *Australian Geomechanics* 36: 25–32, 2001.

23. Applied Mining Technologies, http://www.appliedminingtech.com. Web. Accessed December 2014.

24. Reid, D. C., Hainsworth, D. W., Ralston, J. C., McPhee, R. J. *Shearer Guidance: A Major Advance in Longwall Mining, in Field and Service Robotics: Recent Advances in Research and Applications*, pp. 469–476, Springer, Berlin, Heidelberg, 2006.

25. Kelly, M., Hainsworth, D. W., Reid, D. C., Lever, P., Gurgenci, H. Longwall automation—a new approach, 3rd International Symposium, Aachen, German June 11–12, 2003. 15p.

26. Reid, D. C., Hainsworth, D. W., Ralston, J. C., McPhee, R. J. *Shearer Guidance: A Major Advance in Longwall Mining, in Field and Service Robotics: Recent Advances in Research and Applications*, 2006.

27. Gibson, G. Towards an integrated roadway development system, *Australian, Longwall Conference 2015*, Hunter Valley, New South Wales, Sydney, October, 2015.

28. Ralston, J. C., Hargrave, C. O., Dunn, M. T. *Longwall Automation: Trends, Challenges and Opportunities in 9th International Symposium on Green Mining*, November 28–30, Wollongong, Australia, 2016.

29. Endsley, M. R., Kaber, D. B. Level of automation effects on performance, situation awareness and workload in a dynamic control task, *Ergonomics* 1999, 42(3): 462–449.

30. CSIRO Mining Equipment, Technology and Services Roadmap. Web. https://www.csiro.au/en/Do-business/Futures/Reports/METS-Roadmap. Accessed May 2017.

31. Digital Transformation Initiative: Mining and Metals Industry Report. World Economic Forum. Web. http://reports.weforum.org/digital-transformation/. Accessed January 2017.

32. Technology Hype Cycle for Emerging Technologies 2016. Web. http://www.gartner.com/. Accessed February 2017.

33. Ralston, J. C., Strange, A. D. 3D robotic imaging of coal seams using ground penetrating radar technology, *16th International Conference on Ground Penetrating Radar (GPR)*, 2016, June 13–16, 2016, Hong Kong, China.
34. Industrial Internet Reference Architecture Web. http://www.iiconsortium.org/IIRA.htm. Accessed March 2017.
35. Guardian Mentor Remote Web. https://research.csiro.au/robotics/gmr/. Accessed March 2017.
36. Rovost, F., and Fawcett, T. *Data Science for Business*. O'Reilly Media, Sebastopol, CA (2013).

8

An Optimized Command System for Full Automation of Digitally Controlled Mine and Gas Train Plants and Sites

Brian J. Evans

CONTENTS

8.1 Introduction

During conventional operations of any type of process system, be it a fluid separator, a cooler, a heater, a motor or generator, a turbine, a conveyor belt, a crusher, or a simple storage tank, any unexpected change in operational conditions can cause a subsequent unexpected change in down-stream operations. This unexpected change, in turn, causes disruption to the smooth flow of the process system, resulting in potential risk of production and associated financial losses. Liquefied Natural Gas (LNG) trains are an extremely illustrative example of the continuous monitoring that takes place at mine sites and related infrastructure to mitigate disruptions in process and financial streams. Equipment on LNG trains are mounted with sensors on or within them, which warns of times when disruptive behavior strikes. However, as is highlighted below, command systems for monitoring and decision making in LNG trains and, indeed, in the mining and oil and gas sector more broadly, are much less advanced and streamlined as they should be.

To apply tomorrow's "disruptive technologies" to optimize the operations of a process stream, total change is needed in direction toward automation and intelligent condition prediction. Toward this end, this chapter discusses a hierarchical system for making an automated command and control system that allows operators to make the optimized process control system decisions, while understanding the operational and financial consequences of those decisions, as well as the risk levels attached to the decisions as a function of a positive economic and lean system performance. In summary, output sensor data would be streamed 24/7 to a control room, which then would automatically apply appropriate checks to ensure the arriving data is valid, and then provide appropriate feedback to suggest to the operator what subsequent actions are optimal. The proposed process stream would also consider important (but often neglected) financial linkages to engineering actions, which can be thought of in terms of Newton's Third Law of Motion; that is, for any positive action by the operator, there will be a financial reaction depending on the strength of the action. Existing control rooms at mine and LNG process sites are typically manned by operators who know the basic equipment and how it functions, but they often do not know the costs of equipment, or how those costs are ever-changing on the international currency exchanges, nor do they consider this important. Here, embracing technological innovation, specifically data analytics, could provide operators with a plethora of technical and economic information to make an optimal decision.

8.2 Existing Control Systems

For many years, the use and application of automation and robotics has seemed something that only happened in the car production and food canning industries, where production line robotics were introduced to increase the speed of assembly while reducing the cost of operations. However, this surge in activity never truly translated into the mining or oil and gas industry. In fact, these industries have been typically slow to respond to peripheral technologies. A broad explanation may be that it is very easy to state that the mining industry works on low marginal profits, while anything to do with the oil and gas industry is always expensive. This breeds a conservative mindset wherein technology adoption is only accepted if it has been demonstrated elsewhere. Therefore, unless there is an immediate benefit, expenditure on automation is a secondary event.

As illustrations of this, while automated haulage systems have been developing at mine sites with the use of automated haul trucks (as shown in Figure 8.1) at BHP Billiton's Jumblebar mine, the use of such technology is certainly not pervasive given that, feasibly, the whole operation from drilling to ore transportation can be fully automated with modern techniques.

From the oil and gas viewpoint, it is rare to have any form of automation, particularly offshore. The most advanced of 400 platforms in the North Sea is

FIGURE 8.1
Autonomous trucks on trial at BHP's Jumblebar mine. (http://www.nrec.ri.cmu.edu/projects/ahs/.)

(a)   (b)

FIGURE 8.2
Shell Shamrock monopod (a) and Woodside Angel condensate platform (b).

Shell's Shamrock gas production monopod (shown in Figure 8.2) which is wind and solar powered. In Australia, the only unmanned offshore platform is the Woodside Angel gas condensate platform, which has temporary accommodation for a maintenance crew for 2 to 3 days every 45-day maintenance cycle.

The advent of "big data" suggests that, where enough sensors are in place, it is entirely realistic that condition monitoring data can be transmitted ashore so that someone in a control room can remotely assess and take active decisions when issues occur. However, it appears there is a lot of technology development and convincing to be done before we see such a setup in action. For example, before any sensor is deployed, its output data must be monitored and believed to be correct before anyone will trust it. Moreover, there is no point in deploying sensors when they themselves may be defective.

8.3 Sensors and Monitors

The rapid development in the sophistication of simple deployable sensors containing computation ability and transmitters is a result of the miniaturization of computer integrated circuit chips, which has been greatly accelerated by the growing popularity of portable electronics like smart phones. An example of a typical sensor which may be deployed in an existing operating plant is the condition monitoring sensor (CMS) which contains three accelerometers, an active strain gauge, and a temperature gauge (Figure 8.3). This sensor transmits data to a remote tablet, where all the components of the CMS are built into the glass-fibre pad (drawn for simplicity in Figure 8.3).

The pad is made of specialized carbon fiber construction and has the instruments embedded within it. The pad does not rust and can be made intrinsically safe where gases are a concern. It has an on-board battery which is charged magnetically and a wifi transmitter for operating in air. If needed to operate under water, the wifi is replaced by an infrared (IR) transmitter, which can be seen underwater by an IR receiver.

Monitoring pads can be built to send their readings to a data collection point, with this transmission then being relayed on to a remote monitoring control room via internet or satellite.[1]

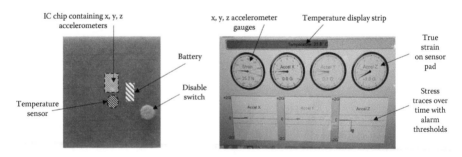

FIGURE 8.3
Example of a condition monitoring sensor (CMS) and tablet display pad.

---

[1] Given the advancement of augmented reality technologies that major corporations are currently developing, in the future it would not be hard to imagine plant maintenance staff having heads-up displays of data on their hard-hat eye guards to receive data directly in front of them.

If a CMS is attached to a tank or other part of a plant, then when plant vibrations or temperature exceed the thresholds set in the receiving mobile tablet, an alarm sounds and draws the operator's attention to the out-of-spec equipment. The next step, all being well, is to have the process control room trained enough to do something automatically with the data and de-risk the management decision-making options for the control room operator.

8.4 Automated Command and Control Systems

Assuming such sensors are mounted on all equipment, they must be cheap enough to make instrumenting financially beneficial. Once this has been achieved, the data coming into a typical control room needs to be handled by operators who are computer savvy. With the great wealth of equipment health data arriving in the control room, the operators no longer need to be fully versed in all of the technologies of plant operation, but they should be there to make the best decision to advise a skeleton maintenance team of the problems and the solutions. Such solutions would be provided by the software in a technical sense as well as procedural and economic sense.

Consider for example a control room which receives data from multiple sources, allowing full operational control of the plant, as well as contact with other areas of the plant for online discussion, as shown in Figure 8.4.

The operator reviews all the ongoing operations of the plant as a function of the plant schematic piping and instrument diagram (P&ID) figures. A failure or a problem deriving from the sensor data would show up as a flashing alarm with an alarm sound. The normal solution now is to call an operator at the site as well as check pressures, flow rates, and so on in preparation for an action with the control mechanisms (e.g., turning on or off valves, reducing pressures, etc.).

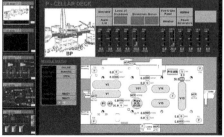

FIGURE 8.4
Example of a remote control room and control system software. (Courtesy of PiSys. www.pisys.co.)

This duty requires the remote operator to have an excellent understanding of the flow process, the mechanics and physics of how equipment operates under different conditions, the consequences of an upstream failure, and, if a wrong action is taken, how it affects the process downstream. Again, the action taken has a potentially severe (Newtonian) financial reaction. This makes plant operators extremely conservative with their decision making, since a poor decision would affect operations, so the operator must discuss with other plant operators at the site to come to a conclusion which can be argued later in a court of law. Both local and remote operators must therefore be at the top of their abilities and knowledge of operating equipment specifications before making a decision and ideally must also possess knowledge of the cost of equipment along with replacement installation timing. Yet this information is often impossible to know, hence, their conservatism in terms of preferring to apply the simplest yet most expensive solution. But with a multitude of sensors and an optimized command and control system, it doesn't have to be like this.

8.5 Addition of an Advisory Menu

A multitude of sensor data can provide the knowledge that components are failing or about to fail. Observation of what is happening downstream can provide further understanding of the impact of failed upstream components. If we can use existing data to make a prediction of a components performance, its expected life, typical failures, its cost to replace in terms of capital cost, along with cost of shipment to site and maintenance crew operations, then a more informed decision which is more optimal than the alternative easy choices of expensive replacement, can be avoided.

For example, consider a typical P&ID diagram as shown below in Figure 8.5. Let's take a hypothetical scenario. A valve V2 fitted with sensors begins to malfunction in some manner, such as developing excessive vibrations or heat when it opens or closes, prompting the sensor to transmit a signal sounding an alarm in the control room. The operator responsible for controlling this valve may have the personal knowledge of it malfunctioning previously, and also may be aware of why this particular valve performs like this. The decision may be that the valve is about to fail and a replacement is needed immediately. If the replacement is not found and replaced soon, its failure will cause multiple issues for production down-stream resulting in the potential loss of millions of dollars. The operator immediately calls stores and the maintenance crew to go to operational stand-by in preparation for isolating the fluid circuit, thereby avoiding shut-down of downstream plant.

However, unfortunately, while the stand-by crew is readied, the operator finds out from the operation's stores that the last valve of this type failed 1 month ago and that replacement valves are on back-order. They won't be

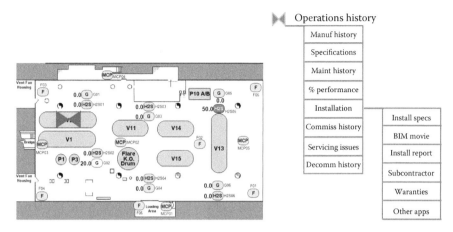

FIGURE 8.5
Control panel alarm of a failing valve, and an example of a control panel response for the operator who touches the valve alarm on the screen.

arriving for at least a week with express delivery to site. The implications of this can be severe and, after exhausting all possibilities of a replacement valve or action (by modifying an existing valve), the decision is made to commence shut-down of the process plant. Once this procedure has begun, it is irreversible and can result in at least 2 weeks lost production. The operator, typically at present, uses personal knowledge and experience to make this decision—that is, to wait until the valve completely fails (accepting a 2 week shut-down and loss of production).

But what if the valve transmits its signal to set off the alarm and instead the operator's software allows commencement of another totally different action. The operator presses the valve alarm and an advisory drop-down menu appears to show the valve's operations history, as detailed in Figure 8.5. The operations history drop-down menu also has submenu applications which provide immediate information on the manufacture history of the valve (where, how, and when it was made), the valve specifications, its maintenance history, its percentage performance (what percentage time it has been fully operational and when it had previously been manually turned off/on), its full installation history, its commissioning history (how it was used and responded before initial commissioning and subsequent shut-down/start-up operations), when it had been serviced over the years of operations, any servicing issues the valve had in the past, and where it had previously been used if at all during any previous decommissioning. All these application menus are linked to the internet and go back to their own individual parts databases stored in the cloud.

Going further into the new system and its possibilities, let us also assume that this operator, who in fact knows little about the valve (as opposed to the walking library of knowledge and experience required of his previous counterpart), would like to now have more information on how this valve was installed 5 years ago. For example, was it a replacement at the time or was this area of the process train a new expansion of the plant? The operator touches the installation application and its drop-down menu provides a new series of data applications. Immediately, installation specifications (detailing peripheral equipment and their connection thread specifications), a built information modeling (BIM) cartoon movie is available if the operator wants to see a cartoon of how the valve is installed or removed, the original installation report, details of the subcontractor who installed the valve (and their present-day information), the valve warranties, and other application buttons are available for further information. By pressing further applications buttons, the operator now has a full range of data to muse over as input before making a decision. Such a procedure can continue but the missing information now is how long the valve may operate successfully until it fails.

Another drop-down menu provides an assessment of life expectancy along with probability statistics data of how long it may take before failure. If the advisory system provides the information that the results of the automated predictive analytics run on the valve data considers that there is an 80% probability that, based on its past performance, the valve will operate for another month before failing again (and a 90% probability it will run for at least 2 weeks before failure), the operator now has the ability and time available to either produce a replacement valve from the company stores or order one if a valve is not on stock.

Now that the pressing mechanical issue of the existing valve has been assessed, the next step is one of economics. Does the operator get a valve ex-stock or maybe there is a recent *oversupply* of valves that suggests it may be more economic to purchase one online now rather than use up stock (after all, this installation may have 7,000 similar valves and this may be popular commodity for replacement)?

8.6 Predictive Analytics: Incorporating Other Data Sets into Decision Making

Continuing our posited example of assessing and replacing the failing valve, what if the available replacement is of dubious origin and not built to the same high-quality specifications? What if the valve is valued in U.S. dollars, and the present exchange rate is poor? Do we dare wait 2 weeks until a new batch is started by the manufacturer for a better price because trends show a better exchange rate, when existing supply may have totally run out?

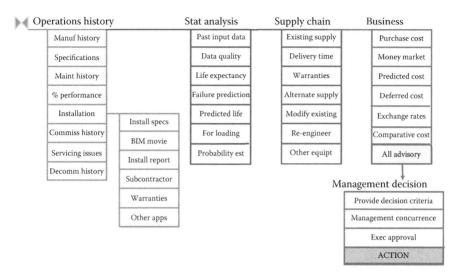

FIGURE 8.6
Other basic data sets needed to arrive at an optimized decision.

Making the most informed decision about such parameters in the above-mentioned case requires drawing upon many other data sets (Figure 8.6). Apart from the statistical analysis data, the supply chain as mentioned above, the economics, finance, and legal position (such as whether partners must be informed of any equipment replacements before they are made) are all components of the advisory menus to ensure that the optimized decisions are made with the associated probabilities of success (or failure). Such knowledge is simply too much for the control room operator who today is aware of little more than how equipment functions and the down-time results of equipment failure. So, we need to introduce the concept of statistical analysis, the supply chain, and business advisory menus, as illustrated in Figure 8.6.

The statistical analysis dropdown advisory system has already been functioning in the background as part of the alarm system; that is, where a component such as a valve changes its operating conditions, its performance has been monitored and predicted with a failure probability statement. However, the operator would now have the full suite available for inspection to review the past data and data quality to ensure the information being displayed is authentic and of adequate resolution for an automated decision. Failure predictions along with life expectancy can now be reviewed and inspected to ensure what is being provided is correct. The operator no long needs to know how the component operates, but should understand the data analytics of the failing component.

Having understood the consequences of component failure and its probability, it is normal to determine if there is a replacement in the stores or not.

Thus, a supply chain inquiry commences in the case when no replacement is at hand. Now the issue becomes whether any purchased replacement can be provided soon (delivery time), if it has a warranty, and, if an alternative valve exists, how it may be re-engineered to replace this component. All this work would be done by a search engine over the internet to provide a database of alternative components and their specification both in material selection and costs.

The money market needs to be checked and predictions it may make, since there is little point in purchasing a component at $40,000 this week, if the exchange rate is predicted to move lower next week making the component a predicted $38,000 if purchased then. So, the business advisory menu automatically provides such data to the operator, who would have a reasonable knowledge of the cost of components and finances involved. Once all this data is reviewed, a final decision would be proposed with a probability statement on the optimized approach to purchase this component. In terms of decision management, all engineering data has been compiled with probability assessments to provide an overall decision with its own probability assessment, as shown in Figure 8.7.

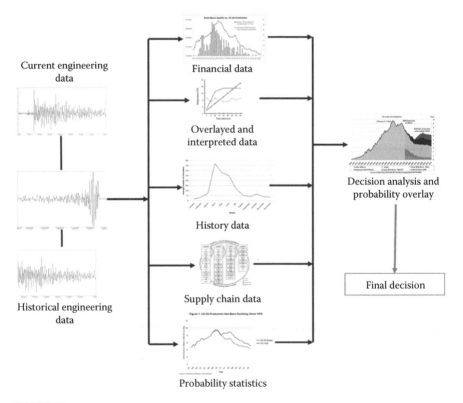

FIGURE 8.7
The decision management system provides input to make an optimized final decision.

The above use of statistical analysis of existing data and linking it with probabilities used to provide all the outstanding data needed for an operator to make the optimum decision will require much greater use of predictive analytics. The operator no longer needs to have an excellent understanding of how basic equipment works upstream and downstream, but is a quite different person who, while having a basic understanding of equipment operation, would have a good grounding in data analysis, supply chain concepts, business, and finance. The operator would be relaxed with interpreting different data sets on the fly using both engineering data and 3D visualization graphics as illustrated in Figure 8.8 to make better decisions. Once the operator is convinced with the decision to be made, it takes a short time to contact the line manager's smart phone or tablet to inform and have concurrence from the manager that the action is approved, which then sets the wheels in motion for the operator to action the final decision.

The immediate power of predictive analytics has now been used along with comprehensive monitoring of plant and internet data to obtain an optimized decision in the shortest possible time frame offering the lowest level of risk. All legal compliance and search efficacy has been instituted to allow documentation in log-form of the decision process for later inspection, if necessary. This system is simply efficient and financially uplifting to a process plant operation.

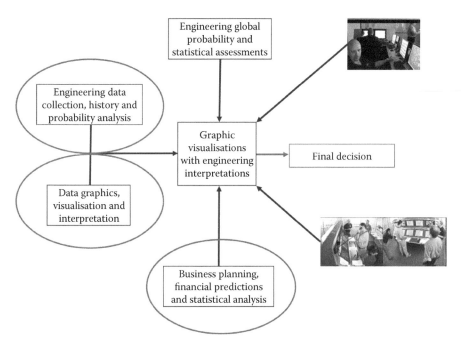

FIGURE 8.8
The data are provided in graphical format to produce a rapid decision scenario.

Furthermore, as explained, the operator is no longer the type of person who must be employed in the process plant work-place for 30 years before being employable in the control room. The operator instead is a computer-savvy engineer who has a knowledge of supply issues and finances.

8.7 Conclusions

There is absolutely no reason that these predictive technologies cannot be implemented today, apart from the need to develop the software applications and hire the engineers who understand them. It is clear that the management decision process of any process plant would be better informed and more agile than at present and any time before, which would require business and financial management to be more computer savvy than they have been in the past. But on this occasion, such people and their decision making are half-way there, because they will have arrived in their management capacity having gone through this sort of engineering scenario in-transit in their earlier careers.

So, it is up to the engineering fraternity to adopt these analytical support systems and follow the lead of the fashion and food industry in relying on them. New accepted innovations are considered to take 30 years before becoming common-place, but the industry cannot wait that long. The science is there and the engineering can be built here and now, so it just requires adoption by a very conservative industry.

9

The Case for "n << all": Why the Big Data Revolution Will Probably Happen Differently in the Mining Sector[1]

Robert K. Perrons and Derek McAuley

CONTENTS

9.1 Introduction

"Big Data" and predictive analytics have received significant attention from the media (e.g., Harford, 2014; Lohr, 2012a, 2012b) and academic literature (e.g., Chen, Chiang, & Storey, 2012; Howe et al., 2008; Lynch, 2008; Perrons & Jensen, 2015) throughout the past few years, and it is likely that these emerging technologies will materially impact the mining sector. In fact, the industry has already made modest inroads into this area (Wilson & Hume, 2013) and, by several measures, the mining sector already has data that could fairly be characterized as "big." Modern seismic data centers like those used by the mining and oil and gas industries can easily contain as much as 20 petabytes of information,[2] which is roughly equivalent to 926 times the size of

---

[1] Reprinted by permission of the publisher, Elsevier, from Robert K. Perrons and Derek McAuley, "The Case for "n << all": Why the Big Data Revolution Will Probably Happen Differently in the Mining Sector," *Resources Policy* 46, no. 2 (2015): 234–238.
[2] This statistic is less impressive when you consider that Walmart, the U.S. retail giant, collects more than 2.5 petabytes of data every hour from customer transactions (McAfee and Brynjolfsson, 2012).

the U.S. Library of Congress (Beckwith, 2011). If this amount of information was copied into books and put on a single continuous bookshelf, it would go around the Earth's equator approximately six times (Beckwith, 2011). And while seismic data sets are notoriously large and cumbersome, many other aspects of the mining industry are also generating significantly more data than they used to (Annavarapu, Kemeny, & Dessureault, 2012; Bascur & Linares, 2006; Dessureault, 2007; Jahedsaravani, Marhaban, & Massinaei, 2014). What is more, there is every reason to believe that this trend toward more digital information will continue to gather pace. Current estimates suggest that the total amount of digital data in the world—including books, images, e-mails, music, and video—is doubling every 2 to 3 years (Lohr, 2012a; Mayer-Schönberger & Cukier, 2013) and it seems unlikely that the mining sector will not be affected by this.

This chapter argues, however, that these technological forces will probably unfold differently in the mining industry than they have in many other sectors. To this end, we begin by offering a brief overview of what Big Data and predictive analytics are, and explain how they are bringing about changes in a broad range of sectors. Then we will discuss the "n = all" approach to data collection and information management—that is, eschewing the sampling of data in favor of using all the data pertaining to a phenomenon—which is frequently advocated by many consultants and technology vendors offering solutions in the Big Data domain (Harford, 2014). Next, based on high-level discussions with other industries and early observations about Big Data's nascent advances into the mining sector, we will make the case for adopting a "n ≪ all" data collection strategy within the mining industry instead of the "N = all" approach.[3] Finally, toward shaping the sector's policies with regards to technology-related investments in this area, we conclude by putting forward a conceptual model for leveraging Big Data tools and analytical techniques that is a more appropriate fit for the mining industry.

9.2 What Is Big Data? and How Can It Deliver Value?

The term "Big Data" is a rather vague label that describes the application of new tools and techniques to digital information on a size and scale well beyond what was possible with traditional approaches (Lohr, 2012b). The idea has grown organically from a broad range of academic disciplines, technology

---

[3] Whereas a single less than symbol ($<$) denotes in mathematics that something is less than something else, the double less than symbol ($\ll$) means that it is much less. Also, the distinction between the upper-case letter N in "N = all" and the lower-case letter n in "n ≪ all" throughout this chapter is intentional. N typically denotes the size of the entire population, while n represents the size of a sample that is smaller than N within that same population (Levin 1987).

providers, and consultants and, as a consequence, a consensus for a single definition has not been reached for this steadily evolving research area. Despite the absence of a precise definition, however, most users of the term broadly agree that it refers to data sets and analytical techniques in applications that are so large and complex that they require advanced data storage, management, analysis, and visualization technologies (Chen et al., 2012).

At its core, the main objective of Big Data and predictive analytics is something to which organizations have been aspiring for a long time: to make better decisions (Regalado, 2014). This objective is achieved by statistically processing large amounts of data—either structured or unstructured—from disparate sources that are often left untapped by conventional business intelligence programs to uncover hidden patterns, unknown correlations, and other types of useful information. These can potentially result in probability-based actionable insights that help human decision-makers to figure out what to do next. These new capabilities have most famously delivered value in the social media sector and retail industries (Harford, 2014; Mayer-Schönberger & Cukier, 2013), but have also led to major breakthroughs in a diverse range of other contexts, including scientific research (Frankel & Reid, 2008; Kluger, 2014; Lynch, 2008), healthcare (Chen et al., 2012), and professional sports (Leahey, 2013). And Big Data also has made significant forays into the heavy equipment sector—an industry that directly supplies the mining community. As Mehta (2013) notes, Caterpillar now uses Big Data in the short-term to anticipate repairs and more efficiently plan component inventories and in the long-term to improve product designs and predict changes in their customers' needs.

By stark contrast, the mining industry has made relatively modest progress in the Big Data domain (e.g., Wilson & Hume, 2013). But these analytical approaches could almost certainly deliver considerable value in that industry. For example, large haul trucks have been a mainstay of the mining industry for a long time (Santos, Porter, & Mayton, 2010) and a significant fraction of the sector's workplace fatalities occur in and around these kinds of large vehicles (Groves, Kecojevic, & Komljenovic, 2007). Large data sets containing geospatial information about the pathways of haul trucks could be compared to synchronized accelerometer data measured from inside each truck to tell road maintenance crews where the most egregious potholes and depressions are appearing (Mednis, Strazdins, Zviedris, Kanonirs, & Selavo, 2011), thereby making it possible to identify potentially hazardous parts of the road before they become more serious.

The Big Data and predictive analytics trend is the convergence of many new ways to extract value from business data. Specifically, these new ways include the following:

- The continued decrease in the cost of technology for capturing, storing, and processing data leading to nearly all business processes becoming digital and generating rich business data.

- The move to make these diverse data sets available in real-time within the business for new and creative uses, together with the increasingly available option of augmenting these data sets with open and public data from other sources.

- The use of computationally intensive analytic techniques to obtain new quantitative insights from previously unexplored combinations of data that, at first, may seem to have no obvious relationship (Mayer-Schönberger & Cukier, 2013). In some circumstances, simply finding such correlations is sufficient to release the value. For many other scenarios, however, these kinds of statistical insights need interpretation by someone who understands the mechanisms behind the phenomena being observed, as correlation is not the same as causality. This is a major issue, for example, in Big Data-based medical research. A team of researchers from Google announced in the high-profile journal *Nature* that, by closely tracking search terms like "flu symptoms" or "pharmacies near me," they were able to predict the spread of influenza through the United States more quickly and accurately than the Centers for Disease Control and Prevention (Harford, 2014). After a few years of providing reliable estimates, however, the model's predictions became much less reliable because "Google did not know—could not begin to know—what linked the search terms with the spread of flu" (Harford, 2014).

To date, many approaches to Big Data and predictive analytics have involved bringing these kinds of digital information together within a single data center, building so called "data lakes," and then processing the data using cluster-computing software such as Hadoop (Shvachko, Kuang, Radia, & Chansler, 2010). For the mining sector, such an environment represents a particularly good vantage point from which to investigate the widest range of possible data interactions, especially since the off-the-shelf software implementing the various analytic techniques has been designed to run efficiently in this type of scenario. Also, with the relevant data in one place, the software is not constrained by arbitrary data partitioning—that is, the process of determining which data subjects, data occurrence groups, and data characteristics are needed at each data site in a geographically distributed system (Wu, Barker, Kim, & Ross, 2013).

9.3 "N = all" and Its Limitations

One of the major tenets of Big Data solutions being prescribed by many consultants and technology vendors is the concept of "N = all"—that is, to collect and analyze all the data potentially pertaining to a phenomenon.

The notion of taking a sampled subset of representative data to identify and measure trends within larger populations has been a cornerstone of many analytical techniques for generations (Cooper & Emory, 1995; Levin, 1987) but as Mayer-Schönberger & Cukier (2013) suggest:

> Sampling is an outgrowth of an era of information-processing constraints, when people were measuring the world but lacked the tools to analyze what they collected… However, the concept of sampling no longer makes as much sense when we can harness large amounts of data. The technical tools for handling data have already changed dramatically, but our methods and mindsets have been slower to adapt.
>
> Yet sampling comes with a cost that has long been acknowledged but shunted aside. It loses detail. In some cases, there is no other way but to sample. In many areas, however, a shift is taking place from collecting some data to gathering as much as possible and if feasible getting everything: $N = all$. (p. 26)[4]
>
> But this perspective fails to take into consideration the uneven costs of gathering data from one scenario or industry to the next. Social media firms such as Facebook or LinkedIn, which are frequently celebrated in the literature as leaders in Big Data and predictive analytics (Mayer-Schönberger & Cukier, 2013), typically enjoy relatively low costs for data acquisition, as much of their information is volunteered for free by their subscribers; the mining sector, by stark contrast, tends to pay more to capture, store, and share digital information from mine sites because of the capital-intensive nature of the sector and the remoteness of the phenomena being measured (Kotey & Rolfe, 2014; Rovig & McArthur, 1998; White, 1998).

The economic costs associated with collecting digital information has historically been a significant factor in the design of data collection strategies and system architectures throughout a broad range of industries (e.g., Benghanem, 2010; van der Veen, Spitzer, Green, & Wild, 2001; Zhu, 2005), and the mining sector is no exception to this. Accordingly, the significantly higher marginal cost of data capture and storage faced by the mining industry creates an environment that is markedly different from those of social media companies and retail-based business models. "N = all" makes much less sense under the conditions of a mine site, where the cost of collecting all the data would clearly be relatively high. Instead, in the next section, we suggest an alternative data collection model for the mining sector that is more suited for the technical and geographical realities of the industry.

---

[4] We argue, however, that the notion of "N = all" is fundamentally flawed since all data is sampled with granularity, and you can consequently never really have all of it. Thus, the "some cases" that Mayer-Schönberger and Cukier (2013) allude to in which there is no alternative to sampling probably includes all cases.

9.4 A Conceptual Model of Big Data and Predictive Analytics for the Mining Industry

A multidisciplinary team consisting of mining engineers and Big Data experts was established to visit the United Kingdom-based offices of four different organizations—specifically, Romaxtech, Rolls Royce, Amantys, and Microsoft—in November 2014 to learn about their respective journeys into the fields of Big Data and high-end data analytics. These organizations were selected because they had already made significant progress with Big Data and advanced analytics and were willing to share their learning from these processes. Moreover, some of them had significant experience in capital-intensive industries (e.g., Rolls Royce), thereby allowing them to shed light on the specific challenges associated with applying Big Data technologies in those kinds of environments. Two important themes emerged from these discussions that are highly relevant to the mining industry.

First, in most cases, less than 1% of the available data contributes to actionable insights. Upon identifying the most statistically important 1% with respect to nature and frequency, these organizations tended to focus their data collection and analysis on only those key variables rather than aggressively capturing a broad range of digital information of questionable utility. In mathematical terms, they had abandoned "N = all" and were instead focusing on "n << all." Many of the same variables that would have been captured with a more widespread "N = all" architecture were still examined—but only in a small number of sampled locations rather than throughout the organization's entire operational domain. This approach made it possible to minimize the costs and disruption associated with applying Big Data types of tools and techniques while still reaping the most benefits. Machine learning and/or statistical analyses were then performed on this smaller subset of data to derive algorithms and discover patterns that the organization could then use to create value in its regular operations. These insights may, in turn, highlight other data that could be collected to further improve the analysis.

Second, even though machine learning plays an important role in the discovery of potential non-obvious trends lurking within their data, most of the visited companies had augmented these automated tools with human supervision. By applying a supervised learning model, they found that they were more able to identify truly useful patterns and relationships that could deliver real value. Other research in Big Data also has taken notice of this trend. Davenport, Barth, and Bean (2012) identify a new breed of professionals known as data scientists who have emerged in a broad range of industries throughout the marketplace. These individuals typically have a skill set that includes data analytics, information technologies, and mathematics while also having reasonably deep domain expertise in their respective industries. Most of the visited companies considered this domain expertise to be an important ingredient in the success of their Big Data and predictive analytics programs.

These perspectives usefully build upon the myriad Big Data and predictive analytics solutions currently being promoted by consultants and vendors to deliver a slightly different model that, as shown in Figure 9.1, essentially consists of four major elements:

1. Gather enough representative data into a "research sandpit"—that is, a stand-alone digital investigation area that is not connected to the operational aspects of the organization—to gain the new statistical insights through machine learning and advanced mathematics, and to define the data analysis to be performed in production.

2. Use domain knowledge to interpret these statistical insights into new models and algorithms with comprehensible mechanisms, for example, physical, chemical, geological, and so on.

3. Scale up the insights from the sandpit by applying the Big Data analysis on a larger scale in the broader production system.

4. Monitor the performance of this highly distributed system against that predicted by the original analysis, and re-analyze if it is divergent.

In this model, predictive analytics are carried out near the data source, and only the actionable insights are relayed to improve the "smartness"

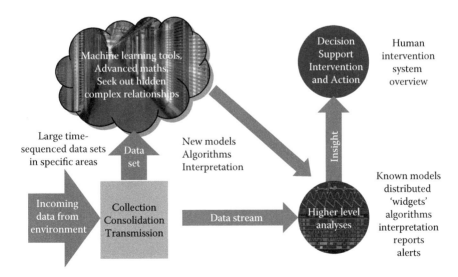

FIGURE 9.1
A conceptual model of Big Data and predictive analytics for the mining industry.

of the overall system. This approach has a lower entry and operating cost structure than more comprehensive "N = all" types of architectures and seems well suited to the mining sector. What is more, with its lower infrastructure requirements, this architecture is easily scalable and relatively "future-proof" since it does not require an inordinate amount of capital investment in equipment that will be expensive to replace as the technology evolves.

9.5 Conclusions and Future Research

While we understand the allure of the "N = all" approach to data collection being promoted by many Big Data consultants and technology vendors in the marketplace, we believe that this logic breaks down in the realities of the mining industry. Instead, we submit that the "n << all" data collection strategy outlined in this paper probably makes more sense for the mining sector and other industries that face relatively higher marginal costs of data acquisition.

We recognize, however, that this argument largely hinges on pronounced differences in these costs from one industry to the next. Sensor technologies of almost every kind—such as accelerometers, global positioning sensors, radio frequency identification (RFID) tags, and temperature sensors—have been declining in price for years (Crassidis, 2006; Dowling, 2004), and there is every reason to believe that this trend will continue unabated into the foreseeable future (Khorram, Koch, van der Wiele, & Nelson, 2012). We therefore acknowledge that the logic underpinning many of our arguments here will become less potent in the years ahead. The marginal cost of acquiring data in remote locations like mine sites will probably fall quite dramatically, and the rise in the performance and reliability of sensors will be equally impressive. Thus, future investigations in Big Data and predictive analytics in the mining industry could examine if the case for different kinds of data collection strategies and system architectures still makes sense at that time.

Acknowledgments

Special thanks to John McGagh, Grant Wellwood, and the rest of the Rio Tinto SM@RT team for their insightful comments and ideas throughout the preparation of this chapter. We also thank the two anonymous reviewers from Resources Policy who provided extremely insightful ideas for improving this work.

References

Annavarapu, S., Kemeny, J., & Dessureault, S. (2012). Joint spacing distributions from oriented core data. *International Journal of Rock Mechanics & Mining Sciences, 52*, 40–45.

Bascur, O. A., & Linares, R. (2006). Grade recovery optimization using data unification and real time gross error detection. *Minerals Engineering, 19*(6), 696–702.

Beckwith, R. (2011). Managing big data: Cloud computing and co-location centers. *Journal of Petroleum Technology, 63*(10), 42–45.

Benghanem, M. (2010). A low cost wireless data acquisition system for weather station monitoring. *Renewable Energy, 35*(4), 862–872.

Chen, H., Chiang, R. H. L., & Storey, V. C. (2012). Business intelligence and analytics: From big data to big impact. *MIS Quarterly, 36*(4), 1165–1188.

Cooper, D. R., & Emory, C. W. (1995). *Business research methods* (5th ed.). Chicago, IL: Irwin.

Crassidis, J. L. (2006). Sigma-point kalman filtering for integrated GPS and inertial navigation. *IEEE TRansactions on Aerospace and Electronic Systems, 42*(2), 750–756.

Davenport, T. H., Barth, P., & Bean, R. (2012). How "Big Data" is different. *MIT Sloan Management Review, 54*(1), 22–24.

Dessureault, S. (2007). Abandoning current faulty budgeting process through information technology and systems re-engineering. *Mining Technology: Transactions of the Institutions of Mining and Metallurgy–Section A, 116*(3), 129–138.

Dowling, A. P. (2004). Development of nanotechnologies. *Materials Today, 7*(12), 30–35.

Frankel, F., & Reid, R. (2008). Distilling meaning from data. *Nature, 455*(7209), 30.

Groves, W., Kecojevic, V., & Komljenovic, D. (2007). Analysis of fatalities and injuries involving mining equipment. *Journal of Safety Research, 38*(4), 461–470.

Harford, T. (2014). Big data: Are we making a big mistake? *Financial Times,* March 28 issue. Retrieved on March 31, 2014 from http://www.ft.com/intl/cms/s/2/21a6e7d8-b479-11e3-a09a-00144feabdc0.html#axzz2xV4TdTmn.

Howe, D., Costanzo, M., Fey, P., Gojobori, T., Hannick, L., Hide, W.,... Rhee, S. Y. (2008). Big data: The future of biocuration. *Nature, 455*(7209), 47–50.

Jahedsaravani, A., Marhaban, M. H., & Massinaei, M. (2014). Prediction of the metallurgical performances of a batch flotation system by image analysis and neural networks. *Minerals Engineering, 69*, 137–145.

Khorram, S., Koch, F. H., van der Wiele, C. F., & Nelson, S. A. C. (2012). *Remote sensing.* New York, NY: Springer.

Kluger, J. (2014). Finding a second earth. *Time, 183*(1), 30–32.

Kotey, B., & Rolfe, J. (2014). Demographic and economic impact of mining on remote communities in Australia. *Resources Policy, 42*, 65–72.

Leahey, C. (2013). The NFL's plan to tackle big data. *Fortune (Asia-Pacific edition), 168*(9), 9.

Levin, R. I. (1987). *Statistics for management.* Englewood Cliffs, NJ: Prentice-Hall.

Lohr, S. (2012a). The age of big data. *New York Times,* February 11 issue. Printed on p. SR1 of the New York edition of the newspaper. Retrieved on May 19, 2014 from http://www.nytimes.com/2012/02/12/sunday-review/big-datas-impact-in-the-world.html.

Lohr, S. (2012b). How big data became so big. *New York Times*, August 11 issue. Also published on p. BU3 of the August 12 issue of the New York Edition with the headline "Amid the Flood, a Catchphrase is Born". Retrieved on May 19, 2014 from http://www.nytimes.com/2012/08/12/business/how-big-data-became-so-big-unboxed.html.

Lynch, C. (2008). How do your data grow? *Nature, 455*(7209), 28–29.

Mayer-Schönberger, V., & Cukier, K. (2013). *Big data: A revolution that will transform how we live, work and think.* London, UK: John Murray Publishers.

McAfee, A., & Brynjolfsson, E. (2012). Big data: The management revolution. *Harvard Business Review, 90*(10), 60–68.

Mednis, A., Strazdins, G., Zviedris, R., Kanonirs, G., & Selavo, L. (2011, June). *Real Time Pothole Detection Using Android Smartphones with Accelerometers.* Paper presented at the 2011 International Conference on Distributed Computing in Sensor Systems and Workshops (DCOSS), Barcelona, Spain.

Mehta, S. N. (2013). Where brains meet brawn. *Fortune (Asia-Pacific edition), 168*(7), 34.

Perrons, R. K., & Jensen, J. W. (2015). Data as an asset: What the oil and gas sector can learn from other industries about "Big Data". *Energy Policy, 81,* 117–121.

Regalado, A. (2014). Data and decision making. *MIT Technology Review, 117*(2), 61–63.

Rovig, A. D., & McArthur, C. K. (1998). Innovation and its management: Some perspectives from the gold industry. In M. C. Kuhn (Ed.), *Managing innovation in the minerals industry* (pp. 37–52). Littleton, CO: Society for Mining, Metallurgy, and Exploration.

Santos, B. R., Porter, W. L., & Mayton, A. G. (2010). *An analysis of injuries to Haul Truck Operators in the U.S. Mining Industry.* Paper presented at the proceedings of the Human Factors and Ergonomics Society Annual Meeting, Santa Monica, CA.

Shvachko, K., Kuang, H., Radia, S., & Chansler, R. (2010). *The Hadoop Distributed File System.* Paper presented at the IEEE Conference on Mass Storage Systems and Technologies, May 3–7, 2010, Incline Village, NV.

van der Veen, M., Spitzer, R., Green, A. G., & Wild, P. (2001). Design and application of a towed land-streamer system for cost-effective 2-D and pseudo-3-D shallow seismic data acquisition. *Geophysics, 66*(2), 482–500.

White, J. W. (1998). Managing innovation from the perspective of a supplier. In M. C. Kuhn (Ed.), *Managing innovation in the minerals industry* (pp. 15–27). Littleton, CO: Society for Mining, Metallurgy, and Exploration.

Wilson, J., & Hume, N. (2013). Efficiency is the quarry as BHP billiton digs deep into its data. *Financial Times*, October 22 issue. Retrieved from http://www.ft.com/intl/cms/s/0/5940335c-3a58-11e3-9243-00144feab7de.html#axzz3M21hziOF.

Wu, L., Barker, R. J., Kim, M. A., & Ross, K. A. (2013). Navigating big data with high-throughput, energy-efficient data partitioning. *ACM SIGARCH Computer Architecture News, 41*(3), 249–260.

Zhu, X. (2005). Cost-constrained data acquisition for intelligent data preparation. *IEEE Transactions on Knowledge and Data Engineering, 17*(11), 1542–1556.

10

Advancement in Digital Oil Field Technology: Maximizing Production and Improving Operational Efficiency through Data-Driven Technologies

Richard Mohan David

CONTENTS

10.1 Introduction

The oil industry is witnessing a period of lower crude prices due to the combination of oversupply and weak global demand, which continues to hamper profit margins. In addressing this challenge, upstream companies are going through a phase of significantly reducing their number of projects, cutting capital expenditures, and downsizing workforces. However, to continue operating in the low oil price and volatile market condition, upstream companies are simultaneously facing intense pressure to improve operational efficiencies and maximizing production from their existing assets. Research by McKinsey & Company estimated that the effective use of digital technologies in the oil and gas sector could make significant inroads toward that end, reducing capital expenditures by up to 20% and operating costs in upstream by 3%–5%.

In addressing this challenging context, leading upstream companies are leveraging next generation technologies, streamlining and automating integrated processes for efficient monitoring, surveillance, and collaborative decision-making and execution. Some examples of strategic initiatives that upstream industry is embarking on are as follows:

- Integrated asset planning
- Real-time production optimization
- Integrated asset modeling
- Integrated reservoir management
- Integrated capacity management
- Predictive maintenance
- Remote operations

Many of these initiatives, often under the terminological umbrella of "digital oil field" (DOF), are nothing new, but have been mainly implemented as standalone, pilot, or "luxury" projects based on the individual digital maturity level of upstream companies. That is, these initiatives are typically not run as part of the company's operating philosophy or operating model. However, with the challenging contemporary market conditions showing no signs of abating, companies are increasingly realizing that DOF initiatives are no longer merely good to have but are must haves for efficient and sustained operations.

Though oil fields generate terabytes of data every day, most companies only utilize a small fraction in decision-making. Yet data is the pulse of the field. Systematically harnessing and analyzing the data can provide vastly improved insights and foresights for effective and efficient oil field management. The next generation DOF technologies simplify data analysis and enable self-servicing analytics for business users.

Data analytics is perhaps the most talked about subject within the upstream industry. Data-driven technologies are transforming the way upstream industry is operating in numerous ways and redefining DOF boundaries. Examples of such technologies include big data analytics, the Industrial Internet of Things (IIoT), machine learning, and data science, often occurring in combination. Data sciences alone encompass innovations relating to decision trees, artificial intelligence, pattern recognition, neural networks, fuzzy logic, regression algorithms, classification, clustering, outlier detection, Bayesian analysis, and nonlinear programming. The following sections discuss some of the most notable trends within upstream oil and gas companies that are adopting data-driven technologies.

10.2 Examples of Data-Driven Technologies

10.2.1 Advancement in Field Automation

Advancements in sophisticated field surveillance, sensors and actuations, connectivity platforms, storage technologies, mobility, big data, and analytics is enabling upstream companies to process and analyze a variety of field data rapidly for understanding reservoir and production potential, improving health and safety, and enhancing operational efficiencies. The industry is moving away from manual field data capture, with more and more sensors being put into the rigs, wells, pipelines, and facilities to measure temperature, pressure, vibrations, chemical injections, and so on. This even includes fiber optics along the well bore path to monitor and observe reservoir condition. Embedded smart sensors in vessels, tanks, compressors, and turbines are also capable of sending real-time data to control rooms, where engineers can monitor processes and provide diagnostics remotely.

10.2.2 The Industrial Internet of Things

Leading upstream companies are taking advantage of IIoT to leverage the full value of their asset information. IIoT solutions generate actionable information by running analytics on data that moves between devices and data in the public/private cloud or in a data center in a manner that is secure, manageable, and user friendly. Industrial networking and gateway technologies are enabling easy integration into existing distributed control systems (DCS) and site control and data acquisition (SCADA) systems. Effective use of IIoT is bridging gaps between operational technology (OT) and information technology (IT) environment, with OT devices increasingly capable of transparently interfacing with the IT systems. IIoT brings together big data analytics, machine learning, data science, and cloud computing, and facilitates optimizing operations in real time which include closed loop control of asset and processes beyond legacy supervisory control. A study by Cisco Consulting Services estimated that by transforming business processes through IIoT, oil and gas companies can capture their share of the U.S.\$600 billion of value at stake between 2016 and 2025.

Edge devices, gateways, and analytics (referring to processing cloud data near to the data source) is an emerging IIoT technology which, once matured, allows upstream companies to securely transfer and analyze important data of critical equipment at the edge of the network to execute and actuate immediate and smart autonomous decisions. Edge computing enables continuous operations even if the network is down and reduces the burden on constrained networks as well as transmission costs. IIoT is pushing the boundaries of DOF, moving toward autonomous assets that enable new business models in upstream companies.

10.2.3 Remote Monitoring and Operations

Enabled by IIoT, remote operations open the possibility to control operations and take decisions remotely with limited physical presence on the platform, well site, or fields. This reduces human footprint onsite, shifting skilled engineers to office-based remote operating centers, thus significantly cutting the number of engineers' trips to the field and improving Health Safety and Environment (HSE) performance. For example, Chevron's Machinery Support Center monitors critical equipment in real time and across global operations to identify abnormal performance, enable preventative maintenance, and avoid unplanned shutdowns.

It is very challenging task to monitor reading the multitude of continuous field measurements such as flow reading, equipment status, and tank levels. A typical offshore production platform can have more than 40,000 data tags. Exception-based surveillance (EBS), however, enables field operations to focus on the key operating parameters that go out of range from its normal operating conditions and to shift the engineers to carryout higher value core activities.

10.2.4 Predictive Analytics

Brownfields need very efficient maintenance schedules to keep production profitable. Utilizing data-driven innovations, upstream companies are moving away from schedule-based maintenance toward condition-based, predictive maintenance. The economic ramifications of preempting mechanical failures are very clear. For instance, according to General Electric, there are more than 130,000 electric submersible pumps (ESPs) installed globally, accounting for 60% of the world's oil production. Extending ESP performance from the average life span of 1.2 years to the designed for 5 years could be a U.S. $10 billion opportunity. As another example, an MIT Sloan case study estimates that an unproductive day in the Liquefied Natural Gas (LNG) facility costs about U.S. $25 million.

Predictive analytics solutions are typically based on pattern recognition and machine learning technology. Deploying predictive analytics techniques enable companies to minimize risk of disastrous failures and process disruptions, while maximizing equipment reliability and production efficiency. Effectively interpreting the massive amounts of manually and machine generated data can predict the asset (e.g., well, reservoir, network facilities, and equipment) behavior, enable efficient forecast, recommend best actions, and optimize operations, automatically or with human intervention, to maximize performance.

As an example of such technology in action, Santos developed a predictive model, using data from SCADA-connected assets and other sources, which can predict asset failure with 87 to nearly 100% accuracy. One use of these models is predictions and early alerts for when its corrosion inhibitor tanks are about to run dry, thus prompting field teams for when they will need to drive out to refill the tank. Such measures have greatly improved pro-active maintenance, boosted production uptime and operational efficiency, and saved millions of dollars with significant HSE benefits. Another illustration of applied predicative analytics is in drilling, where advanced sensors and machine-to-machine connections enable operators to anticipate equipment failures and take proactive measures to minimize drilling downtime. ExxonMobil's highly successful drilling advisory system (DAS) is built on real-time drilling data and analytical algorithms that compute drilling performance measures and generate recommendations for improved parameters.

10.2.5 Cognitive Computing

Cognitive computing involves self-learning systems that use data mining, pattern recognition, and natural language processing to mimic the way the human brain works to continuously learn from previous interactions, thus gaining value and knowledge over time. Energy companies such as Woodside have used another company's (Watson's) cognitive computing platform to analyze more than 60,000 documents covering the corporation's collective knowledge of 30 years of operating experience. This bank of knowledge can provide meaningful insights to their engineers in seconds, aiding data-driven decision-making in a relatable format. Chevron plans to follow Woodside to use Watson cognitive computing in new phase of its DOF strategy.

10.2.6 Modeling

The trend toward data-driven models such as proxy and surrogate models are complimenting their more traditional, physics-based counterparts. New modeling iterations can process unprecedented levels of data, virtually in real time, revealing many previously hidden physical relationships within and between different data sets, thus enabling fast and efficient optimization decisions, something which was much more challenging with legacy simulation models. This modeling has been used to greatly enhance the understanding of operational aspects like production systems and their various nodes. Surrogate reservoir modeling, using machine learning alongside pattern recognition techniques to mimic the behavior of numerical simulation, has also been used to carry out rapid reservoir analysis and fluid flow behavior with high accuracy.

10.3 Conclusions

Early adopters and innovative upstream companies are strategically revisiting corporate strategies toward implementation of digital innovation, increasingly leveraging data-driven technologies as part of a broader approach to compete in a prevailing global context of lower price levels. These market conditions are also likely, however, to force even the most reluctant companies to eventually follow suit to ensure their operations remain sustainable.

Though data-driven technologies are game changer, upstream companies need to invest in business process reengineering efforts to facilitate automation, integration, and realization of the full potential of their digital investments. Operating next generation DOFs requires engineers with cross knowledge on business function, data science, and data analytics. It is important to assess the current team capabilities and develop plans to improve team skills in data sciences and analytics.

Bibliography

Baaziz, A., & Luc Quoniam. 2014. How to use big data technologies to optimize operations in upstream petroleum industry. Paper presented at the *21st World Petroleum Congress,* Moscow, Russia.

Choudhry, H., A. Mohammad, K. Tee Tan, and R. Ward. 2016. The next frontier for digital technologies in the oil and gas sector. *McKinsey and Company.* http://www.mckinsey.com/industries/oil-and-gas/our-insights/the-next-frontier-for-digital-technologies-in-oil-and-gas.

Head, B. 2016. Woodside retains corporate memory using cognitive computing-Interview with Woodside's Shaun Gregory. *Australian Finance Review Magazine,* August.

IBM. 2015. Santos saving millions with a predictive asset monitoring and alert system. http://www-01.ibm.com/common/ssi/cgi-bin/ssialias?subtype=AB&infotype=PM&htmlfid=ASC12379USEN&attachment=ASC12379USEN.PDF.

Miklovic, D. 2017. Oil and gas operational excellence enabled IIoT. https://www.ge.com/digital/blog/oil-gas-operational-excellence-enabled-iiot.

Mohan, R. 2016. Approach towards establishing unified petroleum data analytics environment enabling data driven operations. *Paper Presented at the Annual SPE Technical Conference and Exhibition,* Dubai, United Arab Emirates.

Morgan, T. 2015. Future 2025: Internet of things and the upstream oil and gas industry. http://www.pidx.org/wp-content/uploads/2015/10/9_T-Morgan.pdf.

Moriarty, R., K. O'Connell, N. Smit, A. Noronha, and J. Barbier. 2015. A new reality for oil & gas: Complex market dynamics create urgent need for digital transformation. https://www.cisco.com/c/dam/en/us/solutions/collateral/industry-solutions/oil-gas-digital-transformation.pdf.

Payette, G, D. Pais, B. Spivey, and L. Wang. 2015. Mitigating drilling dysfunctions with a drilling advisory system. Paper presented at the International Petroleum Technology Conference, Doha, Qatar.

Schneider Electric. 2016. Industrial internet of things (IIoT) impact on the oil & gas industry value chain. https://www.schneider-electric.com/en/download/document/998-2095-10-13-16AR0_EN/.

World Economic Forum. 2017. Digital transformation initiative—Oil and gas industry. http://reports.weforum.org/digital-transformation/wp-content/blogs.dir/94/mp/files/pages/files/dti-oil-and-gas-industry-white-paper.pdf.

11

A Statistical Framework for Data-Driven Assessment of Unconventional Oil and Gas Resources

Justin Montgomery and Francis O'Sullivan

CONTENTS

11.1 Introduction

Over the last decade, the combination of horizontal drilling and hydraulic fracturing has greatly expanded North American shale gas and tight oil production, leading to increased estimates of domestic recoverable resources and expectations of a more hydrocarbon-abundant future (Deutch 2011; Joskow 2013). Similarly, in recent years hydraulic fracturing of coal seam gas in Australia has enabled rapid growth of the liquefied natural gas (LNG) export business, with Australia poised to overtake Qatar as the world's largest exporter of LNG by 2019 (Cully et al. 2016). Given this prevalence, the term "unconventional," which has long been used to describe these resources no longer seems entirely appropriate.

Despite this, it is becoming increasingly apparent that unconventional resources present the oil and gas sector with a different type of economic risk than it is used to dealing with (Crooks 2014; McGlade et al. 2013; Inman 2014). Economic risk in "conventional"—reservoirs in which oil and gas freely flows into well bores without the need for techniques like hydraulic fracturing—is driven primarily by uncertainty about the size and presence of hydrocarbon accumulations (Rose 2001b; Suslick and Schiozer 2004). This geological exploration risk is most prevalent early in a field's exploration and appraisal, but geological data are collected, subsurface models are improved, and uncertainty about production is substantially reduced as initial wells are drilled (Dake 1983; Prensky 1994; Rose 2001b; Suslick and Schiozer 2004; Mustafiz and Islam 2008). Economic risk in shale and tight resource plays is more statistical than geological in nature—resources are known to be present continuously throughout the region referred to as a "play" but reservoir rock quality is highly heterogeneous and rates of production from individual wells (and hence their economics) are unpredictable even in mature fields with many existing wells (Hall et al. 2010; Dong et al. 2015; Charpentier and Cook 2013; Cook and Van Wagener 2014).

The ability to characterize the distribution of well productivity in shale gas and tight oil fields is critical to better understanding this inherent variability and to better managing economic risk (McGlade et al. 2013). The statistical characterization discussed in this chapter provides an important step in this direction and a useful framework for the development of predictive analytics tools to help operating companies manage risk and reach more optimal decisions throughout the development process (Harrell 2017).

11.2 Background—Statistical Resource Assessment in Oil and Gas

The petroleum industry has historically preferred to avoid a "casino" analogy to development in favor of a more deterministic perspective (Rose 2001a). However, such a view does not eliminate the large uncertainty associated with petroleum development projects and may lead to inadequate consideration of risk and poor decision making as compared to more probabilistic approaches (Rose 2001a). Fortunately, in recent years, there has been growing recognition that quantification of uncertainty and statistical models can improve estimates and decision making, especially in resource plays, which have more statistical risk than geological risk (Kaufman 1993; Rose 2001a; Hall et al. 2010; McGlade et al. 2013).

One frequently useful probability distribution in petroleum resource evaluation is the lognormal distribution—a distribution which is Gaussian when the natural logarithm is taken of it. Arps and Roberts (1958) first introduced the use of this distribution to describe oil and gas resources and it gained popularity in the 1960s due to further research, most notably by Kaufman on the distribution of conventional oil and gas field sizes

(Kaufman 1963; Drew and Griffith 1965; Kaufman et al. 1975; Schuenemeyer and Drew 1983; Andreatta et al. 1988; Attanasi and Drew 1985).

The extension of these probabilistic models to characterizing production variability in unconventionals is not direct though. With the shift from discrete conventional oil and gas deposits to continuous unconventional plays, uncertainty about field size has been replaced by uncertainty about the in situ and engineered properties of the near-well bore reservoir. Early accounts of unconventional oil and gas resources frequently hypothesized that the resource was geologically homogenous and all variations in production rates should be explainable by well and completion differences. However, in recent years this hypothesis has been shown to be a dangerously inaccurate assumption and as a result the interest in statistical resource approaches has grown (Hall et al. 2010; Hale and Cobb 2010; Haskett and Brown 2011; Breyer 2012; Montgomery and O'Sullivan 2017).

The lognormal distribution has also been applied extensively in the mining industry for assessment of mineral deposits (Krige 1951, 1960, 1966). Lognormality is in fact commonly found in natural systems due to the salient role of multiplicative effects (as opposed to additive effects which generate Gaussian-like distributions) and the lower bounding at zero of values that must be nonnegative (Blackwood 1992; Limpert et al. 2001; Sornette 2006). An example of multiplicative effects is the amount of petroleum in a reservoir, which is the product of bulk rock volume, porosity, and pore petroleum saturation. A normal distribution in each of these parameters leads to a lognormal distribution for the product (Sornette 2006). This is, in effect a variant of the central limit theorem and the lognormal distribution should be thought of as just as fundamental as the normal distribution, despite its derivative name (Aitchison and Brown 1957).

The reasons for making such a distributional assumption were identified by Chambers et al. (1983):

Reason 1. Allows for compact description of data—mean and standard deviation are adequate parameters to describe normal or lognormal distributions.

Reason 2. Enables useful statistical procedures (e.g., analysis-of-variance).

Reason 3. Characterizes sampling distributions for parameters and make statistical inferences from data.

Reason 4. Improves understanding of physical mechanisms generating data.

All these reasons apply to the case of a lognormal distributional assumption for well productivity. Simple, yet accurate, descriptions of well productivity (enabled by reason 1) will lead to better resource assessments. Statistical procedures like analysis-of-variance for lognormal distributions (as in reason 2) can be used in the future to evaluate different completion (hydraulic fracturing) techniques for effectiveness. Making statistical inferences (from reason 3) is a key enabler for deployment of machine-learning techniques and enables operators to assess the productivity of resource areas early on, as demonstrated using a Bayesian approach (Montgomery 2015). A distributional assumption can also direct the efforts of and organize the results from fundamental research into understanding the complex physical processes governing shale gas and tight oil production (leading to future realizations of reason 4).

11.3 Data and Methods

11.3.1 Data in Study

We consider the distribution of individual well productivity in shale gas and tight oil fields, using each well's peak monthly production rate as our response variable. Although there is additional uncertainty in the production decline rate for each well, peak monthly production is a widely used metric for early-life well productivity and is correlated with total lifetime recovery from wells (Lee and Sidle 2010). Shale gas and tight oil resource economics are highly dependent on early-life well production rates; production falls substantially in wells early in their life—by some estimates as much as 50% of the peak production rate after only 1 year (Strickland et al. 2011). In addition to absolute productivity of wells, we analyze the specific productivity—a length-normalized index defined as the absolute productivity divided by the length of the productive lateral section.

The production data in this study was accessed on July 3, 2014 from the "drillinginfo HPDI" online database of U.S. oil and gas production data. This is a service that aggregates production data for oil and gas wells in 33 states in the United States, drawing on publicly available repositories managed by the respective states. The information in these databases has been reported by oil and gas operating companies according to state requirements. To generate the database of wells in each play for the study, we used criteria which are summarized in Table 11.1 for each play.

The plays we selected for inclusion in this study—Barnett, Marcellus, Bakken, Eagle Ford, and Haynesville—are key U.S. shale plays and have had high levels of drilling activity in recent years, thus providing abundant production. They also have been developed similarly enough to warrant comparison, yet have known geological differences. They include a range of reservoir fluid types, from the dry gas of Haynesville to the liquids-heavy gas condensate of the Eagle Ford (we analyzed the gas production rates from Eagle Ford to avoid complicating our analysis with consideration of varying gas-oil-ratios) and the black oil of the Bakken. Some descriptive characteristics of these plays are summarized in Table 11.2.

TABLE 11.1

The Plays Used in the Study Include Barnett, Bakken, Marcellus, Eagle Ford, and Haynesville. In Order to Generate the Database of Wells in Each Play for the Study the Criteria in This Table Was Used for Each Play

Criteria	Barnett	Bakken	Marcellus	Eagle Ford (Gas)	Haynesville
			Plays		
State(s)	Texas	North Dakota	Pennsylvania	Texas	Texas, Louisiana
Number of wells	9138	5483	1584	2159	2221
Production type	Gas	Oil; Oil and Gas	Gas	Gas	Gas
Well type	Horizontal	Horizontal	Horizontal	Horizontal	Horizontal
Status	Active	Active	Active	Active	Active
Earliest start date	2005	2009	2009	2010	2009
Depths	At least 3,000 ft	At least 5,000 ft	At least 4,000 ft	At least 4,500 ft	At least 10,500 ft
Perforation length	At least 1,000 ft	At least 1,000 ft	At least 1,500 ft	At least 1,000 ft	At least 1,000 ft
Peak rate	Highest rate in first 3 months	Highest rate in first 6 months	Highest rate in first 3 months	Highest rate in first 3 months	Highest rate in first 3 months
Production history	At least 3 months of production	At least 6 months of production	At least 3 months of production	At least 3 months of production	At least 3 months of production

TABLE 11.2

General Details about the Plays Included in the Study Are Summarized Here, Including the Location, Geological Age, Extent, Depth, Average Thickness, Total Organic Carbon, Porosity, and Technically Recoverable Resources

	Barnett	Bakken	Plays Marcellus	Eagle Ford	Haynesville
Location	North Texas (Curtis et al. 2012)	North Dakota, South Dakota, and Montana (USGS 2013)	Ohio, Pennsylvania, New York, and West Virginia (Curtis et al. 2012)	South Texas (Curtis et al. 2012)	East Texas and Northwest Louisiana (Curtis et al. 2012)
Geological age	Mississippian (Curtis et al. 2012)	Devonian–Mississippian (Webster 1987)	Devonian (Curtis et al. 2012)	Cretaceous (Curtis et al. 2012)	Jurassic (Curtis et al. 2012)
Extent, sq mi	6500 (EIA 2011)	6500 (EIA 2011)	95000 (EIA 2011)	3300 (EIA 2011)	9000 (EIA 2011)
Depth, ft	7500 (EIA 2011)	4500–7500 (EIA 2011)	4000–8500 (EIA 2011)	7000 (EIA 2011)	10500–13500 (EIA 2011)
Average Thickness, ft	300 (EIA 2011)	22 (EIA 2011)	125 (EIA 2011)	200 (EIA 2011)	250 (EIA 2011)
Total organic carbon (TOC), %	4.2 (Curtis et al. 2012)	11 (Sonnenberg 2011)	4.8 (Curtis et al. 2012)	3.8 (Curtis et al. 2012)	2.4 (Curtis et al. 2012)

(Continued)

TABLE 11.2 (*Continued*)

General Details about the Plays Included in the Study Are Summarized Here, Including the Location, Geological Age, Extent, Depth, Average Thickness, Total Organic Carbon, Porosity, and Technically Recoverable Resources

	Plays				
	Barnett	Bakken	Marcellus	Eagle Ford	Haynesville
Porosity, %	5 (Montgomery et al. 2005)	8 (EIA 2011)	8 (EIA 2011)	9 (EIA 2011)	8.5 (EIA 2011)
Technically recoverable resources	100 MMBO, 26,700 BCFG, 1,100 MMBNGL (USGS 2004)	7,400 MMBO, 6,700 BCFG, 500 MMBNGL (USGS 2013)	84,200 BCFG, 3,400 MMBNGL (USGS 2011b)	900 MMBO, 51,900 BCFG, 2,000 MMBNGL (USGS 2011a)	60,700 BCFG (USGS 2011a)

Source: Montgomery, J. B. et al., *AAPG Bull.*, 89(2), 155–175, 2005; U.S. Energy Information Administration (EIA). 2011. *Review of Emerging Resources: U.S. Shale Gas and Shale Oil Plays.* Washington, DC: U.S. Department of Energy; Bowker, K. A., *AAPG Bull.*, 91(4), 523–533, 2007; Sonnenberg, S. A., TOC and pyrolysis data for the Bakken Shales, Williston Basin, North Dakota and Montana. In *The Bakken-Three Forks Petroleum System in the Williston Basin*, R. Mountain (Ed.), Association of Geologists, pp. 308–331, 2011. Denver, CO; Hammes, U. et al., *AAPG Bull.*, 95(10), 1643–1666, 2011; U.S. Geological Survey (USGS). 2004. *Assessment of Undiscovered Oil and Gas Resources of the Bend Arch-Fort Worth Basin Province of North-Central Texas and Southwestern Oklahoma, 2003.* Washington, DC: U.S. Department of the Interior; USGS. 2011a. *Assessment of Undiscovered Oil and Gas Resources in Jurassic and Cretaceous Strata of the Gulf Coast, 2010.* Washington, DC: U.S. Department of Interior; USGS. 2011b. *Assessment of Undiscovered Oil and Gas Resources of the Devonian Marcellus Shale of the Appalachian Basin Province, 2011.* Washington, DC: U.S. Department of Interior; USGS. 2013. *Assessment of Undiscovered Oil Resources in the Bakken and Three Forks Formations, Williston Basin Province, Montana, North Dakota, and South Dakota, 2013.* Washington, DC: U.S. Department of Interior; Curtis, M. E. et al., *AAPG Bull.*, 96(4), 665–677, 2012; Webster, R. L., Petroleum source rocks and stratigraphy of the Bakken Formation in North Dakota. In *The Rocky Mountain Association of Geologists Guidebook, Williston Basin, Anatomy of a Cratonic Oil Province.* Denver, CO: Rocky Mountain Association of Geologists, 1987.

Notes: This information was assimilated from a range of references and is only intended to be descriptive of the plays. Many of the values still have a high degree of uncertainty associated with them. References for these plays are included.

MMBO, million barrels of oil; BCFG, billion cubic feet of gas; MMBNGL, million barrels of natural gas liquids.

11.3.2 Normalization of Data

The probability density function of a lognormal distribution is:

$$f_{(X)}(x;\mu,\sigma) = \frac{1}{x\sigma\sqrt{2\pi}} e^{-\frac{(\ln x - \mu)}{2\sigma^2}}, x > 0 \tag{11.1}$$

In our case, x is the well's peak production rate, μ is the arithmetic mean of the log-transformed peak production rates in the well ensemble, and σ is the standard deviation of the log-transformed peak production rates in the well ensemble.

Although the shape of the productivity distribution for wells is consistent across all well ensembles, the central tendency and spread may vary. To compare the productivity distribution shape for different well ensembles, we normalized the production rates within well ensembles. We took the log-transformation of the peak production rate for each well and then calculated the standard score for each well relative to the other wells in the ensemble under consideration. The standard score, z, is calculated using the equation,

$$z = \frac{\ln x - \mu}{\sigma} \tag{11.2}$$

In addition to the *absolute* peak production rate, we considered the specific peak production rate. To calculate the specific peak production rate, we used the equation,

$$Q_{spec,peak} - \frac{Q_{peak}}{D_{perf,lower} - D_{perf,upper}} \tag{11.3}$$

where:
$Q_{spec,peak}$ is the monthly specific production rate in the peak month
Q_{peak} is the (absolute) monthly production rate
$D_{perf,lower}$ is the measured depth of the lowest perforation in the well
$D_{perf,upper}$ is the measured depth of the highest perforation in the well

The denominator in this expression represents the productive length of the well that has been perforated.

11.3.3 Graphical Analysis with Probability Plots

To compare the empirical productivity distribution of well ensembles with the ideal lognormal distribution, we use theoretical quantile–quantile plots, also called probability plots. Specifically, we use a normal quantile–quantile plot to graph the log-transformed standard scores of well peak production rates. This test for distribution shape is not affected by the normalization we performed because it is invariant to location and spread. The normalization does allow multiple ensembles to be compared simultaneously on a common scale.

Constructing a probability plot involves rank ordering the samples and plotting each sample's actual value against the theoretical, "ideal" distribution value for the observation. We use the y-axis for the theoretical distribution and the x-axis for the data values. For ease of comparison, the plots are set to a standard horizontal scale of $z = -3$ to $z = 3$. This pushed some left-hand extreme values off the graph but allows for easier visual inspection over a suitably wide-range of the data. All of the data plotting was carried out in MATLAB, using the "probplot" function with the default midpoint probability plotting positions.

The interpretation of probability plots is explored in literature, and we provide only a brief discussion here (Chambers et al. 1983). A good fit of data with a distribution is indicated by straightness of the points in the plot. In addition to checking for goodness of distributional fit, systematic departures from the line may reveal important information. Often, outliers exist at either end of the data. Additionally, there tends to be greater variation in distribution tails with density that gradually tapers to zero at extreme values (such as normal or lognormal distributions). Defined and systematic curvature at the ends may indicate longer or shorter tails in the data than the ideal distribution. With our axis selection, curvature upward at the left tail or downward at the right tail indicates longer tails at those ends of the distribution. The opposite orientations indicate shorter tails at either end. Asymmetry can also be identified. Convexity of the plotted data indicates that the empirical distribution is more left-skewed than the ideal distribution (and contrarily, concavity indicates right-skewness) (Chambers et al. 1983).

11.3.4 Analysis of Blocks of Wells

In order to group wells into 10 square mile sections, we created evenly-spaced divisions within Johnson County in the Barnett. Uniformly-dimensioned "blocks" of 10 square miles were established based on lines of latitude and longitude and the location coordinates for each well in that county were used to associate wells with one of these blocks. Only blocks that had 16 or more wells were included in the analysis. The wells in each block were then treated as a separate well ensemble and normalized prior to comparison in a probability plot.

11.4 Results and Discussion

The Barnett shale play, the most extensively developed (and thus sampled) shale resource play, exemplifies the lognormality of well productivity (Figure 11.1a and b). The same pattern is also apparent for the Marcellus,

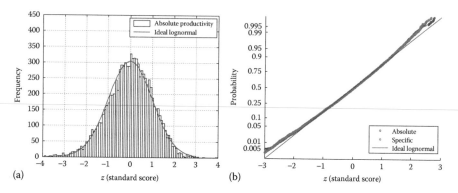

FIGURE 11.1
Lognormal distribution of absolute and specific productivity in Barnett shale. (a) Lognormal histogram of well productivity in the Barnett. (b) Probability plot comparing normalized absolute and specific productivity in the Barnett to an ideal lognormal distribution.

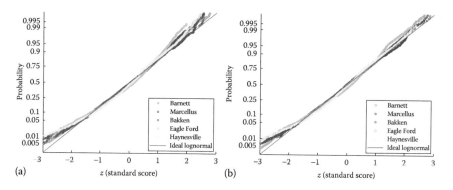

FIGURE 11.2
Lognormal distribution of absolute and specific productivity for different unconventional plays. (a) Probability plot comparing normalized absolute productivity of five major shale plays to an ideal lognormal distribution. (b) Probability plot comparing normalized specific productivity of five major shale plays to an ideal lognormal distribution.

Bakken, Eagle Ford, and Haynesville plays (Figure 11.2a and b). For all these, the correspondence of the empirical distribution to an ideal lognormal distribution is exceptional between the 5th and 95th percentile. The deviations at the extremes of the data are unsurprising given the tapering of a lognormal probability distribution toward zero in the tails, which tends to exaggerate differences in these low probability regions (Chambers et al. 1983). However, there may also be physical and economic factors more prevalent at these ends of the distribution. For instance, high density development in productive sweet-spot areas is known to lead to interference—in which neighboring wells share a pressure drawdown area—reducing individual well output for some of the highest performing wells (Ikonnikova et al. 2014). This distributional assumption may prove useful for identifying such inter-actions. For a further discussion of possible explanations of the deviation at the distribution tails, see Montgomery (2015).

The characterization of production variability as lognormal also applies at various levels within plays. It is a reliable description of variability within individual counties of a play, like the Barnett, including core areas such as Johnson County and lower performing non-core areas like Parker County (Figure 11.3a). Although some counties have higher median well productiv-ity, the shape and spread (measured as the P90-P10 ratio, or 90th percentile value over 10th percentile value) of the distribution is remarkably consistent, as shown in Table 11.3. Even more remarkably, the shape of this distribution is scale-invariant, even down to the scale of 10 square mile sections (Figure 11.3b).

How reliable is this distribution assumption over time with poten-tially evolving technology? To answer this, we look at some additional

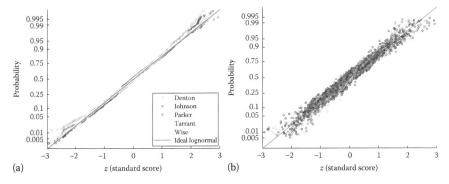

(a)   (b)

FIGURE 11.3

Lognormal distribution of productivity is scale-invariant. (a) Probability plot comparing nor-malized specific productivity in different counties of the Barnett to an ideal lognormal dis-tribution. (b) Probability plot comparing normalized specific productivity of Barnett wells in 10-square-mile blocks having 16 or more wells to an ideal lognormal distribution.

TABLE 11.3

Summary Statistics for Specific Productivity of Counties in Barnett (Figure 11.3a)

County	P90-P10	P50 (mscf/mo/ft)	Mean (mscf/mo/ft)
Denton	3.38	19.63	21.51
Johnson	3.95	23.56	27.25
Parker	3.22	13.93	16.07
Tarrant	3.37	25.90	29.29
Wise	3.92	13.91	16.00

Note: The unit mscf/mo/ft is thousand standard cubic feet per month per foot.

subgroupings of well ensembles in plays. First, we consider each vintage, or development year, as a well ensemble. The probability plot examining different vintages in the Barnett are shown in Figure 11.4a. In addition to the consistency of the distribution shape, it is worth noting that the spread has remained consistent and quite broad over this period of development for both absolute and specific productivity, as shown in Table 11.4.

Furthermore, categorizing wells by the length of the perforated section of well yields the same distribution shape. In general, wells have been increasing in length (and number of frack stages) over time, making this a relevant consideration for understanding how technology evolution might impact the shape of the productivity distribution. Figure 11.4b shows the results of this categorization for well productivity.

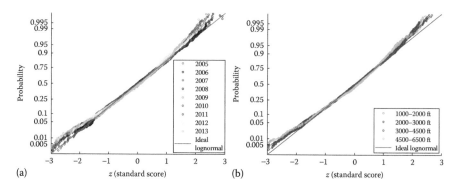

(a)

(b)

FIGURE 11.4
Lognormal distribution of productivity is time (and technology)-invariant. (a) Probability plot comparing normalized specific productivity of different vintages in the Barnett to an ideal lognormal distribution. (b) Probability plot comparing normalized absolute productivity for Barnett wells with different perforated well-lengths to an ideal lognormal distribution.

TABLE 11.4

Spread of Productivity by Vintage in Barnett (Figure 11.4a)

Vintage	Absolute Productivity P90-P10	Specific Productivity P90-P10
2005	4.48	5.08
2006	4.59	4.49
2007	4.33	4.58
2008	3.90	4.43
2009	4.06	3.97
2010	4.00	4.66
2011	4.81	4.46
2012	4.14	3.88
2013	5.22	4.26

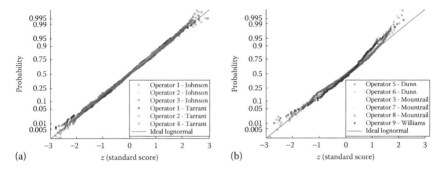

FIGURE 11.5
Lognormal distribution of productivity applies within portfolios of individual operating companies. (a) Probability plot comparing normalized specific productivity for different operators in the Barnett to an ideal lognormal distribution. (b) Probability plot comparing normalized specific productivity for different operators in the Bakken to an ideal lognormal distribution.

Finally, if this distributional assumption is to be used by individual operating companies, it is worth considering the distribution for a specific company's portfolio of wells in each area. Figure 11.5a shows the distribution of well productivity for specific county-level individual company well portfolios in the Barnett and Figure 11.5b shows the same for the tight oil Bakken play. These samples represent the largest individual company well portfolios in each play.

The immense heterogeneity present in these rocks precludes the ability to accurately predict the output of a well deterministically even when many similar wells have been drilled nearby (Hall et al. 2010). However, based on the universality of the distribution we find (for further evidence, see Montgomery [2015]), geostatistical techniques like kriging may be useful for assessing and identifying optimal development locations as has been done in mining to address local variability in ore quality of a similar pattern (Cressie 1990; McGlade et al. 2013). This technique will reduce the number of wells required to access a given quantity of resource, thus improving the economics and reducing the local disturbance and environmental impacts of development. An early example of this kriging approach with tight oil can be found in Montgomery and O'Sullivan (2017).

11.5 Conclusions

The saliency of this broad, skewed distribution for unconventional well productivity has stark implications for resource size, economics, and environmental impact. It is critical to consider the productivity distribution for different areas when assessing the resource size in unconventionals, and this distribution will not be well understood until a number of wells have been drilled. This casts some doubt on the reliability of unconventional oil and gas production forecasts failing to account for this variability with potentially dire consequences for agencies planning long-term energy infrastructure or with budgets reliant on oil and gas revenues. The right-skewed nature of the distribution of well productivity may lead to a tendency to overestimate production. Additionally, it is inadequate to rely on one measure of a field's well productivity, such as the mean, when assessing the environmental footprint of development. Each field will have some high-performing wells and a much larger number of low-performing wells and the expected environmental impact per well depends on how many wells will ultimately be drilled.

Thus far, development of unconventional oil and gas resources has proceeded at a rapid pace without an adequate appreciation for the underlying uncertainty. Claims that these resources can be manufactured vastly understate their economic risk (Forbes et al. 2009). A more apt metaphor may be slot machines, in which it is critical to quickly and accurately understand the table odds. The pressure from lower oil prices and increasing scrutiny on the long-term resource potential marks a changing landscape for companies engaged in unconventional petroleum production. It is now clear that new strategies will be necessary to make this abundant resource more efficient and predictable to extract. Characterization of production variability is a critical step toward this. The next step is using this statistical characterization as a framework for data analytics tools and better decision-making practices which incorporate abundant field data into the planning and engineering of fields. This type of data-driven approach may unlock an even more unconventional-centric resource future.

References

Aitchison, J., and J. A. C. Brown. 1957. *The Lognormal Distribution with Special Reference to Its Uses in Economics.* London, UK: Cambridge University Press.

Andreatta, G., G. M. Kaufman, R. G. McCrossan, and R. M. Procter. 1988. The shape of Lloydminster oil and gas deposit attribute data. In *Qualitative Analysis of Mineral and Energy Resources,* C. F. Chung, A. G. Fabbri, and R. Sinding-Larsen (Eds.), pp. 411–431. New York: Springer.

Arps, J., and T. Roberts. 1958. Economics of drilling for Cretaceous oil and gas on the east flank of the Denver-Julesberg Basin. *American Association of Petroleum Geologists Bulletin* 42(11):2549–2566.

Attanasi, E. D., and L. J. Drew. 1985. Lognormal field size distributions as a consequence of economic truncation. *Mathematical Geology* 17(4):335–351.

Blackwood, L. G. 1992. The lognormal distribution, environmental data, and radiological monitoring. *Environmental Monitoring and Assessment* 21:193–210.

Bowker, K. A. 2007. Barnett Shale gas production, Fort Worth Basin: Issues and discussion. *AAPG Bulletin* 91(4):523–533.

Breyer, J. A. 2012. Shale Reservoirs. In *Shale Reservoirs—Giant Resources for the 21st Century*, J. A. Breyer (Ed.), pp. x–xii. Tulsa, OK: American Association of Petroleum Geologists.

Chambers, J. M., W. S. Cleveland, B. Kleiner, and P. A. Tukey. 1983. *Graphical Methods For Data Analysis*. Boston, MA: Wadsorth International Group and Duxbury Press.

Charpentier, R. R., and T. A. Cook. 2013. Variability of oil and gas well productivities for continuous (unconventional) petroleum accumulations. Washington, DC: U.S. Geological Survey.

Cook, T., and D. Van Wagener. 2014. Improving well productivity based modeling with the incorporation of geologic dependencies. Washington, DC: U.S. Energy Information Administration.

Cressie, N. 1990. The origins of kriging. *Mathematical Geology* 22(3):239–252.

Crooks, E. 2014. U.S. Shale: What lies beneath. *Financial Times*, August 26, 2014. http:// ig-legacy.ft.com/content/e178031e-2cf4-11e4-8105-00144feabdc0#axzz5291 cRiOu.

Cully, M., N. Thomas, and D. Whitelaw. 2016. Factors influencing Australia's gas supply and demand. Canberra, Australia: Office of the Chief Economist, Deptartment of Industry, Innovation, and Science, Australia.

Curtis, M. E., C. H. Sondergeld, R. J. Ambrose, and C. S. Rai. 2012. Microstructural investigation of gas shales in two and three dimensions using nanometer-scale resolution imaging. *AAPG Bulletin* 96(4):665–677.

Dake, L. P. 1983. *Fundamentals of Reservoir Engineering*. New York: Elsevier.

Deutch, J. 2011. The good news about gas-The natural gas revolution and its consequences. *Foreign Affairs* 90:82.

Dong, Z., S. A. Holditch, D. A. McVay, W. B. Ayers, W. J. Lee, and E. Morales. 2015. Probabilistic assessment of world recoverable shale-gas resources. *SPE Economics and Management* 7(2):72–82.

Drew, L. J., and J. C. Griffith. 1965. Size, shape and arrangement of some oilfields in the USA. *Symposium on Computer Applications in the Mineral Industries*.

EIA. 2014. Annual Energy Outlook 2014 with projections to 2040. https://www.eia. gov/outlooks/archive/aeo14/.

Forbes, B., J. Ehlert, and H. Wilczynski. 2009. *The Flexible Factory: The Next Step in Unconventional Gas Development*. Sugar Land, TX: Schlumberger Business Consulting.

Hale, B. W., and W. M. Cobb. 2010. Barnett Shale: A resource play-locally random and regionally complex. In *Proceedings of the Society of Petroleum Engineers Eastern Regional Meeting 2010*, pp. 217–230. Richardson, TX: Society of Petroleum Engineers.

Hall, R., R. Bertram, G. Gonzenbach, J. Gouveia, B. Hale, P. Lupardus, P. McDonald, N. Meehan, B. Vail, and M. Watson. 2010. *Guidelines for the Practical Evaluation of Undeveloped Reserves in Resource Plays*. Richardson, TX: Society of Petroleum Engineers.

Hammes, U., H. S. Hamlin, and T. E. Ewing. 2011. Geologic analysis of the Upper Jurassic Haynesville Shale in East Texas and West Louisiana. *AAPG Bulletin* 95(10):1643–1666.

Harrell, D. R. 2017. *Musings of a Seasoned Petroleum Reserves Evaluator about Improving the Reliability of Our Reserves Estimates from Unconventional Reservoirs*. Houston, TX: 7th Reserves Estimation: Unconventionals.

Haskett, W. J., and P. J. Brown. 2011. Recurrent issues in the evaluation of unconventional resources. *American Association of Petroleum Geologists Search and Discovery* 40674:12–15.

Ikonnikova, S., J. Browning, S. Horvath, and S. Tinker. 2014. Well recovery, drainage area, and future drill-well inventory: Empirical study of the Barnett Shale gas play. *Reservoir Evaluation and Engineering* 17(4):484–496.

Inman, M. 2014. The fracking fallacy. *Nature* 516(7529):28–30.

Joskow, P. L. 2013. Natural gas: From shortages to abundance in the United States. *The American Economic Review* 103(3):338–343.

Kaufman, G. M. 1963. *Statistical Decision and Related Techniques in Oil and Gas Exploration*. Upper Saddle River, NJ: Prentice-Hall.

Kaufman, G. M. 1993. Statistical issues in the assessment of undiscovered oil and gas resources. *The Energy Journal* 14(1):183–215.

Kaufman, G. M., Y. Balcer, and D. Kruyt. 1975. A probabilistic model of oil and gas discovery. In *Methods of Estimating the Volume of Undiscovered Oil and Gas Resources*, pp. 113–142. Tulsa, OK: American Association of Petroleum Geologists

Krige, D. G. 1951. A statistical approach to some basic mine valuation problems on the Witwatersrand. *Journal of the Chemical, Metallurgical and Mining Society of South Africa* 52(6):119–139.

Krige, D. G. 1960. One the departure of ore value distributions from the lognormal model in South African gold mines. *Journal of the South African Institute of Mining and Metallurgy* 61(4):231–244.

Krige, D. G. 1966. A study of gold and uranium distribution patterns in the Klerksdorp gold field. *Geoexploration* 4:43–53.

Lee, J., and R. Sidle. 2010. Gas-reserve estimation in resource plays. *SPE Economics Management* 2:86–91.

Limpert, E., W. A. Stahel, and M. Abbt. 2001. Log-normal distributions across the sciences: Keys and clues. *BioScience* 51(5):341–352.

McGlade, C., J. Speirs, and S. Sorrell. 2013. Methods of estimating shale gas resources—Comparison, evaluation and implications. *Energy* 59:116–125.

Montgomery, J. B. 2015. Characterizing shale gas and tight oil drilling and production performance variability. Master of Science Thesis, MIT.

Montgomery, J. B., and F. M. O'Sullivan. 2017. Spatial variability of tight oil well productivity and the impact of technology. *Applied Energy* 195:344–355.

Montgomery, S. L., D. M. Jarvie, K. A. Bowker, and R. M. Pollastro. 2005. Mississippian Barnett Shale, Fort Worth Basin, North-Central Texas: Gas-shale play with multi–trillion cubic foot potential. *AAPG Bulletin* 89(2):155–175.

Mustafiz, S., and M. R. Islam. 2008. State-of-the-art petroleum reservoir simulation. *Petroleum Science and Technology* 26(10–11):130–329.

Prensky, S. E. 1994. A survey of recent developments and emerging technology in well logging and rock characterization. *The Log Analyst* 35(2):15–45.

Rose, P. R. 2001b. *Risk Analysis and Management of Petroleum Exploration Ventures*. Tulsa, OK: American Association of Petroleum Geologists.

Rose, P. R. 2001a. *Risk Analysis and Management of Petroleum Exploration Ventures, Methods in Exploration*. Tulsa, OK: American Association of Petroleum Geologists.

Schuenemeyer, J. H., and L. J. Drew. 1983. A procedure to estimate the parent population of the size of oil and gas fields as revealed by a study of economic truncation. *Mathematical Geology* 15(1):145–161.

Sonnenberg, S. A. 2011. TOC and pyrolysis data for the Bakken Shales, Williston Basin, North Dakota and Montana. In *The Bakken-Three Forks Petroleum System in the Williston Basin*, R. Mountain (Ed.), pp. 308–331. Denver, CO: Association of Geologists.

Sornette, D. 2006. *Critical Phenomena in Natural Sciences: Chaos, Fractals, Selforganization and Disorder: Concepts and Tools*. 2nd ed. New York: Springer.

Strickland, R., D. Purvis, and T. Blasingame. 2011. Practical aspects of reserves determinations for shale gas. *Paper presented at the Society of Petroleum Engineers North American Unconventional Gas Conference and Exhibition*. Woodlands, TX.

Suslick, S. B., and D. J. Schiozer. 2004. Risk analysis applied to petroleum exploration and production: An overview. *Journal of Petroleum Science and Engineering* 44(1–2):1–9.

U.S. Energy Information Administration (EIA). 2011. *Review of Emerging Resources: U.S. Shale Gas and Shale Oil Plays*. Washington, DC: U.S. Department of Energy.

U.S. Geological Survey (USGS). 2004. *Assessment of Undiscovered Oil and Gas Resources of the Bend Arch-Fort Worth Basin Province of North-Central Texas and Southwestern Oklahoma, 2003*. Washington, DC: U.S. Department of the Interior.

USGS. 2011a. *Assessment of Undiscovered Oil and Gas Resources in Jurassic and Cretaceous Strata of the Gulf Coast, 2010*. Washington, DC: U.S. Department of Interior.

USGS. 2011b. *Assessment of Undiscovered Oil and Gas Resources of the Devonian Marcellus Shale of the Appalachian Basin Province, 2011*. Washington, DC: U.S. Department of Interior.

USGS. 2013. *Assessment of Undiscovered Oil Resources in the Bakken and Three Forks Formations, Williston Basin Province, Montana, North Dakota, and South Dakota, 2013*. Washington, DC: U.S. Department of Interior.

Webster, R. L. 1987. Petroleum source rocks and stratigraphy of the Bakken Formation in North Dakota. In *The Rocky Mountain Association of Geologists Guidebook, Williston Basin, Anatomy of a Cratonic Oil Province*. Denver, CO: Rocky Mountain Association of Geologists.

12

Application of Advanced Data Analytics to Improve Haul Trucks Energy Efficiency in Surface Mines

Ali Soofastaei, Peter Knights, and Mehmet Kizil

CONTENTS

12.1 Introduction

Truck haulage is responsible for a majority of cost in a surface mining operation. Diesel fuel, which is costly and has a significant environmental footprint, is used as a source of energy for haul trucks in surface mines. Accordingly, improving truck energy efficiency would lead to a reduction in fuel consumption and therefore greenhouse gas emissions.

The determination of haul trucks fuel consumption is complex and requires multiple parameters including the mine, fleet, truck, speed, payload, operator inputs, fuel, climate, tire, and road conditions as inputs. Data analytics can be used to simulate the complex relationships between the input parameters affecting haul trucks fuel consumption. The aim of this chapter is to introduce an advanced data analytics model to improve the energy efficiency of haul trucks in surface mines.

The most important controllable parameters affecting fuel consumption are payload, truck speed, and total rolling resistance. From these parameters a comprehensive analytical framework can be developed to determine the opportunities for minimizing truck fuel consumption. The first stage of the analytical framework includes the development of the artificial neural network (ANN) model to determine the relationship between truck fuel consumption and payload, truck speed, and total resistance.

This model can be trained and tested using real data collected from some large surface mines in Australia, the United States, and Canada. A fitness function for the haul truck fuel consumption can be successfully generated. This fitness function is then used in the second stage of the analytical framework to develop a digital learning algorithm based on a novel multiobjective genetic algorithm (GA). The aim of this algorithm is to establish the optimum set points of the three controllable parameters to reduce the diesel fuel consumption, with these set points being specific to individual mines and fleet operations.

12.2 Context: Reducing Energy Consumption via Data Analytics

Energy efficiency has gradually become a more important consideration worldwide since the rise of the cost of fuel in the 1970s. The mining industry annually consumes trillions of British thermal units (BTUs) of energy in operations such as exploration, extraction, transportation, and processing. Mining operations use energy in a variety of ways, including excavation, materials handling, mineral processing, ventilation, and dewatering. It also uses significant quantities of power. The Mining industry consumed 520 petajoules (PJ) of energy in 2014–2015 or 9% of the national energy end use in Australia (Allison et al. 2016). Energy consumption in mining is rising at around 6% annually in Australia due to lower grade ores located deeper underground (EEO 2012), a trend seen in other developed countries (DOE 2012). As well as improving margins through efficiency savings, energy streamlining in the sector can also result in the reduction of millions of tons of gas emissions because the primary energy sources used in the mining industry are petroleum products: electricity, coal, and natural gas.

The potential for energy (and financial) savings has motivated the mining industry and governments to conduct research into the reduction of energy consumption. Consequently, a large number of research studies and industrial projects have been carried out in an attempt to do this in mining operations across the world (Soofastaei 2016). Current investments in the improvement of mining technologies and energy management systems have resulted in a significant reduction of energy consumption. Based on completed industrial projects, significant further opportunities exist within the mining industry to reduce energy consumption. The case study presented here—haulage equipment—is one of these potential areas.

Service trucks, front-end loaders, bulldozers, hydraulic excavators, rear dump trucks, and ancillary equipment, such as pick-up trucks and mobile maintenance equipment, are prominent examples of the diesel equipment and associated energy footprint of mining operations. In surface mines, the most commonly used means of mining and hauling of materials is by a truck and shovel operation. The trucking of overburden constitutes a major portion of energy consumption. However, as will be discussed, the rate of this energy consumption is the result of many different parameters (EEO 2010) which can be analyzed and altered to obtain optimal levels of performance.

Data analytics represents a very appropriate approach to pulling together these disparate data sources since it is the science of examining raw data to draw conclusions about that information. The main advantages of data analytics can be presented by cost reduction, faster and better decision-making, and finally new products and services (Soofastaei and Davis 2016). The uses of data analytics are many and can apply to areas that many might not have thought of before. One area that sees much potential in data analytics is the mining industry. For an industry that does trillions of dollars in business every year, data analytics should be considered a necessity not a luxury. Indeed, there are many phases of the mining process where data analytics can be put to use. The four main phases are the (1) extraction of ore, (2) materials handling, (3) ore comminution and separation, and (4) mineral processing. Of particular focus for some companies is efficiency improvements in the second phase, materials handling. Without data analytics at the heart of this phase, operators are more than likely to be subject to suboptimal functioning of their equipment, including in haulage vehicles and infrastructure.

As Figure 12.1 illustrates, use of data analytics in organizations cover two dimensions: time frame (the past, present, or future) and competitive advantage (value of insight generated). At the lowest level, analytics are routinely used to produce reports and alerts. These are simple, retrospective processing and reporting tools, such as pie graphs, top-ten histograms and trending plots, typically addressing variations of the basic question of "what happened and why?" Increasingly, sophisticated analytical tools, capable of working at or near real-time and providing rapid insights for process improvement, can show the user "what *just* happened" and assist them in understanding "why" as well as the next best action to take. However, at the top of the pictured comparative advantage scale, there are predictive models and optimization tools, aimed at evaluating "what will happen" and identifying the best available responses.

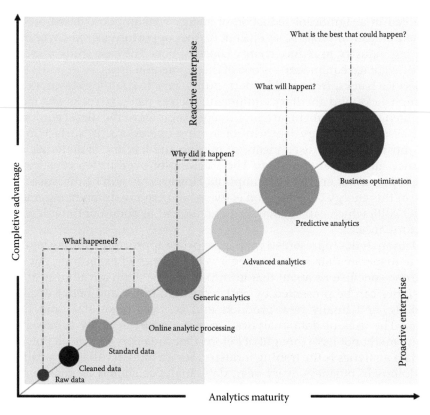

FIGURE 12.1
Competitive advantages of data analytics in organizations. (From Davenport, T. H. et al., *Analytics at Work: Smarter Decisions, Better Results*, Harvard Business Press, New York, 2010.)

12.3 Modeling Haul Trucks' Fuel Consumption

In this chapter, the effects of the three most important and effective parameters on fuel consumption of haul trucks are examined. These parameters are payload (P), truck speed (S), and total rolling resistance (TR). On a real mine site, the correlation between fuel consumption and the three parameters is complex. We use two artificial intelligence methods to create an advanced data analytic model to estimate and reduce haul truck fuel consumption in surface mines. The model can estimate the energy consumption of haul trucks in surface mines using an artificial neural network (ANN) and can also find the optimum values of P, S, and TR using a GA. We analyze each of these in turn and then present the results of our modeling.

12.3.1 Artificial Neural Network

ANNs, also known as neural networks (NNs), simulated neural networks (SNNs), or parallel distributed processing (PDP), are the representation of methods that the brain uses for learning (Hammond 2012). ANNs are a series of mathematical models that imitate a few of the known characteristics of natural nerve systems and sketch on the analogies of adaptive natural learning (Rodriguez et al. 2013). The key component of a ANN paradigm is the unusual structure of the data processing system. ANNs are utilized in various computer applications to solve complex problems. They are fault-tolerant and straightforward models that do not require information to identify the related factors and do not need the mathematical description of the phenomena involved in the process (Beigmoradi et al. 2014).

The main part of a NN structure is a "node." Biological nodes sum the signals received from numerous sources in different ways and then carry out a nonlinear action on the results to create the outputs. NNs typically have an input layer, one or more hidden layers, and an output layer (Figure 12.2).

Each input is multiplied by its connected weight, and in the simplest state, these quantities and biases are combined. They then pass through the activation functions to create the output (Equations 12.1 through 12.3).

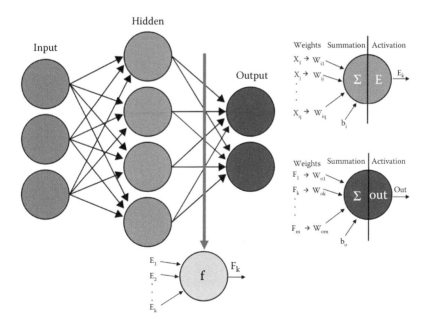

FIGURE 12.2
Artificial neural network structure. (From Soofastaei, A., *Development of an Advanced Data Analytics Model to Improve the Energy Efficiency of Haul Trucks in Surface Mines*, PhD thesis, The University of Queensland, School of Mechanical and Mining Engineering, Brisbane, Australia, 2016.)

$$E_k = \sum_{j=1}^{q} (w_{i,j,k}x_j + b_{i,k}) \qquad k = 1, 2, ..., m \qquad (12.1)$$

where:
  x is the normalized input variable
  w is the weight of that variable
  i is the input
  b is the bias
  q is the number of input variables
  k and m are the counter and number of NN nodes, respectively, in the
     hidden layer

In general, the activation functions consist of linear and nonlinear equations. The coefficients associated with the hidden layer are grouped into matrices $w_{i,j,k}$ and $b_{i,k}$. Equation 12.2 can be used as the activation function between the hidden and the output layers (in this equation, f is the transfer function).

$$F_k = f(E_k) \qquad (12.2)$$

The output layer computes the weighted sum of the signals provided by the hidden layer, and the associated coefficients are grouped into matrices $W_{o,k}$ and b_o. Using the matrix notation, the network output can be given by Equation 12.3.

$$\text{Out} = \left(\sum_{k=1}^{m} w_{o,k}F_k \right) + b_o \qquad (12.3)$$

Network training is the most important part of NN modeling and is carried out using two methods: controllable and uncontrollable training. The most common training algorithm is that of back-propagation. A training algorithm is defined as a procedure that consists of adjusting the coefficients (weights and biases) of a network to minimize the error function between the estimated network outputs and the real outputs.

12.3.2 Optimization of Effective Parameters on Haul Truck Fuel Consumption

Optimization is a part of computational science that represents a very effective way to find the best measurable solution for problems. To solve a given problem, it is important to consider two components: (1) search area and (2) objective function. In the search area, all the possibilities of the solution are considered. The objective function is a mathematical function that associates each point in the search area to a real value, applicable to evaluate all the members of the search area.

Traditional optimization methods are characterized by the stiffness of their mathematical models, making their application limited in representing "real-life" dynamic and complex situations (Selvakumar et al. 2013). Introducing optimization techniques based on artificial intelligence, underpinned by heuristic rulings, have reduced the problem of stiffness. Heuristic rules can be defined as reasonable rules derived from experience and observations of behavior tendencies within a system of analysis.

Using analogies with nature, some heuristic algorithms were proposed during the 1950s by trying to simulate biological phenomena in engineering. Accordingly, these algorithms were termed natural optimization methods. One of the best advantages of using the mentioned algorithms is their random characteristic. Due to their innate flexibility, they have been found to be appropriate to solve all types of problems in engineering (Singh and Rossi 2013; Soleimani et al. 2013; Soofastaei et al. 2016). Rapid advances in computing during the 1980s made the use of these complex algorithms for optimization of functions and processes more practicable when traditional methods were not successful in this area. During the 1990s some new heuristic methods were created by the previously completed algorithms, such as swarm algorithms, simulated annealing, ant colony optimization, and the method used in this study, GAs.

GAs were proposed by Holland (1975) as an abstraction of biological evolution, drawing on ideas from natural evolution and genetics for the design and implementation of robust adaptive systems (Sivanandam and Deepa 2008). Use of the new generation of GAs is comparatively novel in optimization methods. They do not use any derivative information and, therefore, have good chances of escape from local minimums. As a result, their application in practical engineering problems can bring more optimal, or at least more satisfactory, solutions than those obtained by other traditional mathematical methods (Whitley et al. 1990).

GAs are analogous with the evolutionary aspects of natural genetics. From randomly selected "individuals" in any search area, the fitness of the solutions, which is the result of the variable that is to be optimized, is determined subsequently from the "fitness function." The individual that generates the best fitness within the population has the highest chance to return in the next generation with the opportunity to reproduce by crossover with another individual, thus producing decedents with both characteristics. If a GA is developed correctly, the population (a group of possible solutions) will converge to an optimal solution for the proposed problem (Xing and Qu 2013). The processes that have more contribution to the evolution are the crossover, based on the selection and reproduction and the mutation.

GAs have been applied to a diverse range of scientific, engineering, and economic problems due to their potential as optimization techniques for complex functions (Singh and Rossi 2013; Stanković et al. 2013; Tian et al. 2013). There are four significant advantages when using GAs to optimize problems (Yousefi et al. 2013). First GAs do not have many mathematical requirements regarding optimization problems. Second, they can handle many types of objective functions and constraints (i.e., linear or nonlinear) defined in discrete, continuous, or mixed search spaces. Third, the periodicity of evolution operators makes them very efficient at performing global searches (in probability). And finally, they provide us with great flexibility to hybridize with domain-dependent heuristics to allow an efficient implementation for a problem.

It is also important to analyze the influence of certain parameters on the behavior and the performance of the GA, to establish their relationship with the problem necessities and the available resources. The influence of each parameter on algorithm performance depends on the context of the challenge being treated. Thus, determining an optimized group of values to these parameters will depend on a good deal of experimentation and testing. There are a few main parameters in the GA method. Details of these five core parameters are illustrated in Figure 12.3 and tabulated in Table 12.1.

The primary genetic parameters are the size of the population that affects the global performance and the efficiency of the GA, the mutation rate that avoids that a given position remains stationary in value, or that the search becomes essentially random.

12.3.3 The Developed Model

An innovative combined model was introduced to improve the three key effective parameters on the energy consumption of haul trucks. Taking the facets of

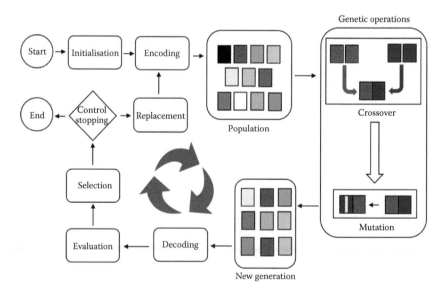

FIGURE 12.3
A simple structure of the genetic algorithm. (From Soofastaei, A., *Development of an Advanced Data Analytics Model to Improve the Energy Efficiency of Haul Trucks in Surface Mines*, PhD thesis, The University of Queensland, School of Mechanical and Mining Engineering, Brisbane, Australia, 2016.)

TABLE 12.1

Genetic Algorithm Parameters

GA Parameter	Details
Fitness function	The primary function for optimization.
Individuals	An individual is any parameter to apply the fitness function. The value of the fitness function for an individual is its score.
Populations and generations	A population is an array of individuals. At each iteration, the GA performs a series of computations on the current population to produce a new population. Each successive population is called a new generation.
Fitness value	The fitness value of an individual is the value of the fitness function calculated for that individual.
Parents and children	To create the next generation, the GA selects certain individuals in the current population, called parents, and uses them to create individuals in the next generation, called children.

the GA approach, in this model P, S, and TR are the individuals and the main function for optimization of the fitness function is fuel consumption. The fitness function was created by an ANN model. This function is a correlation between the fuel consumption of the haul truck, P, S, and TR. After the first step, the completed function goes to the GA phase of the computer code as an input. The developed code starts all GA processes under stopping criteria defined by the model (MSE and R^2). Finally, the improved P, S, and TR will be presented by the model. These optimized parameters can be used to minimize the fuel consumption of haul trucks (Figures 12.4 and 12.5).

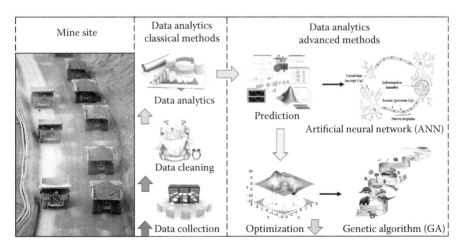

FIGURE 12.4
A schematic of the developed idea to create a combined artificial intelligence model. (From Soofastaei, A., *Development of an Advanced Data Analytics Model to Improve the Energy Efficiency of Haul Trucks in Surface Mines*, PhD thesis, The University of Queensland, School of Mechanical and Mining Engineering, Brisbane, Australia, 2016.)

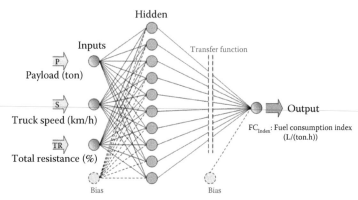

Artificial neural network (prediction) model

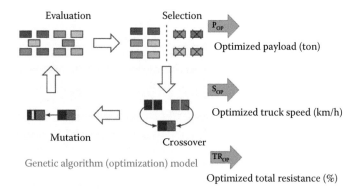

Genetic algorithm (optimization) model

FIGURE 12.5
Details of developed model. (From Soofastaei, A., *Development of an Advanced Data Analytics Model to Improve the Energy Efficiency of Haul Trucks in Surface Mines*, PhD thesis, The University of Queensland, School of Mechanical and Mining Engineering, Brisbane, Australia, 2016.)

12.4 Results

The indicated artificial intelligence model that was developed was then tested against real data taken from some types of popular trucks in four big surface mines in the United States, Canada, and Australia. Some information about these mines and trucks is presented in Table 12.2 (Figures 12.6 through 12.9).

To test the developed networks and validate the developed model, 1,000,000 independent samples collected from four mines were used. As our figures illustrate, the results show good agreement between the actual and estimated values of fuel consumption. Figure 12.10 presents sample values

TABLE 12.2

Case Studies

Case Study	Location	Mine Type	Mine Details	Investigated Truck
Mine 1	Queensland, Australia	Surface coal mine	The mine has coal reserves amounting to 877 million tons of coking coal, one of the largest coal reserves in Asia and the world. It has an annual production capacity of 13 million tons of coal.	CAT 793D
Mine 2	Arizona, United States	Surface copper mine	The mine represents one of the largest copper reserves in the United States and the world, having estimated reserves of 3.2 billion tons of ore grading 0.16% copper.	CAT 777D
Mine 3	Arizona, United States	Surface copper mine	The deposit had estimated reserves (in 2017) of 907 million tons of ore grading 0.26% copper and 0.03% molybdenum.	CAT 775G
Mine 4	Ontario, Canada	Surface gold mine	This mine produced 235,000 ounces of gold in 2016, at the cost of sales of $795 per ounce, and all-in sustaining costs of $839 per ounce. The mine's proven mineral reserves as of December 2016, were 1.6 million ounces of gold.	CAT 785D

for the estimated (using the ANN) and the independent (tested) fuel consumption to highlight the insignificance of the values of the absolute errors in the analysis for the four mines that were studied.

All processes in the developed model, then, certainly work based on the present dataset collected from four large surface mines. The results of using developed model for the selected real-life mines are given in Tables 12.3 through 12.6. They, therefore, could presumably be replicated using the same method for other surface mines.

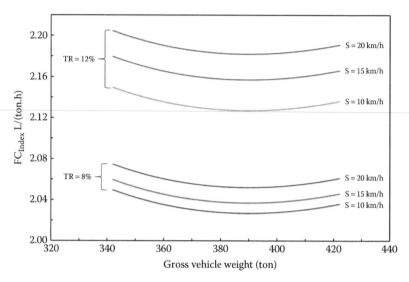

FIGURE 12.6
Correlation between Gross Vehicle Weight, S, TR, and FC_{Index} based on the developed ANN model for CAT 793D. All data were collected from a surface coal mine located in Central Queensland, Australia (Mine 1).

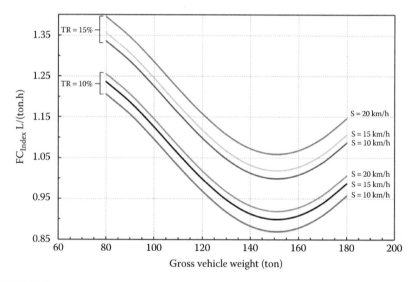

FIGURE 12.7
Correlation between GVW, S, TR, and FC_{Index} based on the developed ANN model for CAT 777D. All data were collected from a surface copper mine located in Arizona, United States (Mine 2).

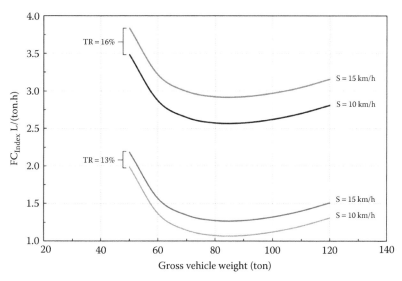

FIGURE 12.8
Correlation between GVW, S, TR, and FC_{Index} based on the developed ANN model for CAT 775G. All data were collected from a surface copper mine located in Arizona, United States (Mine 3).

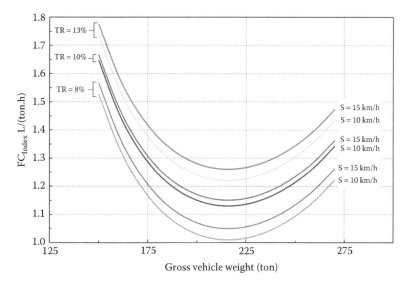

FIGURE 12.9
Correlation between GVW, S, TR, and FC_{Index} based on the developed ANN model for CAT 785D. All data were collected from a surface coal mine located in Ontario, Canada (Mine 4).

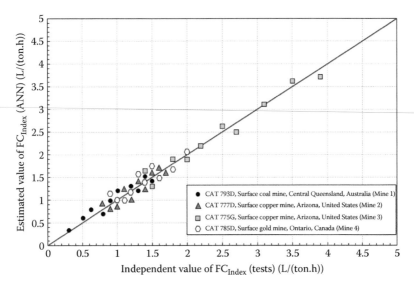

FIGURE 12.10
Sample values for the estimated (using the ANN) and the independent (tested) fuel consumption index.

TABLE 12.3

The Range of Normal Values and Optimized Range of Variables by GA Model to Minimize Fuel Consumption by Haul Trucks (Caterpillar 793D in Mine 1)

	Normal Values		Optimized Values	
Variables	Minimum	Maximum	Minimum	Maximum
Gross vehicle weight (ton)	340	430	380	400
Total resistance (%)	8	12	8	9
Truck speed (km/hr)	10	20	10	15

TABLE 12.4

The Range of Normal Values and Optimized Range of Variables by GA Model to Minimize Fuel Consumption by Haul Trucks (Caterpillar 777D in Mine 2)

	Normal Values		Optimized Values	
Variables	Minimum	Maximum	Minimum	Maximum
Gross vehicle weight (ton)	80	180	140	160
Total resistance (%)	10	15	10	11
Truck speed (km/hr)	10	20	10	12

TABLE 12.5

The Range of Normal Values and Optimized Range of Variables by GA Model to Minimize Fuel Consumption by Haul Trucks (Caterpillar 775G in Mine 3)

Variables	Normal Values		Optimized Values	
	Minimum	Maximum	Minimum	Maximum
Gross vehicle weight (ton)	50	120	70	90
Total resistance (%)	13	26	13	14
Truck speed (km/hr)	10	15	10	13

TABLE 12.6

The Range of Normal Values and Optimized Range of Variables by GA Model to Minimize Fuel Consumption by Haul Trucks (Caterpillar 785D in Mine 4)

Variables	Normal Values		Optimized Values	
	Minimum	Maximum	Minimum	Maximum
Gross vehicle weight (ton)	150	275	200	225
Total resistance (%)	8	13	8	9
Truck speed (km/hr)	10	15	10	12

12.5 Conclusions

The aim of this chapter was to formulate an advanced data analytics model capable of improving haul truck fuel consumption based on the relationship between P, S, and TR. From the available "real-life" datasets obtained from surface mining operations, this relationship is extremely complex to disect using traditional analysis. Therefore, an artificial intelligence method was adopted to create a reliable model to analyze the problem.

The first element of this method was to utilize an ANN model to establish a correlation between P, S, and TR with fuel consumption. The results of this correlation showed that fuel consumption has a nonlinear relationship with the investigated parameters. The ANN was then trained and tested using the collected real mine site dataset, with there being good agreement between the actual and estimated values of fuel consumption. Building upon this material, a GA model was developed for considering the optimization of effective parameters on fuel consumption in haulage trucks, which in turn could maximize the energy efficiency in haulage operations.

From these amalgamated models, the range of all studied effective parameters on fuel consumption of haul trucks were optimized, and the best values of P, S, and TR to minimize fuel consumption index (FC_{Index}) were highlighted. The developed model was applied to analyze data for four big coal, copper, and gold surface mines in the United States, Canada, and Australia.

References

Allison, B., A. Shamim, M. Caitlin, and P. Pha. 2016. *Australian Energy Update 2016*. Canberra, Australia: Office of the chief economist, Department of Industry Innovation and Science.

Beigmoradi, S., H. Hajabdollahi, and A. Ramezani. 2014. Multi-objective aeroacoustic optimisation of rear end in a simplified car model by using hybrid robust parameter design, artificial neural networks and genetic algorithm methods. *Computers and Fluids* 90:123–132.

Davenport, T. H., J. G. Harris, and R. Morison. 2010. *Analytics at Work: Smarter Decisions, Better Results*. Vol. 3. New York: Harvard Business Press, p. 93.

DOE. 2012. *Mining Industry Energy Bandwidth Study*. Washington, DC: Department of Energy, USA Government.

EEO. 2010. *Energy-Mass Balance: Mining*. Canberra, Australia: Australian Government, Department of Resources Energy and Tourism.

EEO. 2012. *First Opportunities in Depth: The Mining Industry*. Canberra, Australia: Australian Government, Department of Resources Energy and Tourism.

Hammond, A. 2012. Development artificial neural network model to study the influence of oxidation process and zinc electroplating on fatigue life of grey cast iron. *International Journal of Mechanical and Mechatronics Engineering* 12(5):128–136.

Holland, J. H. 1975. *Adaptation in Natural and Artificial Systems: An Introductory Analysis with Applications to Biology, Control, and Artificial Intelligence*. Vol. 2. Washington, DC: University of Michigan Press, pp. 183–195.

Rodriguez, J. A., Y. El Hamzaoui, J. A. Hernandez, J. C. García, J. E. Flores, and A. L. Tejeda. 2013. The use of the artificial neural network (ANN) for modelling the useful life of the failure assessment in blades of steam turbines. *Engineering Failure Analysis* 35:562–575.

Selvakumar, S., K. P. Arulshri, and K. P. Padmanaban. 2013. Machining fixture layout optimisation using a genetic algorithm and artificial neural network. *International Journal of Manufacturing Research* 8(2):171–195.

Singh, A., and A. Rossi. 2013. A genetic algorithm based exact approach for lifetime maximisation of directional sensor networks. *Ad Hoc Networks* 11(3):1006–1021.

Sivanandam, S. N., and S. N. Deepa. 2008. *Introduction to Genetic Algorithms*. Vol. 3. New Delhi, India: Springer, p. 453.

Soleimani, H., M. Seyyed-Esfahani, and M. A. Shirazi. 2013. Designing and planning a multi-echelon multi-period multi-product closed-loop supply chain utilising genetic algorithm. *The International Journal of Advanced Manufacturing Technology* 68(1–4):917–931.

Soofastaei, A., S. M. Aminossadati, M. M. Arefi, and M. S. Kizil. 2016. Development of a multi-layer perceptron artificial neural network model to determine haul trucks energy consumption. *International Journal of Mining Science and Technology* 26(2):285–293.

Soofastaei, A., S. M. Aminossadati, M. S. Kizil, and P. Knights. 2016. Reducing fuel consumption of haul trucks in surface mines using genetic algorithm. *Applied Soft Computing*.

Soofastaei, A., and J. Davis. 2016. Advanced data analytic: A new competitive advantage to increase energy efficiency in surface mines. *Australian Resources and Investment* 1(1):68–69.

Stanković, L. J., V. Popović-Bugarin, and F. Radenović. 2013. Genetic algorithm for rigid body reconstruction after micro-Doppler removal in the radar imaging analysis. *Signal Processing* 93(7):1921–1932.

Tian, J., M. Gao, and Y. Li. 2013. Urban logistics demand forecasting based on regression support vector machine optimised by chaos genetic algorithm. *Advances in Information Sciences and Service Sciences* 5(7):471–478.

Whitley, D., T. Starkweather, and C. Bogart. 1990. Genetic algorithms and neural networks: Optimising connections and connectivity. *Parallel Computing* 14(3):347–361.

Xing, H., and R. Qu. 2013. A nondominated sorting genetic algorithm for bi-objective network coding based multicast routeing problems. *Information Sciences* 233(3):36–53.

Yousefi, T., A. Karami, E. Rezaei, and E. Ghanbari. 2013. Optimisation of the free convection from a vertical array of isothermal horizontal elliptic cylinders via a genetic algorithm. *Journal of Engineering Physics and Thermophysics* 86(2):424–430.

13

Augmented Reality in Geological Modeling: Development and Use of the Leapfrog Aspect Viewer

P.J. Hollenbeck

CONTENTS

13.1 Introduction

Leapfrog Aspect represents a shift in the geologic modeling process by allowing the user to view their model as it is in the real world. Rather than bringing the "real world" into the modeling software through data types such as aerial imagery, surface and airborne LiDAR, and photogrammetry, the digital model is brought into the "real world." The user can simply see how their model stacks up with reality to instantly recognize and understand the changes that need to be made. While Leapfrog Aspect is purpose-built to display geologic data, it can be used to superimpose any geo-referenced data in relation to the user's physical space. This chapter discusses the context and considerations behind conceiving the tool, the technological challenges and solutions involved in its development, and the exciting potential future evolution of the package.

13.2 Augmented Reality and the Mining Sector

Augmented reality (AR) is generally defined as a method of superimposing data or digital media onto the real world. It has recently become a "buzz word" in the consumer space, but has existed in one form or another for decades. For example, one of the more well-known versions of AR would be the heads-up display used by military fighter pilots. Likewise, numerous sporting events are now using AR as part of their television feed—it is not uncommon to see scores, play lines, ball trajectories, or even athlete's names digitally overlain on the screen to enhance the viewer's experience.

Leapfrog Aspect is a very similar concept. It superimposes digital model data onto a mobile device's camera feed to allow the user to visualize the model and the physical space simultaneously (Figure 13.1). As will be discussed in more depth in this chapter, Leapfrog Aspect will use the device's coordinate position to accurately place the user within their data, and units from the dataset will automatically convert to real-world size, rendering a realistic first-person representation of anything that is modeled and loaded into the viewer. Typically, these data will be geology-specific (given the data originates in our Leapfrog desktop suite of geo-modeling products), such as lithological domains, mineralization, drilling programs, and grade interpolations, which results in an "X-ray" view of the groundmass in question. However, Leapfrog Aspect's viewer can display anything that is modeled in 3D and (preferably) incorporates georeferenced location information.

FIGURE 13.1
A primary use-case for Leapfrog Aspect is visualizing geologic model data (black and dark gray vein) with the physical outcrop.

The initial idea for an AR viewer for geologic data began in 2011, when smartphones were not only becoming ubiquitous, but also had the requisite sensors to be able to process and display AR data. (Incidentally, there are four sensors that are absolutely required for using AR—an accelerometer to measure overall device orientation, a gyroscope for fine-motion tracking, a magnetometer (compass) to tie the view back to a real-world reference point, and preferably (though somewhat optional) a camera to get the physical view in conjunction with the digital overlay.) The primary use-case for AR was in apps such as Yelp, wherein a person could look for pizza parlors nearby and be able to see flags in real space to show a first-person distribution of their search results (Figure 13.2).

The potential for such an application in the mining and geologic space seemed completely obvious. To do any modeling in the various software packages, the input data already must be georeferenced to "some" grid in space, whether or not a standard coordinate space or a local grid, ensuring that all the data resides in the same general location. As such, the resulting geologic model would also be georeferenced and scaled to the same units originally used in the input data. Because mining data tends to be of much higher positional accuracy than consumer-grade GPS and map positions, the thought was that if we could feed precisely located data to the AR viewer, then it should certainly be a more useful tool than a largely superfluous restaurant-tracking app.

FIGURE 13.2
Yelp's "Monocle" AR viewer option showing tags for restaurants over their physical locations.

From a consulting standpoint, the implementation of an AR viewer was considered instrumental in gaining rapid insight into an operation during a relatively short site visit. To fully obtain perspective on the magnitude of a given project, the consultant may find themselves standing on a ridge and hand-waving with the on-site geologist to orient oneself to the property. Alternatively, the consultant could stand on the same ridge with the Leapfrog Aspect viewer in hand and see the model as it is situated relative to the property, as well as all the drilling, structural measurements, and surface mapping done to contribute to the model. The consultant suddenly goes from having a general idea to having a very detailed understanding of the property in the same timeframe.

Likewise, from an operational sense, the use-cases for AR are endless. A geologist could review their model in the field by viewing it in relation to the actual geology to confirm if it is reasonable and sensical. Annotations and adjustments could be made "on the fly," or completely new structural measurements could be taken with the same mobile device to be automatically integrated into the model upon return to the office. Operations crews could also use AR for a variety of applications. 3D ore control could be projected on a dig face, while haulage routes could be shown on a heads-up display for accurate ore/waste transfer. Safety concerns could likewise be addressed in AR. Rock-fall hazards could be identified and displayed in the viewer, as could underground hazards (old mine workings, etc.) or even active blast patterns to avoid.

Finally, civil projects would also see a benefit from an AR perspective. Building projects could "show" the finished product well before the first shovel started on the foundation. Tunnels could be completely visualized from the portal location or anywhere else to see exactly where the excavation would run, and in conjunction with any geologic model that had been constructed to help drive the design.

13.3 The Platform and Input Data

The first iterations of Leapfrog Aspect were extremely limited in capability. It was initially tested as a native-built Android application, which caused significant problems in attempting to load the relatively large amounts of data in a geologic model (Figure 13.3). The same platform used by Yelp was also investigated, but we quickly found that displaying a dozen location flags in Yelp was not the same as displaying 250,000 triangles on a given lithologic model.

Unity is a gaming platform used across nearly all consumer device types, and is one of, if not the biggest, gaming development platforms available. Unity also has a very extensive virtual reality and AR development library,

FIGURE 13.3
One of the first builds of Leapfrog Aspect as a native android application. It was limited to 25 points displayed before it would freeze.

and is purpose-built to display extremely detailed graphical objects. As such, it represented an ideal new platform for our needs, and we set about rebuilding Leapfrog Aspect from scratch using this development environment. In addition to all the built-in AR tools, one key element to using Unity was the cross-platform functionality. Rather than building a completely different version of Leapfrog Aspect for each device type (Android, iOS, and potentially Windows), we could leverage the existing capability of Unity to build a single version of the app, and deploy it quickly to all the desired device types and platforms.

The Leapfrog suite of software also proved to be ideal for generating content for Leapfrog Aspect. In summary, Leapfrog desktop software uses implicit modeling to build geologic domains, mineralized zones, and grade interpolations. Leapfrog products can also create exported "scenes" which are effectively data packets that can be loaded and viewed on a freely available viewer, and their purpose is to allow the end-user to visualize the model exactly as it was built and manipulated within the original editing software. (It is not unlike a 3D PDF, but with significantly more functionality.)

These scene files can be comprised of any data type used in the modeling software—drill holes, points and polylines, textured wireframe meshes, and block models can all be exported within scenes (Figures 13.4 and 13.5). The exported scene file was an ideal data source for Leapfrog Aspect as well—it is highly portable and self-contained, making it easy to move onto the mobile device using a hard-wire connection or a cloud storage option.

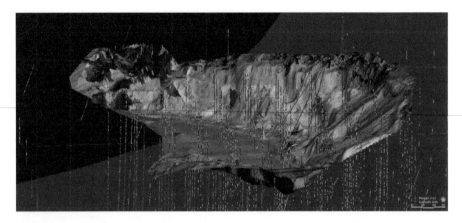

FIGURE 13.4
A "scene" exported from Leapfrog Geo, showing a topographic surface (textured with an aerial photo), a fault plane (gray), and drill holes displaying assay results (multicolored lines).

FIGURE 13.5
The same scene as in Figure 13.4 shown on-site in Leapfrog Aspect.

13.4 Development Challenges and Solutions

AR, particularly in the capacity to which Leapfrog Aspect is built, is surprisingly complex. It is necessary to account for all real-world variables that would affect how the data would display—physical position, camera orientation, compass declination, and data scaling were some of the factors we needed to consider.

Virtual reality would have been significantly easier to build into an app. The key difference between AR and virtual reality is the camera feed—virtual reality can be considered total immersion, wherein the user is completely engulfed in the virtual world. All sense of the outside space is cut off, as there is no visual connection with that world. Once a virtual reality headset is put on, the real world disappears. As such, there are relatively few positional considerations to make—physical location is unnecessary, as is a north direction—with no reference to reality, it doesn't matter which direction or position the view takes for the user.

In contrast, the fact that AR is literal to its name means that it allows the viewer not only to see the real world such that it can be augmented with digital material, but also to accurately position that data so it looks correct relative to the rest of the world.

13.4.1 Accurate Positioning in XYZ

The first factor for accurate positioning was to incorporate GPS to at least start in a reasonable point within the data. However, GPS is not enough. Often the built-in GPS is not sufficient—certainly in situations where GPS is unavailable, such as tunnels or underground mines, it would be pointless. Likewise, many consumer-grade devices (i.e., those utilized by our expected user base) do not have sufficient accuracy with their built-in GPS, or do not have GPS at all, as is the case with most iPad tablets. As such, we needed to consider ways to improve positional accuracy, particularly relative to the data. Perhaps unsurprisingly, the datasets themselves would sometimes conflict with accurately positioned viewer locations, so it was more necessary to set the position relative to the data rather than at an absolute real-world location.

Two additional positioning capabilities were added to address the issue of conflicting GPS or lack thereof. The first was to simply key in coordinate positions into an XYZ series of cells. If the user was standing at a known location, they could plug in those positional values and the viewer would override any external positioning data to locate exactly at that spot. This worked well, except when the survey point was measured at ground-level—it would pull the model up to the level of the viewer, which would typically be off by 1.5 m or so. As such, an option was added to set the viewer's height so that the data would be offset downward to set the "ground level" to the ground.

The second fix to the positioning issue was essentially an automation to the manual coordinate entry method. We incorporated a QR code reader into the application such that the user could build a QR code to represent a known coordinate position, and then post that QR code at the appropriate position in the field. Then, when the QR code reader option was used, it would see the code, read the coordinate values, and set the position of the viewer accordingly.

GPS positioning is also notorious for sub-par elevation readings—consumer-grade devices can typically get the XY position correct to within

1–2 m, but elevation can be off by 50 m or more. To combat this problem, we added an option to override the GPS elevation with a fixed value—this was of particular utility when used in open-pit mines, as the benches are typically identified by their elevation. Setting the elevation meant the user had to be aware if they left a given mine bench, but otherwise would have significantly more accurate positioning results when on that selected bench.

13.4.2 Accurate Direction Adjustments

A second factor to consider was that of the magnetometer and correcting for magnetic versus true north as well as any other directional issues that may arise. Again, due to the variable quality of the target devices, the compass could be off by any number of degrees and would need a way to correct that offset. In most cases we encountered, the data was entered using true north, while the mobile device was feeding back magnetic north directions, so the model would be negatively rotated around the Z axis. This was particularly problematic in locations like New Zealand where the magnetic declination is around 25°.

Automatic adjustment was initially considered for a fix, in that the app would identify the location of the device and pull the declination automatically from an online resource. However, three problems were identified in this approach: (1) the device would need to be connected to the internet to complete this task, which is often an issue in a remote mine, (2) the second issue was that of local rotated grids—the magnetic declination alone may not be representative of other rotation factors applied to the local grid in use on a mine site, so depending upon a solely automatic approach would not solve this problem, (3) likewise, devices with problematic compasses would not be able to be used if a variable correction factor could not be applied. As such, a manual rotation factor was added so the user would be able to "tweak" the adjustment factor to force north to point north, and all data would then display relative to that orientation adjustment.

13.4.3 Viewing Angle

A third factor to consider was that of the viewer's pitch angle. If the camera is pointed down, it needs to display what is directly down from the device, while if it is pointed straight ahead, it must show what is ahead. For the most part, this was not an issue because of the libraries built into Unity. Unity was already prepared to handle the accelerometer and gyroscope readings provided by the device. The bigger issue comes from the devices. While smartphones tend to be equipped with the requisite sensors to run AR, the larger and seemingly better suited Android and Windows tablets often do not have a gyroscope sensor. As such, when attempting to use these devices with Leapfrog Aspect, objects would appear on the display as though the user was looking straight down and would not adjust with motion. It was necessary to come up with a recommended list of devices to counteract this issue. We vetted or tested several common tablets to make sure they would work with the app and had all the required components.

13.4.4 Scaling and Units

An ongoing issue in the mining industry in North America (most often affecting the United States) is that of units and unit conversions, and they proved to be an issue for Leapfrog Aspect during development. The units reported from the device's GPS were always Lat/Long, which were then converted to UTM WGS84 metric units for use in the app. However, since Leapfrog products are unit-agnostic, any models built using Imperial units instead of metric would appear 3.2808x larger than they should have been and in a location not relative to the actual site of interest. As such, we implemented a suite of transformation capabilities to go from simple feet-to-meter conversions all the way to the most obscure local grid system imaginable so that the data could be visualized anywhere. To perform the transform, we also leveraged the same projection file system used by most desktop GIS platforms to be able to read a standard.prj definition file, so the conversion would be as simple as possible.

13.4.5 Form Factor

Certainly the biggest hurdle, and one that we haven't surpassed, is that of the tablet and smartphone form factor. The difficulty in these products is that we are attempting to superimpose 3D data onto a 2D flat screen, which leads to issues of depth perception and skewed reality. For example, the following image shows a lithologic body "floating" in space because it is being viewed against a 2D camera feed. The depth of the 3D model is lost, and it ends up appearing closer than it is, or simply skewed and distorted (Figure 13.6).

Unfortunately, there's not a quick fix for this issue, as it's unlikely that these devices will see true 3D displays in the near future. 3D displays were attempted on a few devices in the past but did not meet with commercial success.

A useful trick for working around the flat-tablet issue is to include a current topographic surface in the scenes to be displayed (at least for surface scenarios; this trick is less useful for underground operations). By including a topography, the scene suddenly has a "real" anchor that can tie the digital data to the physical view and provide a bit of implied depth. If the viewer can see that the digital topography is matching the real topography, then the associated objects in the scene can also be tied to that digital topography and by proxy the physical positioning of those objects looks much more realistic (Figure 13.7).

FIGURE 13.6
A lithologic volume and drill holes in the Leapfrog Aspect viewer. It is a bit difficult to get a proper impression of depth and 3D positioning, particularly of the drill holes, due to the 2D camera feed and device screen.

FIGURE 13.7
Adding topography (white color region) gives a better sense of 3D since the drill holes can now be visualized "behind" the pit wall and below the floor rather than simply floating in space.

13.4.6 Device Capability

Another factor that is a challenge for Leapfrog Aspect is the processing and graphical power of a typical mobile device. We are used to our desktop or laptop computers and their ability to process and display millions of data points quickly and to move about the screen in a fluid and smooth manner.

The intent of those larger devices has always been of "content creation" so it's not unreasonable to expect that they would display large content if that is required.

Mobile devices, on the other hand, are geared more toward "content consumption," such as watching videos, aggregating news feeds, reading e-mail, and so on. They are not intended for creating large amounts of data and their run-time capability of displaying that data is not up-to-par with their larger counterparts. That paradigm is quickly shifting with the popularity of laptop-hybrid tablets such as the Windows Surface Pro series or the iPad Pro, and arguably smartphones such as the Lenovo Phab 2 Pro with its LiDAR-esque capability to map the user's surroundings are bridging that gap between consumption and creation. Additionally, some devices such as the Shield tablets built by Nvidia are equipped with dedicated Kepler graphics processors, predominantly for playing high-resolution games but with the fortunate side-effect of improving the graphical performance of Leapfrog Aspect. However, for the general consumer, the device at hand is still intended for consumption and not large-scale runtime display of data.

For the most part, large data files will load slowly, taking as long as 2 min or more before they will display on a typical device. Text will also impact the load time and viewer smoothness. If a collection of one thousand drill holes are all labeled with their Hole ID in the scene, movement will result in jittery rendering and "tracer" effects. This is not surprising, as similar issues exist on desktops. Text is notoriously graphics-intensive and displaying a lot of text in 3D space will slow down even high-end desktop computers if they don't have a powerful graphics card.

Unity is also not intended to work as we're using it—game environments are typically static, predefined collections of meshes, objects, and textures, and while they can be data-dense and high resolution within the game, they are more quickly read by the system due to their existence within the game code rather than being built on-the-fly. Our option to use that platform rather than to develop stand-alone apps for each operating system has resulted in slower load-times than what we may expect in a platform-specific build, and certainly slower than if each scene was built as a static game environment optimized within Unity. But to do that we would eliminate the ability to dynamically load and change scene files within the viewer, which would significantly reduce the utility of Leapfrog Aspect.

13.4.7 Reception and Criticism

Leapfrog Aspect has been generally very well received by the geologic modeling industry, with over 1,000 installs to date. The numbers are small relative to a consumer-grade app like Instagram, but that's to be expected—it's an early adopter product in an extremely niche market. It has also only been available to consumers on Android, which has limited the user uptake.

The biggest criticism received thus far is that data is slow to load and very large datasets cannot be visualized. As such, it has been necessary to set certain expectations for what sort of data to load into Leapfrog Aspect. Arguably, it would be fantastic to see a complete model with all drilling, meshes, grade shells, and an aerial image simultaneously within the Leapfrog Aspect viewer, but it's not particularly practical. Once in the field, all that data would be messy and would need to be pared down regardless of what resides in the file. For example, it's not meaningful to show a giant volume of host rock when you're really interested in the narrow vein hosting the gold, so is there a reason to include the host rock mesh in the viewer file?

The most productive use-case for Leapfrog Aspect in all scenarios is to have focused scene files, wherein the user is interested in a specific feature or situation in their data. Perhaps the modeler wants to make sure a fault is situated in their model in the same position as it is appearing in the freshly excavated mine wall. A simple scene could be built with the fault surface and topography to reduce the load time and clutter. Likewise, rather than loading all drill holes from the entire district, the user could focus on the latest year's drilling campaign or only show core holes from the past 5 years. Certainly, as the hardware matures and becomes more capable these limitations should disappear (would anyone have expected to be able to build a 40 GB 3D model on their Windows computer 15 years ago?) but for now we must work within the capabilities of the available mobile devices (Figure 13.8).

FIGURE 13.8
The simplest example of a good Aspect scene: a basic structural measurement (gray disk) against the fault from which it was measured.

13.4.8 What Is Next for Leapfrog Aspect

AR and virtual reality have been concepts for a long time (Ivan Sutherland invented the first AR headset in 1966) but they are also considered to be in their infancy relative to modern technology and potential capabilities. Many of the obstacles and shortcomings we are experiencing in Leapfrog Aspect are already becoming obsolete and will be non-existent in the next few years.

Positioning capabilities are improving rapidly due to methodologies like simultaneous location and mapping (SLAM) which uses the device's inertial sensors and camera to accurately map the movement of the device down to centimeters without using GPS or other external input. Additionally, wireless beacons are already in use in urban areas to pinpoint mobile devices down to a floor level in a mall. There is no reason these devices could not also be utilized in a mining environment aside from some ruggedization requirements.

Graphical and processing capabilities will continue to build and improve, largely because Leapfrog Aspect will piggy-back on the massive mobile gaming market. We can already see the improvements brought to market by the Nvidia Kepler graphics processing unit and similar dedicated graphics cards. The processors and RAM in mobile devices are quickly catching up to larger machines as well, since 64-bit processors are becoming commonplace and typically paired with 6–8 GB of RAM. The combination of specs is rivaling laptops made within the past 5 years.

The natural progression of AR is to follow the path of virtual reality and move away from the hand-held tablet form-factor to head-mounted units more appropriate for visualization (Ivan Sutherland's original AR setup was a headset, not a tablet). Smart glass wearables like Google Glass are going to undoubtedly solve the problem of the 2D display, since many are now utilizing stereoscopic eyepieces—glasses that have pass-through displays for each eye—allowing depth perception to be simply integrated into the viewer. Stereoscopic smart glasses will also eliminate the use of the camera feed as the physical world is completely visible through the eyepieces. These glasses are already in use in other industrial applications and are available as ruggedized units with ANSI Z.87 ratings suitable for mining environments.

In a similar sense, mixed reality is another step forward for the digital/physical convergence we are seeking with Leapfrog Aspect. Microsoft has effectively pioneered this concept through their HoloLens headset; it allows the user to see the physical world while data is superimposed around them, not dissimilar to AR. The area sensors of HoloLens go beyond AR. It can detect the physical world through a combination of an infrared scanning camera in conjunction with a fisheye camera lens, effectively providing low-power LiDAR-style surface mapping. This information can be fed back to the system such that it can integrate physical objects into the digital environment. It's not just that the digital feed is superimposed, but rather that it's interacting with the physical space. For instance, a digital ball thrown at a real table top will bounce off the table top due to the digital representation of that table top mapped by the HoloLens' sensors. For Leapfrog Aspect, this can mean several possible scenarios. The geologist can have a fully 3D "model" sitting on their desk in mixed reality while they adjust the model on their desktop. The model can then be blown up to life-size dimensions so the geologist can move around as though they were looking at the model in the field. They could then take that model and physically move to site and see the model against the actual mine, like Leapfrog Aspect's current functionality but with the added benefit of 3D stereoscopic visualization. Finally, the same model could be simulcast to multiple HoloLens devices globally, such that a virtual tour of a property would be instantly accessible to anyone, anywhere. It would no longer be necessary to fly an expert to a site to assist with a project, and key decision makers could gain a more complete understanding of an active project without ever setting foot on the property (Figure 13.9).

The HoloLens and the ability to "virtual tour" have already proven to be quite successful for the National Aeronautics and Space Administration (NASA). They can virtually send geologists to Mars by generating photogrammetric environments constructed from photos taken by the Mars rovers. Through the HoloLens, the geologists can "walk" on the virtual landscape and gain a more complete understanding of the planet's structure and formation while also avoiding the 2-year travel time.

It's impossible to consider mobile devices without also considering cloud computing and remote file access. The ideal scenario for Leapfrog Aspect would be to receive on-the-fly updates from a local server such that when

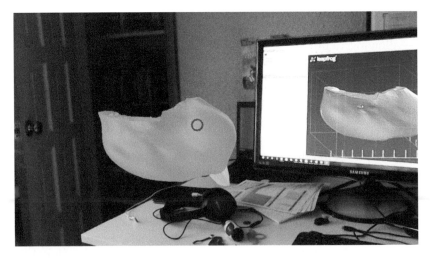

FIGURE 13.9
A scene as viewed from the HoloLens headset "placed" on the physical desktop. Note the same scene in Leapfrog on the computer display.

an edit is made to a model in the office, that change is instantly propagated to the Leapfrog Aspect viewer, so the field geologist sees the update in real time. This does not have to be limited to geologic work either—3D dig faces could be sent out to equipment operators the moment an ore control calculation is completed, reducing any potential down-time for the shovel at the face. Likewise, geotechnical or other safety issues could be instantly sent out to the entire mine site as a viewable "hot spot" or polygonal boundaries to delineate exactly where the hazard exists and what to avoid.

13.5 Conclusion

As other chapters in the book indicate, modern mining operations are extremely data-rich, and companies are increasingly seeking to utilize these reams of collected information to improve efficiency and competitiveness. Leapfrog Aspect is at the forefront of such development, offering the potential to be a paradigm shift in data collection and visualization by moving data out of the traditional plan map and directly into the user's physical space. As the preceding text outlines, it is already a highly functional and cutting-edge tool, presenting complex, multilayered data in an intuitive fashion. However, we are merely scratching the surface regarding its potential and with rapid iterations of hardware improvements the software capabilities will be realized with conjugate speed, such that soon a set of smart glasses will join the Brunton and rock hammer as required tools for any field geologist.

14

Streamlining and Standardizing Data Use in the Extractive Industries

Energistics Consortium

How does your company provide an innovative edge to extractive industry clients?

Energy operators are deploying increasingly sophisticated workflows to solve the challenges of subsurface modeling, drilling, and resource extraction. This deployment invariably involves the use of a wide range of software and data management tools from different vendors. Energistics data standards enable solutions that move data around with the assurance of consistency and integrity. Energistics standards also support innovative tech companies, often start-ups or relatively small enterprises, by providing a well-documented and widely accepted set of tools to address comprehensive data input and output formats and protocols. This support allows them to focus the bulk of their efforts on the original technologies that they are developing, which in turns accelerates their time to market and ensures that their solutions will fit into the information technology (IT) "ecosystem" of their customers.

What have been the largest barriers to entry in developing and incorporating innovation in the oil and gas industry?

A handful of large vendor and service companies develop and support the major IT platforms that support the bulk of the exploration, drilling, and production workflows. While most of them provide some (inter)connection facilities, they are not designed to collaborate with each other or with third-party solutions. As happens in most industries, innovation occurs mainly in the research and development (R&D) laboratories of major and national oil companies, and through the efforts of small independent vendors focused on solving specific problems. Until recently, such advances struggled to gain acceptance because of the efforts required to make these innovative apps interoperate with each of the different major platforms. The amount of work is often quite substantial as it involves a learning curve on how to establish the connectivity processes for each separate platform.

Has the pace of the expansion in the amount of data available to companies typically outstripped their capacity to effectively utilize it?

There are two facets to that question. In terms of the ability of the software to handle terabyte-sized datasets and thousands of discrete and complex data items such as drilled bore holes and their related measurements, the oil and gas industry has managed to keep abreast with software tools and computer architectures that have the necessary bandwidth to cope. There are, however, severe bottlenecks affecting the preparation of the data for a project because the necessary information is often collated from different sources or databases, requiring the laborious involvement of subject matter experts (SMEs) to validate the data prior to initiating computational or analytical processes. Energistics provides not only data standards but also a rich set of metadata that retains information regarding history and provenance for data items, which will significantly reduce the need for SME verification when properly integrated into a workflow. There is an accelerating trend in oil and gas companies to research and deploy automated workflows. For these workflows to be successful, it is essential that the process be able to assume that all the data can be easily assimilated, validated, and trusted.

What have been the most significant recent achievements in terms of innovation?

In terms of general trends, the usage of measurements from the field to predict rather than just monitor events is a game changer for many aspects of the industry. Analytical tools applied to past and present data can detect patterns and foresee critical situations at a point in time when remedial action can be deployed to avert time-consuming and even dangerous outcomes. By making it possible to manage remote facilities with increasing insight and precision, manpower on location can be reduced. This has a favorable impact in terms of health and safety (fewer people in proximity of harsh or hazardous conditions) and an ability to increase on-site automation, which improves efficiencies while driving down costs. What can be expected is a circle where the value of harvesting data from sensors drives an increase in the number and sophistication of measuring devices. The low cost of sensors and the implementation of standardized data streams based on industry-accepted formats make this possible.

What do you see as being the most prominent future trends for innovation in data collection and analysis in the oil and gas sector?

The data collection side of the equation has the necessary momentum to continue to deliver more data and more accurate data going forward, leveraging advances in relatively cheap and durable solid-state measurement devices, increased bandwidth, and other technologies that ride on the extraordinary pace of development in the IT and telecom industries. The more remarkable changes will come at the receiving end of the data, where the data streams will deliver increasingly insightful information to scientists, engineers, and decision makers across the whole company. As well as enabling real-time analytics, the amount and relevance of data will contribute to breaking down the data silos that have existed for so long. It also opens new avenues for collaboration within global companies and between companies and their partners and stakeholders, making information more accessible and useable. An upside to this evolution will be to make the industry more appealing to future generations. Apart from a few areas of intense 3D visualization and characterization related to seismic data interpretation and reservoir simulation, the rest of the oil and gas industry has lagged behind other sectors in deploying modern tools to which young professionals educated with computers and smartphones and playing with 3D games can relate.

15

Vignette: Innovations in Deep Seabed Mining

Lindsey Harris

What have been the key technical innovations that have allowed for deep sea mining to proceed?

Many important technical advances have made deep sea mining (DSM) more feasible and profitable. Though the DSM industry has consciously attempted to distance itself from traditional mining, much of the technology has been appropriated and refurbished for the environmental challenges posed by working at depths over 1500 m. For example, DSM companies extracting seafloor massive sulfides designed and built a bulk head cutter for cutting and pulverizing dormant hydrothermal vent chimneys (Nautilus Minerals 2017). For polymetallic nodule extraction, engineers debated different collection methods, from the long bucket method and trawling to suction or scooping (Nautilus Minerals 2017; Hammond 1974). Recently, engineers have developed and built riser and lifting systems for pumping the slurry of solids and seawater from the seabed up to a waiting production support vessel using rigid pipes and a positive displacement pump (Nautilus Minerals 2017). In both cases, advances in remote sensing technology have been a necessity. This technology has developed in leaps and bounds since interest in DSM peaked in the 1970s and through advances in other industries such as "wet diamonds," offshore oil and gas, and surveying. Now, DSM companies capitalize on these advances to identify sites, map, sample, and manage seafloor and submersible equipment. Scientists are now better able to visualize the topography of the seabed and its content with high definition cameras, lasers, and acoustic equipment (Harris 2016).

What are the social and environmental concerns about adoption of these technologies?

The social and environmental concerns raised by DSM in general are closely linked, yet the impacts of the activity are still largely unknown. DSM companies have released abbreviated profiles on their mining equipment, but many activists, Pacific Island community members, and scientists remain skeptical that the equipment and methods will not seriously harm ecosystems and the communities that depend on them. Moreover, scientists working closely with the International Seabed Authority (ISA) have recently released the first report from the Areas of Particular Environmental Interest in the Clarion Clipperton Zone, reporting new and endemic species (Kaiser et al. 2017). Scientists have also raised concerns about the impacts daily operations of proposed DSM operations will have, including light and sound pollution, temperature changes, and the disposal of sediments, wastes, and other effluents in the layers of the ocean and generated by mining equipment on the seafloor (Borowski and Thiel 1998; Ramirez-Llodra et al. 2011). They are also concerned with the slow regeneration rates of benthic ecosystems after exploitation (Gollner et al. 2017). These concerns coincide with those of community and environmental activists living nearest to potential mine sites. For example, shark callers on the west coast of New Ireland state that exploration projects have scared noise-sensitive sharks away, and increased levels of dust in the ocean have affected local fishermen's catches (Lowrey 2012; Matbob 2013). They cited these as not only environmental and economic disruptions, but also as cultural ones including the disruption of sites and the loss of culturally important species, local and indigenous knowledge, history, and natural resource management (Lowrey 2012; Matbob 2013). Finally, mining has traditionally led to local boom and bust economies. In the long run, these episodes further entrenched communities in development, destroyed local ecosystems, and failed to deliver the long-term benefits promised by mining (Bainton 2010). DSM could continue this debilitating history of exploitation for small island states, because the returns to these states may be low and temporary compared to the risk of or actual environmental harm. This issue raises pressing questions: The seabed beyond national jurisdiction is called the Area, and it is designated by treaty as the common heritage of mankind. This concept is so far uninterpreted judicially. However, it does appear to entail that it be utilized for the benefit of all mankind as a whole. Thus a benefit sharing regime has yet to clarified. (Jaeckel et al. 2016; Noyes 2011). Similarly, in the case of projects with in a state's national jurisdiction the long-term impacts of seabed mining on nearby communities remains problematic and unclear.

What are the regulatory barriers to successful implementation of these technologies?

The regulatory barriers to DSM depend on whether the projects will occur within a state's national jurisdiction or in the Area. A DSM project in a state's national jurisdiction is subject to the laws of that state. There is a general call for DSM to adhere to the "precautionary principle." The problem is how to implement it in a state of high uncertainty. Some argue that DSM should not take place until the industry or academia demonstrates persuasively that DSM projects will not cause harm to the public or the environment. Others insist that DSM projects must proceed slowly until there is scientific consensus that they will not cause serious harm (Schmidt 2015). Also as previously discussed, a controversial principle was written into the Convention on the Law of the Sea: the common heritage of mankind, this principle, which suggests that the seabed beyond national jurisdiction and the resources therein are a kind of inheritance for all peoples, has yet to be interpreted and continues to bedevil legal scholars (Lodge 2012; Shackelford 2009). Finally, and most importantly, as of this writing, there is no adopted, legally binding exploitation code for projects in the Area. Thus, the biggest regulatory barrier, is the absence of a regulatory regime for exploitation. The ISA has released a draft exploitation code to international community and will begin debating it at the ISA's annual meeting in 2018.

Are there any social innovations that communities considering deep sea mining are adopting or could be further developed?

The communities affected by seabed mining are heterogenous and their responses, needs, goals, and concerns do not all align. Innovations among groups opposed to seabed mining include reaffirming their commitment to a collective understanding of land and ocean. For example, a group of concerned environmentalists and activists recently convened at the fourth meeting of the Melanesian Indigenous Land Defense Alliance (MILDA) in Buala, Isabel Province, Solomon Islands, to reaffirm their commitment to organize and defend their resources: cultural traditions, land, seas, seabed, ecosystems, biodiversity, rivers, and air (Cullwick 2017). Environmental non-governmental organizations (NGOs) in areas nearest to potential sites network with each other and partner with natural and social scientists across the globe to raise awareness about deep seabed mining and its impacts.

For local stakeholders in favor of deep seabed mining, the innovation is in how much and in what ways benefits flowing from DSM projects will be apportioned. It is hoped that DSM projects will innovate from the unfulfilled development dreams of other mining projects and bring better infrastructure, education and training, and jobs. Here the social innovation from the community is how they will come to control, manage, and negotiate their demands and the product of these demands with governments and companies.

As is increasingly being witnessed across a broad range of natural resource uses worldwide, participants in debates surrounding DSM are not restricted to local, potentially affected communities; they also include far-flung and unevenly distributed networks: communities in diaspora, indigenous rights groups, academics, and policymakers among them. In these cases, the most important social innovations will be legal (in the case of applications of the common heritage of mankind), ethical (whether to proceed with DSM and who should bear the costs and benefits first), and political (research on who will benefit from DSM projects and how).

Are there any additional innovations that you can identify as being needed in order for deep sea mining to become more mainstream?

Necessary innovations must be legal, technical, and cultural in scope. As previously stated, the common heritage of mankind has yet to be interpreted and implemented. Considering DSM in the Area, more advisory bodies must be added to the ISA. One example may be an environmental committee that evaluates environmental impact assessments submitted by contractors and provides expert opinions and guidance to the ISA Assembly and Council. Similarly, technical innovations that minimize the environmental impact of mining equipment are essential, especially those that address the size and content of plumes and noise pollution. Finally, although states and contractors are quick to highlight legal and technical innovations that improve the feasibility of their projects, social license is much more difficult to procure (Filer and Gabriel 2017). However, as compared with terrestrial mining companies, DSM has the opportunity to innovate approaches to corporate social responsibility, which should facilitate obtaining a social license. DSM must consider the ways it will interact with and affect communities' relationship to their seascapes. Many communities have cultural, epistemological, historical, political, and commercial ties to the seabed and deep ocean, even if no one lives on the seabed. Such considerations move beyond corporate social responsibility and relate to questions of regulation, just and sustainable use and management, and environmental health.

References

Bainton, N. A. 2010. *The Lihir Destiny: Cultural Responses to Mining in Melanesia. Asia-Pacific Environment Monograph; 5*. Canberra, Australia: ANU E Press.

Borowski, C., and H. Thiel. 1998. Deep-sea macrofaunal impacts of a large-scale physical disturbance experiment in the southeast pacific. *Deep-Sea Research Part II* 45(1): 55–81.

Cullwick, J. 2016. The Buala declaration. *Vanuatu Daily Post*.

Filer, C., and J. Gabriel. 2016. How could Nautilus Minerals get a social licence to operate the world's first deep sea mine? *Marine Policy* InPress: 6.

Gollner, K., L. Menzel, D. O. Jones, A. Brown, N. C. Mestre, D. Van Oevelen, L. Menot et al. 2017. Resilience of benthic deep-sea fauna to mining activities. *Marine Environmental Research* 129: 95–97.

Hammond, A. 1974. Manganese nodules (II): Prospects for deep sea mining. *Science.* 183(4125): 644–646.

Harris, L. 2016. *Sustainable Seabed Mining: Corporate Geoscientists' Visions in the Solomon Islands*. Honolulu, HI: Master of Arts, Anthropology, University of Hawai'i at Manoa.

Jaeckel, A., and K. M. Gjerde. 2016. Sharing benefits of the common heritage of mankind—Is the deep seabed mining regime ready? *Marine Policy* 70: 198–204.

Kaiser, S., C. R. Smith, and P. M. Arbizu. 2017. Editorial: Biodiversity of the clarion clipperton fracture zone. *Marine Biodiversity* 47(2): 259–264.

Lodge, M. W. 2012. The common heritage of mankind. *International Journal of Marine and Coastal Law* 27(4): 733–742.

Lowrey, N. 2012. New Ireland locals fear Nautilus destroying shark calling. *Deep Sea Mining Campaign*.

Matbob, N. 2013. Messe shark callers concerned over experimental seabed mining. *Papua New Guinea Mine Watch*.

Nautilus Minerals. 2017. *Status of the Equipment*. Accessed May 22. www.nautilusminerals.com/irm/content/status-of-the-equipment.aspx?RID=424.

New Zealand Parliament. 2003. *The Foreshore and Seabed—Maori Customary Rights and Some Legal Issues*. Wellington, New Zealand: Parliamentary Library.

Noyes, J. E. 2011. The common heritage of mankind: Past, present, and future. *Denver Journal of International Law and Policy* 40(1): 447.

Ramirez-Llodra, E., P. A. Tyler, M. C. Baker, O. Aksel Bergstad, M. R. Clark, E. Escobar, L. A. Levin et al. 2011. Man and the last great wilderness: Human impact on the deep sea (Anthropogenic Impact on the Deep Sea). *PLoS ONE* 6(8): E22588.

Schmidt, C. 2015. Going deep: Cautious steps toward seabed mining. *Environmental Health Perspectives* 123(9): A234.

Shackelford, S. 2009. The tragedy of the common heritage of mankind. *Stanford Environmental Law Journal* 28: 109–577.

16

Extracting Off-Earth Resources

Serkan Saydam

CONTENTS

16.1 Introduction

Mining is one of the oldest human activities. Since the prehistoric time, extracting mineral resources has played a very important role for building our civilizations. As the old mining proverb states "where she be, there she be," where there are minerals to be extracted, humans will rush there, even if they need to battle with extreme conditions.

The main motivations for off-Earth mining are the abundance of mineral resources that are inevitably necessary for our technologically obsessed civilization and our passion of travelling and discovering new places which we can colonize. An additional motivation is that the off-Earth mining research activities have significant potential to generate spin-off technologies which can be applied to terrestrial mining operations. Obtaining mineral resources in space may also have enormous worth in terms of sustainability and gathering materials without placing further ecological stress on the Earth.

Whereas space missions have previously been determined exclusively by government agencies and international scientific bodies, today commercial enterprises are increasingly becoming involved in a range of activities, including launching rockets, satellite deployment, resupplying space stations, and even space tourism. It is expected that in the next 15 to 20 years, commercial activities will also embrace extracting minerals from asteroids, the Moon, and Mars. It is also highly likely that in the near future off-Earth mining activities will be the primary motivation in the space economy.

The bulk of previous work into the subject includes mining on the Moon and asteroids, particularly focusing on volatiles, minerals, and/or possible preliminary mining operational scenarios. The predominant, longstanding focus of the National Aeronautics and Space Administration (NASA) (2003) has been on the Moon due to the previous human exploration programs (i.e., the Space Exploration Initiative 1989–1992 and the Constellation Program 2005–2010). However, in the last 10 years, the target of research has shifted to Mars.

Prior work for extraterrestrial colonies has a relatively long history that began just after the launch of *Sputnik* in 1957. Holbrook (1958) conducted the first unclassified publication in 1958 for investigation of the planetary physical environment, space transportation, mission design, off-Earth human physiology and psychology, exploration methods and equipment design, base design, concept of operations and logistics, and colonization (Shishko et al. 2016).

Craig et al. (2014) provided perhaps the most comprehensive off-Earth mining review. According to their study, O'Leary (1988) conducted one of the first works on potential mineral resources on the Moon, Mars, Mars' two moons, and an asteroid called 1982 DB. O'Leary's analysis in 1988 focused on the ramifications for the feasibility of human missions or settlement; however, since then much more literature has been produced on the subject of commercial mining in off-Earth environments. This chapter will provide a review of past and current research status of off-Earth mining and briefly comment on some of the main challenges confronting the sector.

16.2 Literature Review

16.2.1 Previous Work on Asteroid Mining

Near-Earth objects, asteroids, and comets that circle the Sun at around a similar speed and distance as the Earth orbits the Sun are now considered to be comprised of an abundance of various mineral resources. Prior research indicated that asteroids have numerous commodities which would be of substantial interest to commercial organizations: water and volatiles, precious metals, rare earth minerals, refractory materials, iron, and nickel.

In the molten stages of formation of planets like the Earth, the heavier metals remained closer to the core while lightweight elements formed the crust through rising to the surface. However, it is assumed that the minerals expected to be in asteroids would not have been exposed to this type of differentiation due to their smaller size and low gravity compared to the planets. Consequently, it is expected that the heavy metals are more evenly distributed all over the asteroids. Between Mars and Jupiter, there are more than two million near Earth asteroids.

The study by Sonter (1997) and his subsequent work are some of the first studies conducted into the economic viability of asteroid mining. Using net present value (NPV) technique and introducing parameters such as orbital mechanics, fuel requirements and associated costs, mining methods and processing technologies, and transport cost, he developed a basic framework for feasibility studies of off-Earth mining operations. He followed this work by another study that established a new terminology called the mass payback ratio which is used in the calculation to demonstrate the requirement to pay the mass in the form of propellant, rocket, and operational consumables to return the product mass to the market (Sonter 2001). In the same study, he also developed an asteroid mining scenario in which mined minerals were moved into the low Earth orbit (LEO) and marketed there to construct a space infrastructure. The materials mined included water for energy, semi-conductors to build solar cells, and iron ore and nickel for constructing the infrastructure.

In other earlier works, Ross (2001) proposed a feasibility study for asteroid mining. He also used NPV analysis and focused on extracting water, volatiles, precious metals, rare earth minerals, iron ore, and nickel. He also conducted a preliminary market analysis for these commodities. Geological uncertainty was not, however, factored into his work, something which Yoshikawa et al. (2007) addressed in their later study. Building on the works of Sonter (1997, 2001) and Ross (2001), Erickson (2006) conducted an economic study with a focus on return-on-investment and, additionally, included operational risks and cost efficiency in his calculations. He also mentioned that such an operation would heavily rely on the technological developments in robotics.

In more recent economic works, Sonter (2013) further developed his proposed asteroid mining scenarios by assessing a typical target asteroid and estimating its geological characteristics. The posited case study was a metallic type (M-type) asteroid that typically containing high-grade nickel and iron which could be used for construction materials. In their similarly orientated work, Craig et al. (2014) conducted a feasibility study through an NPV analysis using a near Earth asteroid, 1986 DA. This asteroid is expected to be 0.5 astronomical units (AU) away from the Earth and consist of huge reserves of naturally occurring stainless steel[1] (Ingebretsen, 2001; Ross, 2001), making it an ideal asteroid to mine. The authors calculated that the selected asteroid has a reserve of 20×10^9 ton of raw stainless steel which would be mined through strip or auger mining methods. The assumed production rate is 20 Mt of stainless steel per year over a mine lifespan of more than 100 years. It is assumed that the steel would be smelted and transported by a spaceship to the Earth orbit and sold for off-Earth construction. The NPV analysis conducted showed that the operation would achieve a very high NPV of US$658 $\times 10^9$. But, due to the a very large capital cost of US 1.3×10^{15}, the project would not be considered feasible.

---

[1] 88% Fe, 10% Ni, and 0.5% Co.

Buet et al. (2013), who designed a mining system for capturing an asteroid, contributed more depth to the potential practicalities of asteroid mining. Their aim was to capture an asteroid and bring back to the Earth's surface. Prado (2013), who compared terrestrial mining systems for use in off-Earth mining activities, utilized strip mining as an example, but did not consider the low gravity environment. Ata et al. (2015) discussed potential separation and extraction techniques that could be used for processing of asteroids. They mentioned that water would be the main element required for processing minerals in space, and a reduction in the amount of water needed for mineral processing may be the key to making space mineral processing ventures viable.

16.2.2 Previous Work on the Moon

In one of the earliest assessments of exploiting the Moon's natural resources, Duke et al. (1997) focused on designing a series of operational scenarios for extracting ice water from the lunar polar regions. Their work focused on using microwave energy to melt the ice to produce water through thermal processing. They also suggested using steam pipes to transport the water. However, their approach was only at a scoping level and did not include any economic feasibility.

A series of feasibility studies have been formulated, however. For example, Charania and DePasquale (2007) considered economic uncertainties associated with resource exploitation on the Moon and developed three business cases: (1) selling directly to a government on the Moon, (2) selling to a government in low lunar orbit (LLO), and (3) selling to another customer in LEO. They considered that the uncertainties were clearly combined by probability distributions through using Monte Carlo simulation techniques. Zacharias et al. (2011) also compared potential mining projects on the Moon, as well as asteroids and Mars, based on each of their dynamic locations. They also conducted NPV analysis only for 10-year mining operations for a random mineral. Their calculations were in favor of the Moon compared to Mars and two posited asteroids. In fact, both the Moon and Mars provided positive NPV, while asteroids received a negative valuing. Furthermore, Pelech (2013) conducted an economic assessment for comets and the Moon for producing water rather than minerals. He established four operation scenarios to provide H_2/LO_x propellant market in LEO. He further developed the "propellant payback ratio," proposed by Sonter (1997, 2001), which examines the economic return on the predetermined prospect to launch the propellant straight from the Earth. He also used the opportunity cost concept as infrastructure launched from the Earth.

The availability of various studies on the subject demonstrates that mining the lunar (and Martian; discussion following) landscape is seriously being considered. Almost 15 years ago, Gertsch and Gertsch (2003) used a typical surface mine design and planning technique to produce a plan to mine the Moon's regolith for gases to support life for 100 people at a hypothetical lunar town. Not long after, NASA conducted a study (Muff et al. 2004) including a prototype design of bucket-wheel excavating equipment for mining the Moon and Mars' surface to extract regolith. Schimitt et al. (2008), reviewing previous work and recommended ideas for exploration techniques for space, emphasized the possibility of mining life sustaining elements such as H, He, C, N, and O on the lunar surface. More recently, Balla et al. (2012) developed a laser fabrication technology to extract minerals from off-Earth bodies and conducted a series of tests using a model of lunar regolith. The tests showed that the method could indeed be used to extract large amounts of materials through the melting and solidification process of regolith. Bernold (2013) designed and developed mining equipment to draw lunar regolith to collect construction materials from the Moon.

Perhaps unsurprisingly, NASA's Kennedy Space Center has been at the forefront of developing equipment for extracting the Moon regolith—the RASSOR (Regolith Advanced Surface Systems Operational Robot). The RASSOR design provided adequate traction and reaction force to the regolith surface by using the counterrotating bladed drums to collect and transport the regolith (Mueller et al. 2013, 2015). Significantly, RASSOR could also be used for Mars. The team also developed another equipment named VIPER (Vibratory Impacting Percussive Excavator for Regolith). Lucas and Hagan (2014) worked on series of tests to compare conventional pick cutting systems and pneumatic excavation systems for an off-Earth mining situation. Figure 16.1 provides a flowchart explaining the space resource processing

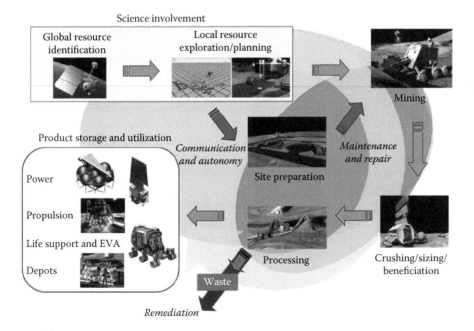

FIGURE 16.1
Space resource processing cycle and integration with surface elements. (From Sanders, G.B. and Larson, W.E., *Adv. Space Res.*, 55, 2381–2404, 2015.)

cycle and integration with surface elements designed by Sanders and Larson (2015). According to their study, RASSOR and VIPER could mine the regolith at tested relative densities.

Research conducted by Ishimatsu et al. (2015) investigated the capability to produce resources on the Moon to reduce the initial mass to LEO. Dorrington et al. (2015) conducted a study on combining a detailed trajectory design technique using economic parameters to improve the economic assessment for asteroid mining missions. The study used the 2014 EK24 asteroid as a case study. According to their assessment the multiple launch opportunities will be existing in the next 10–20 years. Lindley et al. (2015) studied developing block modeling techniques to form meaningful ore reserves for asteroids. Their conclusion highlighted the importance of the orbital data for assessing the mission cost. Dello-Iacovo et al. (2017) conducted research on developing a new approach to measure the seismic velocity of regolith on lunar regolith.

16.2.3 Previous Work on Mars

Many scientists agree that the red planet, Mars, is the most rational destination for the next manned visit to interplanetary space. The earliest NASA investigations on Mars were conducted in the 1960s. In these investigations, the exploration for natural resources was always a focus to assess the feasibility of human colonization of Mars in the future. According to many researchers, Mars could be considered as an ambitious goal; however, human colonization of Mars could be possible within this century. Such a colony will certainly require *in situ* resources to survive and if there is to be any growth of the colony. For this reason, viable business ventures must be developed through satisfying the demand for *in situ* resources from and on Mars.

Badescu (2009) examined mining on Mars extensively. He explored the potentials and limitations of numerous systems which could be useful for manned missions on Mars. The author investigated possible energy sources on Mars and then proposed techniques for surveying, exploration, and excavating various resources. His work also included possible Mars colonization scenarios.

In the last 15 years, researchers and scientists established and documented that massive quantities of water/ice in numerous form within Mars. The availability of water is vital to support human existence on Mars and would also be significant in reducing the demand for resources from the Earth. Water must, in fact, be considered more broadly as perhaps the most vital commodity for space economy, not only through conventional human use (drinking, agriculture, etc.) but also for its use in industrial processes (i.e., hydrolyzing the water can harvest hydrogen and oxygen, which can be used as an energy source).

Research on Mars has concentrated almost exclusively on harvesting water rather than focusing on extraction of more traditional mineral resource.

Saydam et al. (2015) identified the possible water resources on Mars: the equatorial region and the south and north pole regions. However, these regions have dissimilar conditions and operational risks. The equatorial region is the most logical location for colonization considering the environment (temperature −87°C to +26°C) and transportation can be achieved easier compared to the polar regions (Figure 16.2) even though ice formation is uneven on the surface. In the polar regions the extreme weather conditions would be challenging, but the polar regions have smoother places which would be easier for landing and launching.

Saydam et al. (2015b) also recommended three different mining systems with similar configurations and approaches to the truck and shovel operations in terrestrial open pit mines. As a first option, the first mining equipment they used in their model is the Mars *In-Situ* Water Extractor (MISWE), developed by Zacny et al. (2012) (Figure 16.3), which consists of a surface mining rover (a small car size) fitted with an auger-like drill that can be used for selective excavation and extraction of water among the ice regolith. Zacny et al. (2012) tested the model with 200 g of icy soil (12 wt% water) and managed to mine an average of 85% of the water. However, it should be noted that MISWE was only developed for sampling not for production. Therefore, Saydam et al. (2015) recommended as a second option altering the design

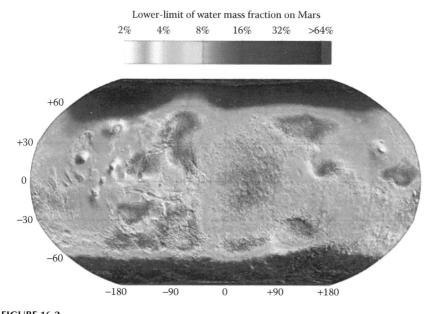

FIGURE 16.2
Water mass map from the neutron spectrometer. (From National Aeronautics and Space Administration—Jet Propulsion Laboratory (NASA-JPL), PIA04907: Water mass map from neutron spectrometer (online), photojournal, December 8, http://photojournal.jpl.nasa.gov/catalog/PIA04907 [accessed November 20, 2016].)

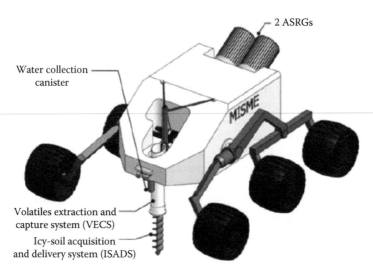

FIGURE 16.3
Mobile in-situ water extractor (MISWE) equipment designed. (From Zacny, K. et al., Mobile in-situ water extractor (MISWE) for mars, moon, and asteroids in situ resource utilization, *Paper Presented at the AIAA SPACE 2012 Conference & Exposition*, 2012.)

and separating the processing section from MISWE. With this alteration, the equipment would be more productive. The third option was to use different equipment in the polar regions, such as micro-tunneling equipment (which was found to have the highest production rates) together with a hauling system with several truck-like rovers.

Publications by Shishko et al. (2015, 2016) and Saydam et al. (2015a) also provided extensive research on water extraction models for Mars. These works included developing collaborative specialized models to investigate the commercial potential of extracting water and ice on Mars in support of a Mars Colony. Their models begin with a "systems architecting framework" to define a Mars Colony and to capture the parameters and technical attributes of such objects. The subsequent database is then connected to a variety of "downstream" analytic models. They specifically combine an extraction process model, a simulation of the colony's environmental control and life support infrastructure known as HabNet,[2] and a risk-based economics model. The developed mining model emphasizes machineries and mining approaches associated with *in situ* resource extraction, processing, storage, handling, and delivery (Saydam et al. 2015). This model calculates the rate of water production and provides optimized equipment selection for possible

---

[2] HabNet is an environmental control and life support (ECLS) system simulation based on an open-source software package named BioSim that was originally developed by TracLabs in the early 2000s under contract to the NASA's Johnson Space Center to support integrated ECLS controls research. HabNet simulates the fundamental sustainability relationships associated with establishing and maintaining the colony's population.

mining scenarios. Tapia Cortez et al. (2017) further developed the mining optimization model proposed by Saydam et al. (2015, 2016) and Shishko et al. (2015, 2016) and formulated a model called Water Extraction Mars Mining Model (WEM³) which simulates, optimizes, and assesses the performance of Martian mining operations from a conceptual engineering point of view. This work is based on typical open pit mining production and fleet optimization techniques.

16.3 Major Technology Challenges in Off-Earth Mining

Terrestrial mining operations have specific operational risks and uncertainties, mainly due to geology and economics. Off-Earth mining operations will have the similar risks and uncertainties with even more extreme environmental conditions. The following section will provide information about some of the challenges which an off-Earth mining operation would face.

Landing and launching of infrastructure and equipment fleet: Current terrestrial mining operations have the capability to produce millions of ton of minerals in a year using large, heavy equipment and substantial machinery fleets. Even with the technological developments and involvement of commercial entities into space missions (e.g., SpaceX) that has reduced cost from around $20,000/kg–$4,000/kg, the costs of launching such a weight of equipment and materials into space is still extremely high. Indeed, it is currently almost impossible. Therefore, different mining systems and equipment will need to be developed.

In-space propulsion: It is important to develop suitable propulsion technologies that can meet future space science and exploration needs. These propulsion technologies will be the key to access the targeted off-Earth objects in a reliable and safe fashion. Trustworthy technologies must focus on developing technical solutions to improve thrust levels, power, system mass, volume, operational complexity, manufacturability, and cost. As well as reducing extensive transportation time, they will increase payload mass, improve spacecraft safety, and decrease manufacturing and operation costs.

Life support system: A life support system is a group of devices that allow the crew to survive in space. The life support system includes supplying air, water, and food. Components of the life support system are life-critical and must be designed and built by safety engineers.

Radiation effects and shielding: Shielding from harmful external impacts (e.g., radiation and micrometeorites) would be necessary to operate any off-Earth mining activity. Space radiation is one of the most dangerous features of a Mars voyage for humans. The Earth's atmosphere acts as a magnetic shield that provides protection from harmful cosmic radiation. Therefore, a spaceship used to travel to Mars and a colony on Mars would requires substantial protective shielding.

Low gravity effects and crew health in space: The human body is weightless in space. Crews traveling to a planet such as Mars, which has only one-third of the Earth's gravity, would be exposed to a comparative change in their body weight on Mars' surface. When crew members return to the Earth, their body weight will be normal again. This transition can affect the human body severely. According to NASA's years of work on the effects of zero gravity on the human body, the bones lose minerals eventually (and sometimes irreversibly), with density dropping at over 1% per month.

Power systems: Electrical power is an essential facet in the ability to live comfortably, safely, and continue domestic, industrial, and scientific activities in an off-Earth habitat. Solar power systems can be considered as a well-sourced alternative. However, it is still unclear whether current standards of solar power technology will be adequate to support mining activities. Improved or entirely new and reliable power systems need to be developed for this purpose.

In situ resource utilization: *In situ* resource utilization (ISRU) refers to the practice of converting indigenous resources into various products that are needed by a space mission. By utilizing indigenous resources, the amount of material that must be brought from the Earth can be reduced. Thus, reducing the initial mass in LEO is typically used as a measure of the mission scope and cost.

16.4 Conclusions

It is likely that the first off-Earth mining operation will take the form of water extraction from asteroids to replace energy on commercial satellites, or extracting water, construction materials, or a combination of both for building a habitat on the Moon and/or on Mars. It is worth noting that there is a serious effort, even a "race" among commercial companies, to start an asteroid mining operation within the next 10 years. If this goal is successful, it will surely catalyze efforts to venture out and begin projects on the Moon or Mars.

One can envisage a typical operation within our near solar system without too much stretch of the imagination with current technological advances: a mining site operating autonomously either from a space station or a space craft in the orbit, communicating with a headquarters on Earth, with the heavy, dangerous work being conducted by robots and managed by humans. As this article has highlighted, a multitude of variables will be at play when considering and planning such operations: the use of resource estimation techniques, the use of current mine planning and systems methodologies, the applicability of the current mining equipment and machinery, and the size of the equipment we want to launch beyond our atmosphere.

The uncertainties and challenges are undeniably substantial, however, and will necessitate the creation of a great deal of further knowledge to bring the plan to life. Cutting-edge mining engineering and interplanetary geology education should be adopted and interwoven toward this end accordingly, with multidisciplinary skills and degrees being essential. Engineering programs will need to expand and enhance their curriculum to offer relevant skills such as automation, robotics, information technology, space orbital mechanics, systems engineering, mining engineering, alongside interplanetary physics, chemistry, and geology.

Encouragingly, the space and extractive resources industries share commonalities in that they both operate in extreme environments and are subject to excessive risks and uncertainties. This commonality provides some synergies in establishing collaborations between research institutions, governments, and resource and space industries that will be the key if off-Earth mining is to be a reality. Moreover, there will be some common benefits for such a relationship between these industries. The space industry could improve their knowledge in operational excellence, safety, production efficiency valuation of the commodities, and market creation, for example. The natural resource industry could benefit in their operations from state-of-the-art systems engineering and applications of information technology, such as autonomous systems and robotics, and large data management.

References

Ata, S., Bournival, G., and Manefield, M. 2015. Resource recovery in space. *The Third International Future Mining Conference*, November 4–6. Saydam, S. and Mitra, R. (Eds.), AusIMM, Sydney, Australia, pp. 275–279.

Badescu, V. 2009. *Mars-Prospective Energy and Material Resources*. Springer-Verlag: Berlin, Germany.

Balla, V.K., Roberson, L.B., O'Connor, G.W., Trigwell, S., Bose, S., and Bandyopadhyay, A. 2012. First demonstration on direct laser fabrication of lunar regolith parts. *Rapid Prototyping Journal*, 18(6): 451–457.

Bernold, L.E. 2013. *An Australian Lunar Soil Simulant to Study Lunar Mining and Construction*. Off Earth Mining Forum delivered at University of New South Wales, Sydney, Australia, February 19–21. www.youtube.com/watch?v=yvBVuewGuxs (accessed May 17, 2013).

Buet, M., Pearson, J., Bennett, D.S., and Komerath, N. 2013. Cornucopia Mission robotic mining system for rapid earth orbit capture of asteroid resources. http://www.csc.caltech.edu/stuff/VoyagerFinalReport.pdf (accessed November 17, 2013).

Charania, A.C. and DePasquale, D. 2007. Economic analysis of a lunar in-situ resource utilization (ISRU) propellant services market, IAC-07-A5.1.03. *58th International Astronautical Conference*, Hyderabad, India, September 24–28, 2007.

Craig, G., Saydam, S., and Dempster, A. 2014. Mining off-earth minerals: A long term play? *Journal of the South African Institute of Mining and Metallurgy (SAIMM)*, 114(12): 1039–1047.

Dello-Iacovo, M., Anderson, R.C., and Saydam, S. 2017. A novel method of measuring seismic velocity in off-Earth conditions: Implications for future research. *48th Lunar and Planetary Science Conference*, The Woodlands, TX.

Dorrington, S., Kinkaid, N., and Olsen, J. 2015. Trajectory design and economic analysis of asteroid mining missions to Asteroid 2014 EK24. *The Third International Future Mining Conference*, November 4–6. Saydam, S. and Mitra, R. (Eds.), AusIMM, Sydney, Australia, pp. 281–287.

Duke, M.B., Gustafson, R.J., and Rice, E.E. 1998. Mining lunar polar ice. American Institute of Aeronautics and Astronautics, AIAA-98-1069, pp. 1–10, Reno, NV.

Erickson, K.R. 2006. Optimal architecture for an asteroid mining mission: Equipment details and integration. *Proceedings of Space 2006*, San Jose, CA, September 19–21, pp. 1–16.

Gertsch, L.S. and Gertsch, R.E. 2003. Surface mine design and planning for lunar regolith production. *Space Technology and Applications International Forum*, 654: 1108.

Holbrook, R. 1958. *Outline of a Study of Extraterrestrial Base Design*. RM-2161, The RAND Corporation, Santa Monica, CA. http://www.rand.org/pubs/research_memoranda/RM2161.

Ingebretsen, M. 2001. Mining asteroids. *Proceedings of IEEE Spectrum SPACE*, August 2011, pp. 34–39.

Ishimatsu, T., DeWeck, O., Hoffman, J., Ohkami, Y., and Shishko, R. 2015. A generalized multi-commodity network flow model for the earth-moon-mars logistics system. *Journal of Spacecraft and Rockets*, 53(1): 28–38.

Lindley, C.A., Sennersten, C., Davie, A., Goldstein, O., Lyu, R., Grace, A., Evans, B. et al. 2015. A multiplayer 3D index tool for recursive block models supporting terrestrial and extra-terrestrial mine planning. *The Third International Future Mining Conference*, November 4–6. Saydam, S. and Mitra, R. (Eds.), AusIMM, Sydney, Australia, pp. 289–296.

Lucas, M.T. and Hagan, P.C. 2014. Comparison of two excavation systems for the mining of lunar regolith. *Mining Education Australia Journal of Research Projects Review*, 3(1): 39–44.

Mueller, R.P., Sibille, L., Leucht, K., Smith, J.D., Townsend, I.I., Nick., J., and Schuler, J.M. 2015. Swamp works—A new approach to develop space mining and resource extraction technologies at NASA's Kennedy Space Center: Prospecting, excavation, load, haul and dump for in situ resource utilisation. *The Third International Future Mining Conference*, November 4–6. Saydam, S. and Mitra, R. (Eds.), AusIMM, Sydney, Australia, pp. 297–303.

Mueller, R.P., Smith, J.D., Ebert, T., Cox, R., Rahmatian, L., and Wood, J. 2013. Regolith advanced surface systems operations robot excavator (RASSOR), *Nasa Tech Briefs*, Article 15471, 1.

Muff, T., Johnson, L., King, R., and Duke, M.B. 2004. A prototype bucket wheel excavator for the moon, mars and Phobos. *Proceedings of the 2004 Space Technology and Applications International Forum (STAIF-004)*, Albuquerque, NM, February 8–11.

National Aeronautics and Space Administration—Jet Propulsion Laboratory (NASA-JPL). 2003. PIA04907: Water mass map from neutron spectrometer (online), photojournal, December 8. http://photojournal.jpl.nasa.gov/catalog/PIA04907 (accessed November 20, 2016).

O'Leary, B. 1988. Asteroid mining and the moons of Mars. *Acta Astronautica*, 17(4): 457–462.

Pelech, T.M. 2013. Technical and economical evaluation of mining comets and the moon for an off-earth water market. *Bachelor of Engineering (Mining) Thesis*. UNSW Australia, Sydney, Australia.

Prado, M.E. 2013. Asteroid mining. *Projects to Employ Resources of the Moon and Asteroids Near Earth in the Near Term*. www.permanent.com/asteroid-geologies. html (accessed May 10, 2013).

Ross, S.D. 2001. Near-earth asteroid mining. www2.esm.vt.edu/~sdross/papers/ ross-asteroid-mining-2001.pdf (accessed May 16, 2013).

Sanders, G.B. and Larson, W.E. 2015. Final review of analog field campaigns for in situ resource utilization technology ad capability maturation. *Advances in Space Research*, 55: 2381–2404.

Saydam, S., Shishko, R., Tapia Cortez, C.A., de Roche, T., Dempster, A., Fradet, R., and Coulton, J. 2015a. An integrated risk-based financial and technical models to evaluate possible mining scenarios on mars' surface. *The Third International Future Mining Conference*, November 4–6. Saydam, S. and Mitra, R. (Eds.), AusIMM, Sydney, Australia, pp. 305–311.

Saydam, S., Tapia Cortez, C., de Roche, T., and Dempster, A. 2015b. An evaluation of mars water extraction mission. *26th Annual General Meeting & Conference*, Freiberg, Germany, June 21–26. Mischo, H. and Drebenstedt, C. (Eds.), WIRmachenDRUCK GmbH, Backnang, Germany, pp. 75–82.

Schimitt, H.H., Farrely, C.T., and Franklin, D.C. 2008. Mining and the future of space exploration. *Proceedings of First International Future Mining Conference and Exhibition 2008*, Sydney, Australia, Saydam, S. (Ed.), The Australian Institute of Mining and Metallurgy, Melbourne, Australia, pp. 91–97.

Shishko, R., Fradet, R., Saydam, S., Dempster, A., and Coulton, J. 2015. *An Integrated Economics Model for ISRU in Support of a Mars Colony AIAA Space 2015*, Pasadena, CA, August 31 to September 2.

Shishko, R., Fradet, R., Saydam, S., Tapia Cortez, C.A., Dempster, A., Coulton, J., and Do, S. 2016. An integrated economics model for ISRU in support of a mars colony—Initial results report. *Space Resources Roundtable (SRR)/Planetary and Terrestrial Mining Sciences Symposium (PTMSS)*, Colorado School of Mines, Golden, CO, June 7–9.

Sonter, M. 2001. Near earth objects as resources for space industrialization. *Solar System Development Journal*, 1(1): 1–31.

Sonter, M. 2013. Project concepts for near-term commercial asteroid mining. Off Earth Mining Forum delivered at University of New South Wales, Sydney, Australia, February 19–21. www.youtube.com/watch?v=yvBVuewGuxs (accessed May 17, 2013).

Sonter, M.J. 1997. The technical and economic feasibility of mining the near-earth asteroids. *Acta Astronautica*, 41(4–10): 637–647.

Tapia Cortez, C., Saydam, S., Coulton, J., and Shishko, R. 2017. WEM3 Off Earth Mining Model. *Presented at the 2017 SME Annual Meeting*, Denver, CO.

Yoshikawa, M., Fujiwara, A., and Kawaguchi, J. 2007. The nature of asteroid Itokawa revealed by Hayabusa. *Proceedings of International Astronomical Union No 236 2006*, Milani, A., Valsecchi, G.B., and Vokrouhlicky, D. (Eds.), Kanagawa, Japan, pp. 401–416.

Zacharias, M., Gertsch, L., Abbud-Madrid, A., Blair, B., and Zacny, K. 2011. Real-world mining feasibility studies applied to asteroids, the moon and mars. *AIAA SPACE 2011 Conference & Exposition*, Long Beach, CA, September 27–29.

Zacny, K., Chu, P., Paulsen, G., Avanesyan, A., Craft, J., and Osborne, L. 2012. Mobile in-situ water extractor (MISWE) for mars, moon, and asteroids in situ resource utilization. *Paper Presented at the AIAA SPACE 2012 Conference & Exposition*, Pasadena, CA.

Section III

Social Responsibility and Environmental Stewardship

17

Vignette: Land Access and Social Consensus

Sharon Flynn

What barriers to entry exist to develop or incorporate innovation in your industry sector?

A significant challenge facing the mining sector is its ability to secure land access for mine development and expansion. Companies have typically relied on market-based negotiations with landowners or government support to secure access. As most exploration geologists will tell you, the high grade ore bodies located under land with easy access are increasingly more challenging to find. Future mine development and expansion is likely to happen in geographies where people living on the land or that have high levels of biodiversity. To overcome the challenges associated with gaining land access, mining companies will have to consider innovation internally and externally, and rethink their approach to project development, scheduling, budgeting, and engagement.

Internally, companies will need to rethink how to consider the time and costs related to land access—for example, how to reduce the footprint to reduce land take and how to get the right data to incorporate land access costs into mine design and the financial model. Externally, companies will need to bring their development timeline forward and be willing to engage earlier with communities and landowners to agree how land access could happen.

In addition, companies will need to get better at recognizing that for many communities, especially poorer, rural communities typically living in areas with mineral resources a transactional, cash-based market negotiation for land access will not work in the long run. Although cash may incentivize people to move in the short term so that projects can maintain their schedules, a land access strategy that does not include a well-designed approach to transitioning livelihoods in the long term may result in communities suffering and becoming poorer. This, in turn, is likely to result in higher costs to the company. Consequently, companies should start their land access planning early, bring in the right people to do it, and engage, consult, and listen to the communities and landowners.

What have been the most significant recent achievements in terms of innovation?

The multi-lateral banks, especially the World Bank and the International Finance Corporation (IFC), have been leading the way over the last 10 years to set standards and approaches for responsible land access and the resettlement of communities. The IFC issued its first Sustainability Framework and set of Performance Standards for environmental and social impacts and risks in 2006. While these standards are only a requirement for IFC clients, they have become the most widely accepted for benchmarking best practice over the last 10 years. The standards and their related tools have pushed integrated approaches to assess impacts and risks to communities and people from large-scale infrastructure and mine development. In addition, in part spurred by the need to comply with the standards, the number of experienced professionals and deep thinkers working in resettlement and livelihoods has grown. There is a wealth of knowledge and lessons learned that companies should recognize and leverage.

What do you see as being the most prominent future trends for innovation in your industry sector?

Three inputs are critical to the ongoing operation of the mining industry: (1) access to land to extract minerals, (2) access to water to process and produce metals, and (3) attaining social and political consensus to operate. Gaining and keeping access to these three inputs will require the mining industry to innovate and think differently.

> *Water*: Water is a public good and a public concern. Many governments and communities around the world worry about the quantity and quality of their water resources. Climate change, growing populations, and increasing demand for water from industry and agriculture will add to the challenge. Mining companies must continue to innovate to reduce water consumption from their own operations and work collaboratively with other stakeholders to improve landscape level approaches to water resources management. Trust and collaboration across sectors—public, private, and civil society—will underpin the ability to innovate and find new solutions

> *Land access*: The challenges of retaining their existing access to land and securing access to new land will drive mining companies to think more broadly and holistically not only about how to address resettlement and livelihood restoration, but also about how to build durable and mutually beneficial agreements with communities and landowners. Transparent, early engagement between communities, government, and companies is the path to understanding the scope and scale of land access impacts and finding approaches that work.

Social consensus: Retaining community and government support for mine development and operations is not a finite end goal in the issuance of a "license." Rather, it is an ongoing process. Gaining public support for developing new mines and operating old ones in a world dominated by social media is getting harder. Most people have little connection to how they use minerals every day and where the raw materials come from. Opinions, views, and "truths" are shared lightning fast across the globe. Unfortunately, the mining sector can be too slow to react to these developments and be too inward looking. Mining companies often have limited responses to events in real time and can fail to communicate messages that resonate with a public who see mining as fundamentally destructive to the environment and to people. As such, the sector needs to embrace these shortcomings by providing agile responses to communities' concerns and being more proactive in their engagement and clearer in their communication about the role and value of their business.

18

Leveraging Social Investment

John D. Moore, Andry Nowosiwsky,
Valentina Kaman, and Gary Krieger

CONTENTS

18.1 Introduction

There is a rich debate and dialog between companies, governments, donor organizations, civil society, academics, and various policymakers regarding how best to harness private sector capacity to achieve positive development outcomes. Public-private partnerships (PPP) are a tool that the international development community, including international financial institutions (IFIs), are increasingly using to enhance critical public infrastructure assets and services.

With this diverse mix of stakeholders looking to better capitalize on a mix of industry investments, there are multiple lessons to learn from the oil and gas sector's experience to date. While the oil and gas industry is renowned for its efforts to identify innovative opportunities to improve exploration, development, production, and downstream technical and commercial performance, significant effort to learn from and implement innovative approaches in the social investment space only began to increase in the 1990s. As the debate moved forward, discussion within the industry—as with other sectors—is now focused on going beyond traditional corporate social responsibility (CSR) paradigms to create longer term business and social value (Kramer and Pfitzer 2016).

One example of how the industry is attempting to create greater value from social investments is ExxonMobil's Papua New Guinea Liquefied Natural Gas Project (referred to in this chapter as "Project") engagement in the community health space. Using a PPP model, the Project first aligned with national priorities and then took advantage of workforce-focused health investments to achieve higher value with respect to business priorities, community health programs, and long-term national health policy and implementation. The Project's Partnership in Health Program (PiHP) is not based on physical infrastructure[1] or traditional clinical services,[2] but on impact mitigation in parallel with higher end scientific and medical services that can provide crucial data for overall countrywide public health policy development as well as planning and program execution. A separate PPP among ExxonMobil, the Baylor College of Medicine-Texas Children's Hospital (BCM-TCH), and the Papua New Guinea (PNG) National Department of Health (NDoH), was designed and implemented to build capacity at the University of PNG (UPNG) School of Medicine and Health Sciences (SMHS) focused on improving national-level pediatric, maternal, and public health capacity.

18.2 Building Shared Value: Evolving Approaches to Industry's Role in Development

Oil, gas, and mining infrastructure developments typically demand significant investment of financial, intellectual, and human resources, as well as the application of an array of technologies and skill sets to achieve planned outcomes. Given the potential political and socio-economic impacts related to large projects, these investments can present multiple risks and benefits. The nature of the risks and benefits presented vary across local contexts, with challenges posed by limited community health infrastructure, education systems, political stability, socio-economic resilience, and economic size and the distribution of wealth being but a few factors of importance. Large projects may improve economic growth opportunities for a country

---

[1] Infrastructure in this instance includes clinics, hospitals, and related facilities.
[2] Services in this instance refer to outpatient and inpatient care.

and its people in addition to creating value for project partners and investors. However, poor management and stewardship of such projects and related revenues can result in increased political as well as economic instability, deteriorating community health, environmental damage, and negative socio-economic outcomes in parallel with increased project cost, schedule, and reputational risks. The core challenge in any large-scale development is to optimize project-related benefits and shared value while identifying, avoiding, and mitigating risks and costs (Nelson and Valikai 2014).

As noted by Nelson and Valikai (2014) at Harvard's John F. Kennedy School of Government, the "primary responsibility for governing project-related benefits and risks rests with government, but investors and project operators play a crucial role in achieving responsible and economically viable natural resource development." Managing risks requires the proactive identification and management of technical as well as non-technical risks inherent in any large-scale resource development. The ability to effectively address fluid risk environments, including the interdependencies between various risks, at a strategic and tactical level is essential. Managing diverse risks, positive engagement with stakeholders, and creating shared value with local government and communities depends in turn on the project operator's ability to execute effectively not only project management elements, but also relationship management as well as the application of international good practices and standards in a manner that is responsive to and realistic about domestic capacity (Nelson and Valikai 2014).

Critical to any effort at creating shared value is for all involved stakeholders to realize that there is no guarantee of a positive outcome for any single program or initiative, but that absent alignment between industry, government, community and related stakeholders sub-optimal or outright negative outcomes are almost certain. Expectation management and transparency in program design and implementation are key, as is the need to refrain from dis-incentivizing primary development actors—the state, local government, communities, and donors—from playing the development roles they should play within a given socio-economic context. Any company that allows the responsibilities of government or other actors to be absorbed into a project may be delaying capacity building and thereby doing a disservice not only to the proposed project, but to the people and communities such efforts are meant to assist.

18.3 Community Health in Focus: Industry and the Global Context

Over the last 20 years, the analysis of project-related health impacts, positive or negative, has become a consistent feature of overall environmental and social impact assessment efforts. In contrast to legislative requirements for environmental and social impact assessment, a formal dedicated health analysis is typically not a mandated legal requirement for most countries (Castro et al. 2016). However, through the International Finance Corporation (IFC)[3] Performance Standards (PSs), "community health, safety and security" (IFC PS No. 4) is a required element of the overall impact assessment process. The IFC does not mandate a stand-alone health impact assessment (HIA), but it does require an analysis of potential health impacts associated with a proposed project. The IFC PSs are widely adopted by IFIs, hence "health" has achieved improved prominence within the overall impact assessment hierarchy.

The analysis of potential impacts and benefits is different than the management and mitigation of identified potential negative impacts, and the promotion and extension of positive benefits to the host communities and the overall country. Mitigation of negative effects and promotion of positive benefits are two sides of the same coin. A project proponent can mitigate potential negative health effects while promoting and accentuating the positive extended benefits of the project. One strategy to simultaneously mitigate and promote is to engage local host resources in a long-term collaborative and transparent process, using the PPP approach.

PPPs are used in numerous countries across a spectrum of public sectors, and often include health, power, water sanitation, transport, telecommunications, and energy. The World Bank Group (World Bank 2014) defines a PPP as "A long-term contractual arrangement between a public entity or authority and a private entity for providing a public asset or service in which the private party bears significant risk and management responsibility." A PPP is a relationship between a government partner and a private sector entity to deliver a public service while sharing the risks and rewards. According to the IFC (2013), a PPP, while not the answer to every infrastructure and social service need, is a tool that can help meet development challenges that are aligned with existing national priorities, resources, and capacity as well as community and private sector interests. Recently IFC reviewed more than 60 IFC projects that involved some type of PPP exercise. As the IFC noted, "...The sheer volume of advice and information on PPPs is intimidating and unwieldy, with a simple Google search of the term "public private partnership" yielding over 20 million hits."

Genuine partnerships are more than a mere slogan. A PPP must have a defined program with objectives and defined key performance indicators. Crucial to a health-related PPP must be a clear understanding of and alignment with the host countries national health strategies. Without proper alignment, the health PPP runs a risk of turning into a project proponent driven effort that is ultimately unsustainable and likely to result in lower shared value. Moreover, industry engagement in a PPP or other program cannot be a substitute for active participation by and coordination with the

---

[3] The IFC is a member of the World Bank Group.

host government. Indeed, government leadership in, or at least support for, the partnership is essential. In turn, industry-driven programs should not—either real or perceived—become a "shadow" or mini-ministry of health for core services that should be provided by government.

In addition to the corporate–community–government relationship, a further consideration in developing a sustainable health PPP, is linkage to the local and international academic health community. In many countries, academic institutions may be weak but they are rarely non-existent. In most developing countries particularly in the tropics, there is a robust relationship between the international public health community and host country academic institutions. There are frequently longstanding collaborative training and research relationships between prominent international academic centers of host country institutions whose academic and research relationships can become an essential contributor to the health PPP. It can also provide a greater opportunity for sustainable extended benefits and enhanced international profile. Adding the academic "leg" to the corporate–community–academic structure adds rigor and credibility to the overall effort.

18.4 PNG in Context: The Development Challenge

PNG[4] is an island nation with a rugged topography and an estimated population of 7.7 million people (World Bank 2017). PNG is in the southwestern Pacific Ocean, just north of Australia and east of Indonesia. The eastern part of the country consists of the island of New Guinea and its offshore islands. The largest offshore islands are New Britain, New Ireland, and Bougainville. The country is one of the most diverse countries in the world, with approximately 87% of the population living in rural areas. The PNG socioeconomic landscape is both diverse and challenging, with the population composed of a complex social and cultural structure that has over 1000 ethnic and tribal groups and more than 800 languages.

Located near fast growing Asian markets, PNG has significant natural resources, including gas reserves. The PNG economy is divided between the agricultural, forestry, and fishing sectors which are the largest employers, and the minerals and oil and gas sectors which account for most of the PNG export earnings and gross domestic product (GDP). The country is a significant producer of gold, copper, nickel, silver, cobalt, oil, and gas. Oil exploration in PNG commenced in the 1920s, with the oil and gas industry emerging in the late 1980s as the first commercial oil exports began in 1992 from the

---

[4] The country gained its independence in 1975 from Australia. PNG has a Westminster parliamentary system of government.

Kutubu field.[5] Following multiple gas discoveries, including the large Hides gas field, liquefied natural gas (LNG) exports began in 2014, as the Project came on-line (McWalter 2011).

PNG faces significant development challenges. Perception of corruption and fiscal instability are concerns expressed across PNG citizenry, and PNG continues to suffer from fiscal woes brought about by lower revenue receipts caused by a decline in oil and mineral prices. In the last quarter of 2016, the government amended its year-end GDP growth projection to 2% because of low revenues from the mining sector. As noted by the World Bank (2016), "Translating revenues from resource development projects and other economic activity into strong, tangible improvements to living standards for all Papua New Guineans remains a key challenge for the government, as is the need to improve public financial management and efficiency of public spending." While PNG's fiscal challenges are certainly real, there are efforts underway aimed at improving governance as well as PNG's attractiveness as an investment destination.[6]

Underlying the broader governance and socio-economic issues, challenges across the community health sector remain a critical impediment to the PNG efforts to move toward a brighter economic future. As PNG moves forward, a renewed focus on addressing existing and emergent community health challenges will prove essential. The pattern and distribution of disease is extremely uneven and some areas of the country are likely undergoing an epidemiological transition, specifically a movement from infectious to non-communicable diseases. Communicable diseases are a major issue in rural PNG where the burden of disease is dominated by respiratory diseases, particularly pneumonia. Vector-borne diseases, such as malaria are important, but cases of malaria have dramatically fallen over the last 10 years (Hetzel 2014) due to efforts by the PNG Ministry of Health, in concert with international health groups such as Global Fund and bilateral donors such as the Australian government. Infectious diseases are still responsible for most of the morbidity and mortality in rural PNG, but research indicates a parallel rise in non-communicable diseases as lifestyles and diets shift.

Tuberculosis (TB) has a major impact on overall morbidity, mortality, and remains a public health issue across PNG. Current research indicates that PNG has significant levels of TB and multi-drug resistant TB (MDR-TB) unrelated to HIV co-infection. In sub-Saharan Africa, TB is often driven by

---

[5] Early development of the Kutubu oil fields was done by Chevron Nuigini.

[6] While challenges are significant, there are efforts underway aimed at improving the long-term attractiveness of PNG as an investment destination. One such initiative is the country's commitment to moving forward with the Extractive Industry Transparency Initiative (EITI). The PNG government is working with the country's mining, oil and gas industry, and civil society to promote revenue transparency and accountability in the country's mining and petroleum sectors. The EITI is a global standard to promote the open and accountable management of oil, gas, and mineral resources. For additional detail please see www.eiti.org. For specific detail on the PNG EITI program please see www.pngeiti.org.pg.

HIV co-infection; however, this does not appear to be the situation in rural PNG (Cross et al. 2014). Underscoring the threat of TB, PNG has been added to the World Health Organization (WHO) High Burden TB country list for TB, TB-HIV co-infection, and MDR-TB (Aia et al. 2016).

The national malnutrition rate in PNG among children under age five attending maternal and child health clinics has continued to decline since 2011. As of 2015, the national rate is at 23%. The level of low birth weight children (birth weight less than 2,500 g) is at 7% (NDoH 2015).

The engagement of industry in PNG socioeconomic development has varied over time, with different concepts and programs applied by mining, oil and gas, forestry, fisheries as well as non-natural resource-focused companies with investments in the country. The oil and gas sector has also had different approaches to the conduct of impact mitigation and social investment in the country. For example, Chevron Nuigini's Community Development Initiative (CDI) Foundation Trust Fund in 2001, was established as a nongovernmental organization (NGO) aimed at generating sustainable benefits to community (Community Development Initiative Foundation 2017). With the acquisition of Chevron assets by Oil Search Ltd. (OSH), this program is now embedded under Oil Search activities in country. Oil Search Limited is the current operator of the PNG oil fields, and is the second largest investor in PNG LNG. In addition to its CDI legacy, a more recent example is the OSH health foundation. Begun in 2011, and rebranded as the Oil Search Foundation in 2015 (Oil Search Foundation 2017), this initiative focus has evolved to focus more on multiple general development sectors (Oil Search Foundation 2015).

18.5 The Project Overview

The Project[7] is an integrated development that has commercialized gas resources in the PNG highlands. The Project is producing approximately 7.4 million tons of LNG each year, which is exported to four major customers in the Asia region (PNG LNG 2017). PNG LNG facilities are connected by over 700 km of onshore and offshore pipeline and include a gas conditioning

---

[7] In 2008, the initial Project partners signed the joint operating agreement and an independent Economic Impact Study was commissioned that showed that the Project would have a significant impact on the PNG economy over time, while providing other benefits such as employment and business opportunities. On 22 May 2008 Project venture participants and the PNG government formally signed the Gas Agreement. The Gas Agreement established the fiscal regime and legal framework by which the Project is to be regulated throughout its lifetime and set the terms and mechanism for government equity participation in the Project. Following this, it was announced that the Project would enter Front End Engineering and Design (FEED). Between December 2009 and March 2010, sales and marketing agreements for the gas were signed with four major customers.

plant in Hides and a liquefaction and storage facility near Port Moresby. ExxonMobil PNG Limited (EMPNG)[8] operates PNG LNG on behalf of six coventure partners.[9] Over the next 30 years, it is estimated that PNG will produce more than 11 trillion cubic feet of LNG. The long-term revenue stream of the Project from PNG LPN has the potential to help transform PNG's macroeconomic position over time.

An environmental impact statement (EIS), which draws upon 26 supporting studies and took 2 years to complete, was approved by the PNG government in October 2009. The EIS documented many rigorous commitments and measures the Project would take to manage the environment. On December 8, 2009 the Project venture participants approved the Project, paving the way for construction to begin. This was supported by the completion of financing arrangements with lenders in March 2010. Engineering, procurement, and construction contracts were approved in late 2009 and in early 2010, construction work began, with LNG production beginning in April 2014. Since then the Project has been reliably supplying LNG to four long-term major customers in the Asia region, namely China Petroleum and Chemical Corporation (Sinopec), Osaka Gas Company Limited, JERA,[10] and the CPC Corporation.

Over 55,000 workers were involved in the construction of the Project, with 21,220 employed at its peak in 2012, with more than 10,000 Papua New Guineans trained for construction and operation roles and more than 2.17 million hours of training through 13,000 training programs having been delivered.

18.6 Case Study—Partnering to Address Community Health Challenges

EMPNG,[11] as operator of the Project, recognized the essential need to develop and strengthen over time partnerships with PNG's national as well provincial government institutions, local and international universities working in

---

[8] ExxonMobil is also participating in PNG's second LNG project, Papua LNG, following the acquisition of InterOil Corporation in February 2017. Papua LNG is a joint venture which includes affiliates of Total SA (operator), ExxonMobil, and Oil Search. The project seeks to commercialize gas associated with the Elk-Antelope fields located in Gulf Province. ExxonMobil is also working on development plans for gas discovered in Western province's P'nyang field. ExxonMobil has been marketing petroleum fuels and other refined products in PNG since 1922.

[9] The equity partners in the Project are ExxonMobil (33.2% and operator), Oil Search (29.0%), Santos (13.5%), National Petroleum Company of PNG (PNG government) (16.8%), JX Nippon Oil and Gas Exploration Company (4.7%), Mineral Resources Development Company (PNG government, on behalf of landowners) (2.8%).

[10] JERA is composed of the former Tokyo Electric Power Company Inc. and Chubu Electric Power.

[11] EMPNG is the local affiliate of ExxonMobil Corporation in PNG. ExxonMobil has had a presence in PNG for over 80 years.

PNG, and international and local NGOs capable of helping implement sustainable community health programs.

Given its global experience, as a PNG LNG operator ExxonMobil recognized that developing a contextually appropriate and sustainable community health program required deep understanding of PNG's complex political economy and its unique sociocultural characteristics. Based on IFC and World Bank published guidance, key points considered were:

- *Political*: The need for a national champion to help develop, refine, and implement the Project's vision, provide context and local level knowledge, and help guide progress and advocate support at the national and local levels.

- *Economic*: Understand the level of financial, intellectual, technical, and other investment required and, more importantly, how much direct Project team monitoring and stewardship was required to achieve success. Early recognition that accounting/financial capacity building was required proved critical along with a realization that business objectives and public policy goals needed to be aligned.

- *Execution*: The need for clear technical "Terms of Reference" covering specific scientific activities. Detailed budgets would be required and needed to be constantly reviewed. In general, "scope creep" and constant change orders were to be strongly discouraged, that is, fiscal discipline and execution excellence were set as the norm not the exception.

From the onset the Project team goals were to (1) accurately characterize and monitor Project area (PA) socioeconomic health indicators and compare to similar control locations, (2) effectively diagnose and monitor disease incidence in key communities, (3) implement specific intervention efforts to encourage health sector improvements and limit adverse health impacts in PA communities, and (4) establish an integrated health and demographics surveillance system (iHDSS) to collect and assess community health information to inform the Project team, provincial, and national governments as well as non-governmental and civil society organizations regarding the burden of disease. Data collected is publicly available online, where it can be used to help inform the efforts of the PNG NDoH and other stakeholders to determine future community health priorities and support sound policy making (IPIECA 2005, 2016).[12]

As part of the analytical process, the Project assessed community health using a HIA tool. The use of HIAs has increased over time, either within an integrated environmental and social impact analysis or as a stand-alone

---

[12] The Project's PiHP was featured as a case study in the IPIECA (2016) HIA guidelines.

analytical effort. For the Project, the intent of the HIA was crucially viewed as helping set a management framework rather than simply being a compliance exercise. The strengths and weaknesses of the PNG health care framework were identified. The assessments included in the Project's HIA showed PNG had an unusually high burden of infectious diseases and a health care system with limited capacity. This system had historically struggled in many locations adjacent to planned project infrastructure to provide basic health services. The HIA also determined that PNG had an existing and highly respected health research unit, the PNG Institute of Medical Research (PNG IMR).

During and after completion of the HIA, the Project engaged with other private sector companies, PNG national research organizations, NGOs, international development agencies, and PNG church medical services agency to understand current capabilities, capacity, and geographical footprint of both disease and related health infrastructure. A joint workshop with the full spectrum of stakeholders was then held, which proved essential to identifying key "lessons learned" that would further inform PiHP implementation:

1. The geographical limitations of NGO outreach across the large Project footprint.
2. The weakness of management structures, particularly financial controls, of PNG government, research, and NGO groups.
3. The need to begin transition planning at the beginning of any project.
4. The difficulty of finding and retaining sufficient national staff given there was a clear shortage/misallocation of health professionals across the proposed Project geography.

The result of the HIA effort was the structuring of a novel PPP to support health risk mitigation, known as the PiHP. The PiHP involved two core agencies, namely the PNG IMR,[13] and a lead NGO contractor, Population Services International (PSI),[14] tasked with identifying local

---

[13] PNG IMR is the statutory but independent research arm of the PNG Ministry of Health. For additional information please see http://www.pngimr.org.pg/.

[14] PSI is a global health network of more than 50 local organizations dedicated to improving the health of people in the developing world by focusing on serious challenges like a lack of family planning, HIV and AIDS, barriers to maternal health, and the greatest threats to children under five, including malaria, diarrhea, pneumonia, and malnutrition. PSI was founded in 1970 to improve reproductive health using commercial marketing strategies. For its first 15 years, PSI worked mostly in family planning (hence the name PSI). For additional information see http://www.psi.org/.

NGO partners and delivering critical services to mitigate potentially negative health impacts identified by the Project HIA and PNG IMR. In addition, ExxonMobil PNG brought in NewFields LLC[15] to assist with program design and implementation as well as quality control. This approach allowed for a clear delineation of authority and the roles and responsibilities of each agency involved. For instance, all monitoring and evaluation was performed by PNG IMR, while service delivery[16] was overseen by PSI. The goals of the PiHP are as follows:

1. Protect the health of the PNG LNG workforce through communicable disease prevention measures.
2. Accurately characterize and track Project Area (PA) socioeconomic indicators of health status and compare to similar control communities.
3. Reliably diagnose and track disease occurrence in communities of interest.
4. Implement specific intervention programs to promote health sector improvements and prevent adverse health outcomes in PA communities.
5. Create a robust and sustainable iHDSS to collect and analyze community health data to inform Project team, provincial and national governments as well as NGO regarding priorities for future health improvement objectives and investments.

Both impact mitigation and extended countrywide benefits for the broader community health sector were integrated into the partnership, with a Memorandum of Understanding (MoU) and Sponsorship Agreements with key stakeholders developed and signed. Scientific, financial, and capacity development objectives were specified as was a set of key performance indicators (Nowosiwsky et al. 2016b). Significant EMPNG stewardship was provided early on, to include financial accountability training and monitoring of partners in relation to their PNG LNG-related activities (Nowosiwsky et al. 2016a, 2016b, 2016c).

---

[15] NewFields is a consultancy founded in 1995. NewFields provides quality individual expert advice, to include specialist support in impact assessment and analysis. For additional information see http://www.newfields.com/.

[16] Service deliver included educational outreach, condom distribution, and water sanitation packages.

18.7 Scaling Up: PiHP and Strengthening National Capacity

In 2011 a 5-year program was designed and implemented. More than 50,000 persons were covered and monitored by the iHDSS, which was accepted into the 20-country international INDEPTH Network.[17] This was the first PPP sponsored site accepted by INDEPTH in its 18-year history, and potentially the first INDEPTH PPP implemented in the oil and gas industry.

Underneath the PiHP, EMPNG, in partnership with PSI PNG-based NGO partners, implemented the Community Health Impact Management Program to work within community areas to mitigate potential adverse health events related to PNG LNG activity. Community-based impact mitigation programs were focused on

1. Water sanitation and hygiene related disease prevention.
2. Sexual and reproductive health education.
3. Communicable disease prevention and control.
4. Health education scholarships and upgrading of medical skills for community health care workers.

From 2011 to 2014, more than 25,500 Papua New Guineans were reached through these programs. More than 13,000 individuals living in villages near the Project participated in interpersonal communication sessions about WASH program (focus on water and sanitation disease prevention) and 9,709 wash kits were distributed. Gender-based violence training was provided to 3,258 people and 1,155 received marital relationship training. Thirty-nine out of 40 scholarship recipients completed their health studies, with most of them returning to their villages to apply their skills (Nowosiwsky 2014).

Critical to the community health program is the iHDSS. The iHDSS provides a means to systematically collect objective health and population data at key Project locations and matched control sites. It also provides timely response to disease outbreaks that could potentially disrupt Project operations through the posting of trained clinicians at designated health facilities throughout the Project area. Information from the iHDSS baseline survey and follow-up health and social demographic data show the absence of negative community impacts while demonstrating the presence of positive community health trends and improvements potentially linked to Project activity.

---

[17] INDEPTH is a global network of health and demographic surveillance systems (HDSSs) that provide a more complete picture of the health status of communities. As of March 2017, INDEPTH membership is composed of 42 member health research centers that observe through 47 HDSS field sites the life events of over 3 million, 800 people in 18 LMICs in Africa, Asia, and Oceania. For additional information on INDEPTH please see http://www.indepth-network.org/.

The lack of a long-term monitoring and surveillance system has often been highlighted in other developments by the World Bank as a major concern in the EIS process (Myorga 2010). The iHDSS is a positive solution to this concern and has significant extended benefits to the host country, particularly regarding providing objective data for public health policy formulation (IPIECA 2005, 2016). The review of iHDSS health study results confirmed Project impact mitigation strategies were effective in preventing negative impacts such as disease increases (as compared to control sites) during the Project.

This unique PPP also saw partnership grow between PNG IMR and the UPNG SMHS, and led to EMPNG providing funding for the construction and outfitting of the National Infectious Disease Diagnostic and Research Laboratory. This modern Biosafety Level 2+ (BSL) infectious diseases research laboratory[18] was developed under the PiHP and colocated within the PNG SMHS campus, with PNG IMR providing management of the facility. The lab is used to advance important biomedical research in tropical medicine and emerging and neglected infectious diseases such as cholera and TB, and stands ready to support other industry and government partners should they want to join in expanding disease surveillance programming to other sites not related to PNG LNG. Important scientific discoveries involving TB, human papilloma virus, febrile surveillance, and noncommunicable diseases have been documented with significant public health policy implications (Badman et al. 2016; PNG IMR 2017).

18.8 Support to the University of PNG School of Medicine and Health Sciences

While community health and human resource challenges create a major challenge to long term development, and setting up the PPP that is delivering the PiHP, ExxonMobil simultaneously looked for innovative ways to bolster the future medical intellectual capacity of PNG. Again, a PPP model was used that brought in a long time ExxonMobil partner, the BCM-TCH. This PPP designed and created a program to improve one of the most valuable resources in the country, the PNG health professional.

With funding and facilitation support from ExxonMobil PNG, initial discussions in 2013 between BCM-TCH, the PNG NDoH, and the UPNG SMHS led to the establishment of a multiyear PPP focused on improving national-level pediatric, maternal, and public health capacity. The PPP was developed

---

[18] A BSL 2 laboratory is suitable for work involving agents that pose moderate hazards to personnel and the environment. The "+" means that mycobacterium can be handled in this facility under certain safety conditions.

as an outgrowth of the HIA process and following a comprehensive needs assessment that identified gaps in the community health system relating to maternal and child health. Key areas for capacity strengthening identified included increasing the quality and availability of SMHS faculty, assistance in efforts to improve research, and patient care. There is also a focus on addressing the emerging crisis of malnutrition and pediatric TB.

Since 2014, three experienced international health experts, a pediatric specialist, an obstetrician and gynecologist, as well as a doctor in public health have been integrated into the UPNG SMHS to support Port Moresby General Hospital. Additional efforts have included outreach to Tari General Hospital in Hela province.

Efforts by the senior pediatric TB specialist at with Port Moresby General Hospital, in collaboration with local stakeholders, have since seen the care of children with TB improved. Not only has the quality of daily care of children with TB been enhanced, but also the management of the children's MDR-TB ward. Other early achievements include updating the national pediatric TB guidelines, and leading a team to implement a newly formulated TB drug for children. Through this partnership, HIV testing for all children with TB is increasing toward its goal of 100%.

While there are many longstanding health programs working in PNG, few have engaged in the core area of medical education at the only medical school in PNG. The BCM-TCH partnership is the only program providing full-time faculty support to the UPNG SMHS. As part of BCM-TCH's commitment to PNG, in 2014 a NGO—Baylor College of Medicine Children's Foundation–PNG—was formed to both host the program's staff and support engagement with potential new sources of funding and collaboration.[19] The BCM-TCH partnership will leave a lasting impact and future PNG health workers will ultimately shape healthcare and public health policies in the country for years to come.

___

18.9 Assessing Outcomes

Governments and communities will not support development projects if there is no investment by industry in supporting positive development outcomes. In turn, industry will not strategically invest in development programs if there is no substantive contribution by such programs to its bottom line. The mere intent of doing good, or even the presence of a program or set

___

[19] To develop future leaders in public health, ExxonMobil is also funding a young PNG candidate, Margaret Yagas, to pursue a Master's in Public Health at The University of Texas. Margaret started her studies in the United States in July 2015 and upon completion will join the UPNG in the NDoH. The skills and experience she will bring to PNG will have a lasting positive impact on the public health education capacity of PNG in the future.

of programs, does not guarantee positive outcomes. It also takes an alignment of interests between the stakeholders involved to create opportunity. No one actor alone can ensure that desired development outcomes will occur, as it requires the application of sound thinking and skilled human resources alongside financial, physical and intellectual capital to create the potential for lasting positive change.

There was a clear business interest in PNG to support community health because that in turn has a direct positive impact on workforce health. The PPP approach adopted by ExxonMobil, suggests that alignment between the various actors involved was, and remains, strong going forward even when industry funding reduces over time. The approach has also demonstrated what is possible in PNG, with the PPP model—and specifically the PiHP and BCM-TCH initiatives—serving as a platform upon which the State, donors and existing as well as new industrial developments can use going forward.

The PiHP community health program is assessed to have helped improve village level capacity to accurately diagnose disease, with PNG IMR medical staff visiting community clinics to share good clinical procedures and diagnostic tools to analyze various illness incidence rates. From 2011 to 2014, more than 70,000 Papua New Guineans participated in iHDSS with approximately 14,000 patients assisted in community clinics by the PNG IMR health team. Several studies were completed including passive and active TB surveillance with over 2000 TB study participants and 1,322 sputum samples collected for active TB surveillance (PNG IMR 2015). Fever studies were completed using rapid diagnostic testing (RDT) for malaria diagnosis to introduce basic diagnostic testing devices and enhance accuracy of diagnosis. A total of 765 women participated in the Healthy Pregnancy Study in the diagnosis of sexually transmitted infections (STIs), and more than 900 people participated in the first non-communicable diseases studies covering issues such as cancer and diabetes. Provincial and national health authorities can now draw upon the data collected and apply their findings to support more effective planning and implement health service delivery.

Following the publication of the first two PNG IMR reports, EMPNG and NDoH co-hosted a workshop in April 2014 to review key findings and how data can better inform public policy development and implementation. A workshop entitled, "Translating Science into Action," was designed to review key findings and how data can better inform public policy development and implementation.[20]

The PPP associated with BCM-TCH continues to show positive results now and for the future. In 2015, the program's team delivered more than 450 hours of lectures and mentored projects at UPNG SMHS. The Public Health Specialist attached to UPNG helped create a Masters in Public Health

---

[20] For additional information refer to IPIECA Health Impact Assessment guidelines, the documentation is included in Annex 4 starting on page 72. (http://www.ipieca.org/ resources/good-practice/health-impact-assessment-a-guide-for-the-oil-and-gas-industry/).

Program and a Diploma of Public Health Program, in addition to actively working with NDoH to support their programs. Educational curriculum now includes didactic teaching support for undergraduate students (medical students years 2–5); the provision of basic and clinical science lectures; problem-based learning curriculum mentoring; direct mentoring of medical students in their fourth and fifth years of study (including case discussions and supervision during clinical rotations); lecturing postgraduate trainees in Diploma of Child Health (mid-level provider) and Masters of Medicine program (Senior Pediatrician Level) programs; practical clinical instruction in general pediatrics, malnutrition, TB, HIV, and inpatient care; and involvement in student, resident, and registrar evaluations. BCM-TCH doctors were also responsible for updating the PNG Standard Treatment Book (STB) sections on TB and malnutrition, and for co-editing and compilation of edits for an updated version of the PNG STB anticipated published in 2016. The STB is the most widely used pediatric reference book for healthcare workers in the country.

Collaboration between the PNG Pediatric Society, WHO, NDoH, and the BCM-TCH team enabled the pilot and rollout of a new and improved pediatric TB medication in Port Moresby in the second half of 2016. It is expected that approximately 600 children will benefit from this treatment. The BCM-TCH team in 2017 aims to provide ongoing support and strengthening the current pediatric, maternal and public health programs to improve education of health care workers and reduce morbidity and mortality rates of children and women.

18.10 Learning Lessons

The PPP model can be successfully adapted to meet both impact mitigation and extended benefit needs for a large oil and gas project. There are significant lessons learned when undertaking PPPs, whether of large-scale initiatives such as PiHP or smaller and specialized and partnerships such as the BCM-TCH.

In the past 5 years, the PiHP has achieved significant results at both the local and national level at the NDoH. In a 2016 paper submitted to the Society of Petroleum Engineers, ExxonMobil (Nowosiwsky et al. 2016a, 2016b, 2016c) detailed many of the critical findings and scientific achievements of the PiHP, with an Independent Science Advisory Board composed of internationally recognized tropical disease and demographic experts, and validated the results of the PiHP. Based on the scientific results, the PiHP has begun to attract and develop additional funding sources particularly in infectious diseases intervention and control, particularly TB and sexually transmitted infections. Similarly, the partnership with BCM-TCH and NDoH has had direct impact on patient care at Port Moresby General Hospital and enhanced capacity around pediatric TB management while strengthening the curriculum and teaching delivery at the UPNG SMHS. These successes have resulted from an alignment between a specific business interest, sustaining a healthy workforce, and achieving national development objectives regarding community health.

Accurately assessing the capability of potential partners and including capacity building as part of program execution has been important in capacity building as well as promoting mutual learning. A recognized concern from many developing world contexts is the tendency by donors, civil society, and the private sector to criticize government for lacking capacity, then demand that government services are provided without addressing gaps in capacity. Providing technical support, mentoring, and strong financial controls, enabled the PiHP to deliver outcomes while ensuring that funding support accountability for program resources were used effectively. From a technical perspective, EMPNG brought in partners with proven performance in similar contexts to assist PNG IMR in implementation of the program. Emphasizing transparency through making data publicly available through the PNG IMR website has ensured that government, donors, NGOs, and broader civil society can access and use the data to implement effective community health policy options. Encouraging all players in the development sector to work under a government-sponsored framework can avoid silos of activity and reduce duplication, enabling programs to create higher value with respect to the cost (of investment) and return (positive development outcomes and improved business performance) curve.

The rigorous financial controls that were established reflect those used in industry, to include correct invoicing by suppliers, proof of payment, proof of labor, and proof of work completed. This level of control is generally more rigorous than a typical "scientific grant" that often is awarded by either a scientific or donor agency. To implement this level of financial and execution control, the Project conducted numerous training sessions for the lead agencies of the PiHP. These efforts were part of the sponsorship agreement between the Project and the main PiHP partners. The PiHP PPPs, BCM-TCH PPPs, partners, and vendors were held to the same contractual and PSs as core business contractors.

A key takeaway is the potential applicability of the PiHP and BCM-TCH models to other resource developments in PNG and elsewhere. EMPNG made significant effort to ensure that relevant donors involved in PNG, including the Australian government, the United Nations, the WHO, and the Global Fund were made aware of the methodology and results. The Australian government is a major source of international assistance for PNG and provides significant support to the health infrastructure of PNG, and Australian aid personnel were also valuable sources of information and perspective to the PiHP and BCM-TCH programs. The willingness of EMPNG to remain engaged, learn from others, and share lessons has served the Project's community health-related PPPs well.

In addition to building awareness among and understanding the donor experience, EMPNG provided briefings to other industry members within industry and other sectors regarding the PiHP and BCM-TCH initiatives. These programs are not viewed as proprietary or a form of competitive advantage, but a business interest to have the broadest set of actors involved in supporting positive development outcomes in PNG. Industry should not assume the role and responsibilities of a government or donor community, but act as a good corporate citizen and look for opportunities that are aligned with business objectives.

18.11 Conclusions: Looking Ahead

The debate on how best to integrate private sector capacity into broader government, donor, and civil society efforts to achieve positive development outcomes is ongoing. As this debate continues, governments, donors, and civil society should not solely view industry as a source of funding support, but as a genuine partner that can bring additional critical thinking necessary to address business and development challenges. Companies should identify opportunities that go beyond CSR paradigms emphasizing transactional external engagement and tactical reputation management, and look for strategic opportunities to build lasting shared value that contributes to business performance over time. The Project's community health case has shown that well-resourced and executed PPPs can be effective tools in managing business risk while creating positive multiplier effects for socio-economic progress and community resilience. As noted by Kania and Kramer (2016), the use of a "collective impact" framework for approaching social investment can guide industry and other stakeholders efforts within a socioeconomic "ecosystem" to identify joint opportunities, then drive them forward to achieve change that benefits both business and the broader society.

The greatest role of any business in promoting positive development outcomes is the conduct of its business. Business conduct encompasses how a business acts, its ethics, standards, management, and approach to achieving profitability. As stated, the role that industry can play in the development space outside of jobs and revenue creation role should not always be seen in terms of dollars and cents. It is the ability of industry to join with other stakeholders who understand both business and development priorities to apply its intellectual and available financial and technical capacity to assist in the success of development initiatives.

References

Aia, P., M. Kal, E. Lavu et al. 2016. The burden of drug-resistant tuberculosis in papua new guinea: Results of a large population-based study. *PLoS ONE* 11(3), e0149806. doi:10.1371/journal.pone.0149806.

Badman, S. G., L. M. Vallely, P. Toliman et al. 2016. Prevalence and risk factors of Chlamydia trachomatis, Neisseria gonorrhoeae, Trichomonas vaginalis and other sexually transmissible infections among women attending antenatal clinics in three provinces in Papua New Guinea: A cross-sectional survey. *Sex Health* 13, 420–427.

Community Development Initiative Foundation. 2017. https://www.chevron.com/-/media/chevron/investors/documents/png_factsheet.pdf and http://www.oilsearch.com/how-we-work/sustainable-development/supporting-community-organisations.

Cross, G. B., K. Coles, M. Nikpour et al. 2014. TB incidence and characteristics in the remote gulf province of Papua New Guinea: A prospective study. *BMC Infectious Diseases* 14(1), 93.

Hetzel, M. W., J. Pulford, H. Guda, A. Hodge, P. M. Siba, and I. Mueller. 2014. *The PNG National Malaria Control Program: Primary Outcome and Impact Indicators, 2009–2014.* Goroka, Papua New Guinea: PNG IMR.

International Finance Corporation (IFC). 2013. *A Winning Framework for Public-Private Partnerships: Lessons from 60-Plus IFC Projects.* Washington, DC: IFC.

IPIECA. 2005. *Health Impact Assessment. A Guide for the Oil and Gas Industry.* London, UK: IPIECA.

IPIECA. 2016. *Health Impact Assessment. A Guide for the Oil and Gas Industry.* London, UK: IPIECA.

Kramer, M. R. and M. W. Pfitzer. 2016. The ecosystem of shared value. *Harvard Business Review* 94(10), 81–89.

Mayorga, A. E. 2010. *Environmental Governance in Oil-Producing Developing Countries: Findings from a Survey of 32 Countries.* Washington, DC: World Bank. worldbank.org/curated/en/284551468163738491/Environmental-governance-in-oil-producing-developing-countries-findings-from-a-survey-of-32-countries (accessed December 20, 2016).

McWalter, M. 2011. *PNG Gas Finds Push LNG Plans.* Tulsa, OK: American Association of Petroleum Geologists. http://www.aapg.org/publications/news/explorer/column/articleid/2257/png-gas-finds-push-lng-plans#sthash.fARVMzb3.dpuf (accessed February 21, 2017).

National Department of Health. 2015. *Sector Performance Annual Review (SPAR).* Port Moresby, Papua New Guinea: Government of Papua New Guinea.

Nelson, J. and K. Valikai. 2014. Building the foundations for a long-term development partnership: The construction phase of the PNG LNG Project, 2014. The CSR Initiative at the Harvard Kennedy School. Cambridge, MA. https://www.hks.harvard.edu/centers/mrcbg/programs/cri.

Nowosiwsky, A. 2014. Partnerships in health. *Paper Presented at the13th PNG Mining and Petroleum Investment Conference,* Sydney, Australia.

Nowosiwsky, A., N. Burke, J. D. Moore, and G. R. Krieger. 2016a. TB or Not TB: Development of an integrated occupational and community health TB surveillance program. *Paper Presented at the Society of Petroleum Engineers International Conference and Exhibition on Health, Safety, Security, Environment, and Social Responsibility*, Stavanger, Norway.

Nowosiwsky, A., N. Burke, D. Prest, J. D. Moore, and G. R. Krieger. 2016b. Public-private partnership: Lessons learned from a large project in a developing country setting. *Paper Presented at the Society of Petroleum Engineers International Conference and Exhibition on Health, Safety, Security, Environment, and Social Responsibility*, Stavanger, Norway.

Nowosiwsky, A., J. D. Moore, D. Hancock, M. Z. Balge, and G. R. Krieger. 2016c. Development of an integrated community health program for a large oil and gas project. *Paper Presented at the Society of Petroleum Engineers International Conference and Exhibition on Health, Safety, Security, Environment, and Social Responsibility*, Stavanger, Norway.

Oil Search Foundation. 2015. Annual Report, 14. www.oilsearchfoundation.org (accessed March 26, 2017).

Oil Search Foundation. 2017. Main webpage. www.oilsearchfoundation.org (accessed March 26, 2017).

PNG IMR. 2015. March 2015 PNG IMR PiHP progress report. http://pngimr.org.pg (accessed March 26, 2017).

PNG IMR. 2017. IMR PiHP progress reports 2012–2017. www.pngimr.org.pg (accessed March 26, 2017).

PNG LNG. 2017. Main webpage. www.pnglng.com (accessed March 26, 2017).

World Bank. 2014. Public-private partnerships. *Reference Guide*. Version 2.0. Washington, DC.

World Bank. 2016. September 2016 PNG country overview. http://www.worldbank.org/en/country/png/overview (accessed March 26, 2017).

World Bank. 2017. PNG country overview. http://www.worldbank.org/en/country/png/overview (accessed March 26, 2017).

19

Mobile Money and Financial Inclusion in the Papua New Guinea Resource Regions

Tim A. Grice and Saleem H. Ali

CONTENTS

19.1 Introduction to Financial Inclusion

Technological innovations in communication and electronic storage technologies have also had a remarkable impact on how even remote communities in the extractive sector manage their finances. There is an emerging consensus that "financial inclusion"—increasing access of poor households to financial services—is linked to overcoming poverty, reducing income disparities, and increasing economic growth.[1] In contrast to wealthy households who live their financial lives with regular access to electronic financial systems such as ATMs and online payment systems, it is more costly and riskier for poor households living outside of the formal economy to perform basic financial activities. To compound the problem, transacting with poor households whose financial lives are outside of the formal economy is often prohibitively costly for utility companies, banks, insurance companies, and other institutions.

Financial inclusion, on the other hand, is "...a state in which all people who can use them have access to a full suite of quality financial services, provided at affordable prices, in a convenient manner, and with dignity for the clients."[2] The Centre for Financial Inclusion offers four core dimensions in their financial inclusion framework:

1. *What is provided*: A full range of services, including a basic product in each of the four main areas: savings, credit, insurance, and payments.
2. *How it is provided*: With quality—for example, convenience, affordability, safety, and dignity of treatment—and with client protections operating.
3. *Who receives*: Everyone who can use the services, including the poor, rural, informal, and groups who are often discriminated against (women, ethnic minorities, and disabled).
4. *Who provides*: A range of providers led by mainstream financial.

A significant multilateral and cross-sectoral effort is underway to "bank the unbanked."[3] The World Bank developed "Findex," a global financial inclusion database that measures the use of financial services worldwide.[3] G20 member countries established the Global Partnership for Financial Inclusion, and the G20s Financial Inclusion Experts Group (FIEG) released the "Principles for Innovative Financial Inclusion."[4] The Alliance for Financial Inclusion (AFI) developed the "Maya Declaration," a global and measurable set of commitments by developing and emerging country governments to "...unlock the economic and social potential of the 2 billion "unbanked" people through greater financial inclusion."[5] The Better than Cash Alliance—founded by the United Nations Capital Development Fund (UNCDF), United States Agency for International Development (USAID), The Bill and Melinda Gates Foundation, Citi, Ford Foundation, Omidyar Network, and Visa—was established to provide expertise on the transition to digital payments to bring about greater financial empowerment in emerging economies.

Collectively, these efforts are designed to increase access and lower the cost of financial services to unserved or underserviced markets through fit for-purpose legislative regimes; savings and credit products; payment and money transfer services; and new mobile money technologies. And recent data from the World Bank 2014 Global Findex database suggests that these multistakeholder efforts are working. Three years ago, 2.5 billion adults were unbanked. Today, 2 billion adults—38% of the world's adult population—remain without an account.

Progress toward global financial inclusion has its foundations in AFI's Maya Declaration, the first global and measurable set of commitments by developing and emerging countries to unlock the economic and social potential of the 2 billion "unbanked." Signatories to the Maya Declaration agree to make measurable commitments in four key areas, which are also aligned with the G20 Principles for Innovative Financial Inclusion:[6]

1. *Creating an enabling environment*: To harness new technology that increases access and lowers the cost of financial services.

2. *Implementing a proportional framework*: That achieves the complementary goals of financial inclusion and financial stability.

3. *Integrating consumer protection and financial literacy*: As key pillars of financial inclusion.

4. *Collecting and utilizing data*: To promote evidence-based policymaking and measurable progress in monitoring and evaluation.

In 2013 the AFI network adopted the Sasana Accord to which AFI members committed to setting quantifiable national goals as well as measuring and reporting progress based on the core set of indicators identified by AFI's Financial Inclusion Data Working Group (FIDWG).[7]

19.2 Financial Inclusion in the Pacific

The Pacific is one of the least-banked regions in the world. In some Pacific countries, fewer than 10% of the population is estimated to have access to basic financial services.[8] Barriers to financial inclusion in the Pacific include challenging geography, poor infrastructure, natural disasters, persistent poverty, subsistence livelihoods, and relatively low levels of financial competency.

The Pacific Financial Inclusion Program (PFIP) is a Pacific-wide program helping low-income households gain access to quality and affordable financial services and financial education.[9] Managed by the UNCDF and the United Nations Development Program (UNDP)—with funding from the Australian Government, the European Union, and the New Zealand Government—PFIP currently covers Fiji, Papua New Guinea (PNG), Samoa, Solomon Islands (SOI), Tonga and Vanuatu, with Kiribati and Tuvalu potentially covered before the end of July 2019. By 2019 PFIP plans to have one million low-income people in the Pacific—with at least 50% women—gain access to appropriate and affordable financial services (687,000 achieved by 2014). Key achievements in Phase 1 include the following:[10]

- Providing financial access to nearly 600,000 previously unbanked people, including 250,000 women and 360,000 people who opened bank accounts with regulated institutions.

- National financial inclusion and financial literacy strategies developed for PNG, Fiji, Solomon Islands, and Vanuatu and a national financial literacy strategy developed for Samoa.
- Establishing national finance inclusion task force in PNG, Solomon Islands, Vanuatu, and Fiji.
- Established financial competency baselines for PNG, Fiji, Samoa, and Solomon Islands.
- Mainstreaming financial education program into the core curriculum in schools (Fiji and Samoa).
- Effective dissemination of information to scale through training, workshops, and publications.

The expected outcomes in Phase Two are as follows:[10]

- Additional 500,000 people, including at least 50% women gain access to an appropriate, affordable financial service.
- Additional 150,000 previously unbanked people with at least 50% women gain access to a formal savings account.
- 15% of PFIP-supported branchless banking clients are active.
- An average savings balance of USD$10 in active savings accounts/ mobile accounts.
- Four additional Pacific Islands Countries (PICs) have national financial inclusion strategies.
- 3 additional PICs offer financial education through core curricula in schools.

19.3 Financial Inclusion in Papua New Guinea

PNG has a population of approximately 7 million people, most of whom live in rural areas. It is estimated that 85% of the total population of PNG do not have bank accounts. Barriers to financial inclusion in PNG are like the barriers experienced in other Pacific nations. Additional challenges that are unique to PNG—or at least experienced to a greater degree than other Pacific nations—include PNG's cultural and language diversity[11] and the high costs of building a distribution system in a country that is geographically diverse with under-developed infrastrucutre.[12]

The Center for Financial Inclusion (CEFI)was launched in 2013 to promote PNG's financial inclusion agenda; to facilitate improvement of financial services delivery; to establish mechanisms for enhanced information exchange; and to promote gender equity in financial services and financial education. CEFI is endorsed by the PNG National Executive Council (NEC) as the industry apex organization for coordinating, advocating, and monitoring all financial inclusion activities in PNG. PNG's multi-stakeholder financial inclusion efforts have resulted in much tangible progress toward improving financial inclusion in the country:[13]

- Together, the five mobile money providers claim over 300,000 registered users since the opening of the mobile money market in 2011. Growing at an annual rate of 202% since 2008, the number of financial services agents now surpasses bank branches in the country.
- Established financial inclusion and financial literacy working groups on Inclusive Insurance, Electronic Banking, FinEd and Financial Literacy, Consumer Protection, Data and Measurement, and Government Coordination in 2014.
- Member of the Alliance for Financial Inclusion's Pacific Island Working Group.
- Mainstreamed financial inclusion in GoPNG priorities, resulting in the launch of the informal Economy Policy 2011–2015 in which access to finance is one of the two pillars for those in the informal economy.
- Facilitated Government to People (G2P) learning visit for GoPNG, including from Department of Treasury, Department of Finance, and Department of Planning and Monitoring on digitizing G2P payments. Financial Diaries research study showed that respondents typically travelled large distances to interact with banks and deposited and withdrew large sums of money.

PNG's progress toward financial inclusion is gaining momentum. Recent data provided by CEFI indicates that an additional 441,396 accounts were opened in 2014; 91,436 (35%) of which were opened by women, taking the total number of bank accounts in PNG to over 1.5 million.[14] However there is much work to be done for a country whose citizens still live in some of the highest unbanked regions in the world.

19.4 Mobile Money

Mobile money services are perhaps the most promising way to deliver financial services profitably and at scale to the poor.[15] The definition of mobile money varies across the industry as it covers a wide scope of overlapping products, applications, and services.[16] Mobile money is used in this report as an overarching term that describes any service in which a mobile phone is used to access financial services.[17] This includes "mobile financial services," "mobile banking," "e-money," "mobile wallet," and "mobile payments." With 255 mobile money services in operation across 89 countries, mobile money services are now available in over 60% of developing markets.[18] This system provides significant potential advantages to the world's poor including[19]

- Increased access to basic financial services leading to enhanced economic activity.
- Reduced risk of money theft and increased personal control over financial resources.
- Increased speed of payments to and from consumers, businesses, and government.
- Improved convenience and reducing the cash in the economy.
- Lower transaction costs and improved transparency and auditability.
- Improved competition through reduced barriers to entry for fee-for-service business models.

Although mobile money is often described as a product or service, the broader network of actors, infrastructure, and services required to deliver mobile money services at like an ecosystem.[20]

With respect to PNG, in the early 2000s, telecommunication services in PNG were limited to urban centers under the then monopoly operator, Telikom PNG.[21] At that time, Bemobile, a Telikom subsidiary, had 160,000 mobile phone subscribers, mostly based in the capital, Port Moresby.[22] PNG's mobile phone landscape dramatically changed in 2007 when the Independent Consumer and Competition Commission (ICCC) opened up the market to additional providers.[23] Following Digicel's entry to the PNG market, signal coverage was rapidly extended across the country—particularly to towns and rural areas. There are now three mobile phone operators in PNG—Digicel PNG, Bemobile, and Citifon (Telikom PNG subsidiaries)—resulting in an increase of mobile phone subscribers to 1,800,000 subscribers in 2012.[24] Average prices for mobile phone calls have also dropped by 60%.[25] As of 2014, the mobile penetration rate in PNG was reported as 41%[26] representing a dramatic increase in usage rates, although still relatively low by world standards. Most Papua New Guinean's in rural areas use basic mobile phone handsets which are easy to use and have a long battery life, although the availability of smartphones with Internet access is increasing.[27] PNG has four formal banks which service major centers within the country. However, with approximately 80% of the total population unbanked or without reasonable access to formal banking and financial services,[27] and with mobile penetration increasing, mobile money represents a strong opportunity to increase financial inclusion in Papua New Guinea.

As identified by van der Vlies and Watson[23] there are two mobile money service models in PNG: the bank-led model and the non-bank-led model. The focus of the bank-led model is on improving banking efficiency and including moving customers from branches to mobile money, and increasing mobile money transaction volumes in order to increase economies of scale, thereby driving down costs. The focus of the non-bank led effort is domestic remittances and increasing the volume of mobile transactions.

19.5 Financial Inclusion and the Resources Sector

Resource projects are often in geographically isolated regions—so are many of the world's unbanked. Moreover, the unbanked poor and the world's undeveloped oil, gas, and mineral deposits have many things in common: they are often in the same locale and they are often "unreached" by mobile money service providers and resources companies for similar reasons—including geographical isolation and sovereign risks. Yet there is a distinct lack of research on the role that resource projects can play as a catalyst for improving financial inclusion within resource regions. This is particularly surprising given that payments made from resources companies to local communities and landowners are often significant and diverse, including compensation for land, royalties, resettlement payments, community investment funds, and payments to local businesses. Payments from resource projects have the potential to penetrate deeply into local economies, often resulting in an in-flux of cash and electronic payments in resource regions.

Moreover, although the global resources industry still has significant social and environmental performance challenges,[28] there is an increasing emphasis on social and environmental performance across the sector.[29] The increased emphasis on social performance in particular means that, in many cases, resources companies and those working on the global financial inclusion agenda will have shared objectives in a resource region. However, little is known about the specific contextual issues that influence financial inclusion in the resources sector (such as land rights, the "politics of distribution" and legislative regimes); the extent to which resources companies promote financial inclusion through the use of digital payment systems (branch and mobile); the legislative, institutional, organizational, attitudinal, and cultural barriers and enablers for strengthening mobile money ecosystems in resource regions; or the extent to which mobile money can improve the efficiency and transparency of company payments and therefore enhance social license and development outcomes.

19.6 Research Questions and Methods

The primary research question of this study was whether strengthening mobile money ecosystems around the PNG resource regions can

- Improve the distribution of compensation and benefits payments for local communities.
- Enhance social license for resources companies.
- Catalyze inclusion efforts in PNG's resource regions.

Secondary research questions included

- What payment methods do resources companies currently use to make compensation and benefits payments to local landowners and community members?
- What are the perceptions around current payments from resources companies to local, provincial, and national stakeholders?
- What are the broader perceptions around the resources industry in PNG?

This study used a mixed-method research design including an online survey, company surveys, and semistructured interviews. A quantitative survey ($n = 88$) was conducted to examine stakeholder attitudes toward financial inclusion, mobile money, and the resources sector in PNG. The survey was hosted online and measured participants' responses to a range of questions measuring perceptions about financial inclusion, mobile money, and the resources industry. The survey was advertised through social networks, mailing lists, and in "public spaces" in PNG. Participation was voluntary and anonymous. A company survey ($n = 9$) was also conducted to assess the payment methods used by resources companies to make payments to local communities and landowners. All mining and gas companies in PNG were contacted to participate in the survey on a voluntary and anonymous basis. Finally, semistructured interviews ($n = 17$) were conducted to explore stakeholder views toward financial inclusion, mobile money, and the Papua New Guinea's resources sector in more depth. Participants included representatives from government regulators, industry bodies, banks, telcos, civil society, and the resources industry. The demographic information of respondents who participated in the online survey is shown in Figure 19.1.

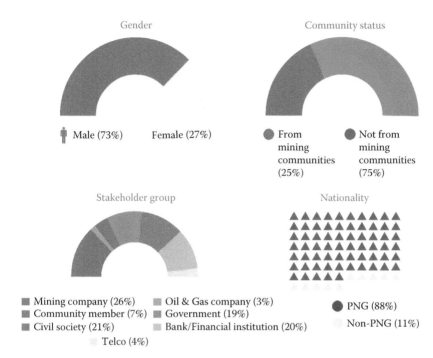

FIGURE 19.1
Online survey sample demographics.

19.7 Results

Finding 1: Payments from PNG's resources companies are a mix of cash, check, and electronic funds transfer with differences across project cycle (cash is more common in exploration) and project size (larger companies are more likely to have more advanced digital payment systems). As seen in Figure 19.2, the three payment methods used by resources companies to make payments to local communities and landowners were cash, check, and electronic funds transfer (companies used multiple payment methods and not all companies made certain payment types). Companies who were in the exploration stage of the project and smaller companies were more likely to use cash than larger companies with developed projects.

Also detailed here is that cash was used for land use payments, compensation payments, investment funds, and meeting sitting fees; the reasons given for using cash was that the recipient didn't have a bank account (57%); the recipient asks for cash (29%); and convenience (14%); and that respondents also reported that trust was an issue for community members and landowners—trust in the

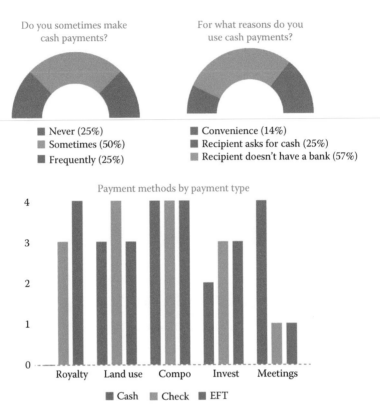

FIGURE 19.2
Payment methods used by resources companies.

financial system (specifically opening bank accounts) and trust that they would receive payments "indirectly" from their community and landowner representatives.

In terms of financial inclusion, none of the companies surveyed reported having a financial inclusion strategy for their project impact area (or being part of a shared stakeholder strategy for the region or province in which they operated). None of the companies surveyed were partnering with mobile money providers to provide mobile phone notifications of payments received and only two of the companies surveyed reported that they "actively promoted mobile money as a way for community members to access payments from the company." However, 50% of companies surveyed reported partnering with financial institutions to provide financial literacy programs.

Finding 2: Overall, there was a perception that the distribution of payments from resources companies is not effective. For some respondents the distributional breakdown of funds seemed "unfair" whereas other respondents felt that payment distribution systems

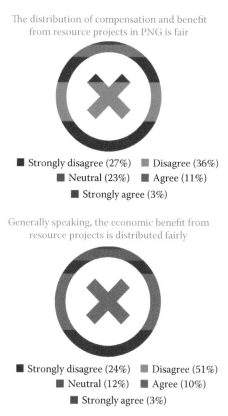

The distribution of compensation and benefit from resource projects in PNG is fair

■ Strongly disagree (27%)   ■ Disagree (36%)
■ Neutral (23%)   ■ Agree (11%)
■ Strongly agree (3%)

Generally speaking, the economic benefit from resource projects is distributed fairly

■ Strongly disagree (24%)   ■ Disagree (51%)
■ Neutral (12%)   ■ Agree (10%)
■ Strongly agree (3%)

FIGURE 19.3
Distributional fairness of compensation payments.

are not effective. As seen in Figure 19.3, only 11% of respondents agreed that compensation and benefits are distributed fairly, while 63% of respondents strongly disagreed or disagreed with this statement. When asked specifically about economic benefits a similar pattern was observed: 75% of respondents strongly disagreed or disagreed that the economic benefits from resource projects are distributed fairly (also Figure 19.3). Thematic analysis of open questions and interviews also highlighted a perception that payment distribution systems are not effective for various institutional and sociocultural reasons, including corruption, politics, incorrect or incomplete identification of rightful recipients, and intra- and inter-clan conflict.

Finding 3: There was also a perception that "grassroots people" do not receive their fair share of the benefits from resource projects. The National Government, men, the Provincial Government, and landowners were the groups perceived to benefit the most from

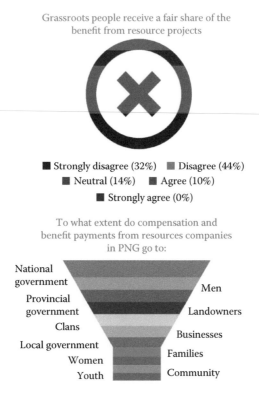

FIGURE 19.4
Benefits distribution to recipients.

resource projects; whereas youth, the broader community, women, and families were perceived to benefit the least. As seen in Figure 19.4, 76% of people either strongly disagreed or agreed that grassroots people receive a fair share of the benefits from resource projects; whereas 14% of respondents were neutral and 10% of respondents agreed with this statement (no respondents strongly agreed). The figure ranks the responses of participants when asked to assess the extent to which various groups receive compensation and benefits payments from resources companies. The categories are not "mutually exclusive" but rather are overlapping groups within the PNG social structure.

Finding 4: There was a perception that compensation payments from resources companies are not paid directly to the bank accounts of the rightful recipients. Respondents did not feel that local people use the payments that they receive from resource projects "effectively" (in a way that improves their lives) and the opinion about whether resources companies encourage local people to open bank accounts was relatively mixed. Of the respondents, 75% disagreed that compensation payment from resources companies are paid directly to the

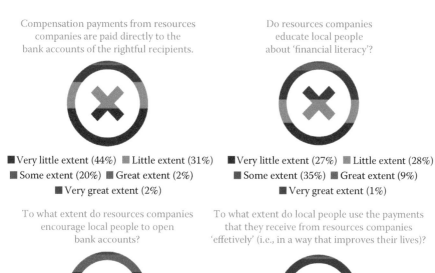

Compensation payments from resources companies are paid directly to the bank accounts of the rightful recipients.

■ Very little extent (44%) ■ Little extent (31%)
■ Some extent (20%) ■ Great extent (2%)
■ Very great extent (2%)

Do resources companies educate local people about 'financial literacy'?

■ Very little extent (27%) ■ Little extent (28%)
■ Some extent (35%) ■ Great extent (9%)
■ Very great extent (1%)

To what extent do resources companies encourage local people to open bank accounts?

■ Very little extent (18%) ■ Little extent (28%)
■ Some extent (31%) ■ Great extent (21%)
■ Very great extent (3%)

To what extent do local people use the payments that they receive from resources companies 'effetively' (i.e., in a way that improves their lives)?

■ Very little extent (44%) ■ Little extent (31%)
■ Some extent (20%) ■ Great extent (2%)
■ Very great extent (2%)

FIGURE 19.5
Mobile banking, financial literacy and resource companies.

bank accounts of the rightful recipients (see Figure 19.5). This figure also shows that respondents were somewhat mixed or negative about whether resources companies encourage local people to open bank accounts and educate people about financial literacy. There was also a perception that local people (landowners and community members) do not use payments that they receive from resources companies "effectively."

Finding 5: Participants agreed that greater financial literacy would improve development outcomes in resource communities and a greater use of mobile money would improve the distribution of payments from resources companies. Participants almost unanimously agreed that greater financial literacy would improve development outcomes in resource communities in PNG and that a greater use of mobile money would improve the distribution of payments from resource projects (Figure 19.6). Responses from the interviews and open questions were mixed concerning the "readiness to adopt" mobile money technology in mining communities. For some, literacy and education levels—together with concerns about security and ability to adopt

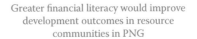

Greater financial literacy would improve
development outcomes in resource
communities in PNG

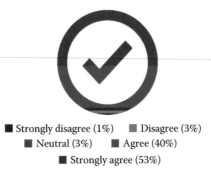

■ Strongly disagree (1%)   ■ Disagree (3%)
■ Neutral (3%)   ■ Agree (40%)
■ Strongly agree (53%)

A greater use of mobile money (using a phone or
mobile device to perform banking transactions)
would improve the distribution of payments
from resources companies

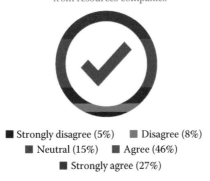

■ Strongly disagree (5%)   ■ Disagree (8%)
■ Neutral (15%)   ■ Agree (46%)
■ Strongly agree (27%)

FIGURE 19.6
Financial literacy, mobile banking and development outcomes in resource regions.

new technologies—were a significant concern. Others, on the other hand, viewed the active use of mobile money technologies as a current reality and well within the capability of their impacted communities.

Finding 6: Overall, the resources industry in PNG received a mixed scorecard when it came to general acceptance, support, and trust. However, the more respondents felt that resources companies encouraged local people to open bank accounts, the greater were their levels of support and acceptance for the resources industry in PNG. As illustrated in Figure 19.7, when asked about the extent to which they accepted and trusted the resources industry in PNG, respondents gave mixed responses. Thematic analysis of interviews revealed that for many respondents the sector is viewed as a mixed blessing—with an acute understanding of the social and economic impacts of resource projects, both positive and negative. Multiple

To what extent do you accept the resources industry in Papua New Guinea?

■ Very little extent (4%) ■ Little extent (23%)
■ Some extent (45%) ■ Great extent (22%)
■ Very great extent (7%)

To what extent do you trust the resources industry in PNG to act in the best interests of society?

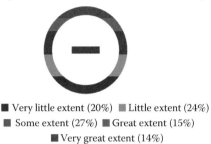

■ Very little extent (20%) ■ Little extent (24%)
■ Some extent (27%) ■ Great extent (15%)
■ Very great extent (14%)

To what extent do you support the resources industry in Papua New Guinea?

■Very little extent (8%) ■ Little extent (14%)
■ Some extent (50%) ■ Great extent (19%)
■ Very great extent (9%)

To what extent do you trust the resources industry in PNG to act responsibly?

■ Very little extent (21%) ■ Little extent (21%)
■ Some extent (36%) ■ Great extent (11%)
■ Very great extent (12%)

FIGURE 19.7
Support and Trust of PNG Resource Sector.

regression analyses were conducted to assess whether perceptions about financial inclusion behaviors from resources companies predicted acceptance and support for the industry. The results revealed that—above and beyond the variance that can be accounted for by general perceptions about development effectiveness—perceptions about whether resources companies encourage local people to open bank accounts were positively related to acceptance of and support for the resources industry in PNG.

19.8 Discussion and Recommendations

19.8.1 Mobile Money Alliances for Resource Regions

Collectively, the results of this study suggest that greater collaboration between resources companies, financial regulators, banks, telcos, mobile phone service providers, and civil society can strengthen the distribution of compensation and benefits payments from resource projects, improve social license for resources companies, and more broadly enhance financial inclusion efforts and development outcomes in PNG. Alliances involving the resources industry in financial inclusion efforts can be structured at both the national level as well as around specific resource projects.

Existing financial inclusion efforts at the national level led by CEFI and the Bank of PNG can be strengthened by enhancing links to the resources industry. For instance, a strategic working group for the resources sector could be considered in the program roll-out of PNG's National Financial Inclusion Strategy 2014–2015. Stakeholders already working on financial inclusion programs could also usefully integrate into relevant programs underway in the resources sector, including the Extractive Industries Transparency Initiative and Chamber of Mines led initiatives such as community relations conferences and resource guides.

At a project level it is suggested that financial inclusion working groups are established for all major resource projects. Working groups should include representatives from the resources company, local stakeholders, and those representing financial inclusion and mobile money efforts (including telcos and banks with commercial interests). For new projects, working groups should be established during the early stages of the project life-cycle so that financial inclusion strategies can be developed as part of the compensation and benefits agreements required by the PNG Mining Act. For existing projects, working groups should be established to collate and share data to assess the state of play for financial inclusion in the impact area. Strategies can then be developed to improve financial inclusion and the distribution of company payments. Project working groups could report back to the Mineral Resource Authority (MRA) through the quarterly reporting process and to CEFI as part of the planning and monitoring meetings for the National Financial Inclusion Strategy.

19.8.2 Financial Inclusion for Resource Projects

This study found that none of the resource projects in PNG had a financial inclusion strategy (or were part of a shared stakeholder strategy for the region or province in which they operated). This is not particularly surprising, given that the financial inclusion and mobile money sector is still relatively new (particularly in PNG). For major resource projects, a structured financial inclusion strategy can help to improve the distribution of company payments over the life of the project thereby reducing social risk and improving social performance. Financial inclusion strategies for resource projects should:

- Be governed by a multistakeholder working group with the roles of resources companies, banks, telcos, and regulators clearly defined.
- Consider the project's social context and the results of the social and economic impact assessment, which itself should adopt a financial inclusion lens.

- Map banking and mobile money infrastructure and agents—and identify commercial partners who may seek to invest in additional infrastructure.
- Comprehensively map the different types of project benefits and compensation payments at the local level including royalties land use payments, other compensation, trust funds, investment, and business development grants.
- Select payment methods based on existing infrastructure and an informed understanding of the social and cultural context of the project.
- Match broader financial inclusion strategies to the project size and project life-cycle, with a plan to equip local people for the transition from cash to digital payments.
- Incorporate a human rights lens with a focus on gender equality.
- Implement stakeholder-led financial literacy programs that manage expectations—the objective is not merely to open bank accounts but to educate local people about the financial opportunities and risks that the resource project will bring and the financial tools and support that is available to them.
- Articulate clear project management, governance, and reporting structures.

19.8.3 A New Payment Architecture

Finally, the study found much discontent regarding the distribution and payment of money from resource projects. With the financial inclusion programs and mobile money services that are now available in PNG, there is an opportunity for resources companies, the MRA as the regulator, and the Chamber of Mines as the industry body to work with the financial inclusion and mobile money industry to develop a new "payment architecture" for resource projects. Mobile money technology offers new possibilities for payment transparency, in the way that:

- Once landowner groups are identified, payments can be made to the lowest level of the social structure with minimal transaction costs.
- Payment receipt notifications can be received on mobile phones in the village so that the recipient doesn't have to travel to town.
- Cash in the economy is minimized, which brings security benefits and potentially increases peoples' ability to save.
- There are transparent records of all payments.
- Mobile money can be a vehicle for greater financial literacy for landowners and community members who are impacted by resource projects.

These direct, efficient, and low-cost payment mechanisms can be established not only for "primary payments" direct from resources companies, but also for "secondary payments" from trust funds, business dividends, and investments. However, clearly no technology solution is a panacea for the social issues and development challenges that come with a resource project in a developing context. The design of payment disbursement architecture begins with the social context in which the resource project finds itself. This includes conducting detailed landowner identification as is currently practiced in many resources projects in PNG. The key principle is that landowners and community members are given the opportunity to choose the level within the social structure to which resource-derived payments are made. Technology or convenience should not be a limiter but rather—with the support given by landowner associations, MRA representatives, and financial institutions—people can make their own financial choices about how they wish to manage compensation and benefits funds.

19.9 Limitations and Future Research

This study was limited to stakeholder interviews and online survey methodology. It was not possible to make site visits or to conduct in-depth assessments of financial inclusion and mobile money ecosystems at resource project sites. Case study methodologies would serve to better identify the project-level issues that influence financial inclusion and mobile money in the resources context. Longitudinal research would also help to inform understanding on the impact of financial inclusion and mobile money in the resources context over time.

Furthermore, the study took a broad-brush approach to the intersection of financial inclusion, mobile money, and the resources sector in PNG. There is a rich anthropological literature on money in PNG and an equally rich literature on the social context of the resources industry. These approaches could usefully be applied to examine the positive and negative social impacts of mobile money in PNG—where commercial interests are intertwined with the development agenda.

It is also important to acknowledge the broader cultural and national context in which the study was conducted: Papua New Guinea. Future research in other developing countries with resource projects would help to create a better understanding of the potentially mutually beneficial alliances that can be formed between resources companies and mobile money providers.

Finally, a more practically oriented toolkit for financial inclusion and mobile money in the resources sector would help practitioners in resources companies and mobile money providers alike.

References

1. Banerjee, A. V. 2013. Microcredit under the microscope: What have we learned in the past two decades, and what do we need to know? *Annual Review of Economics* 5(1):487–519. doi:10.1146/ annurev-economics-082912-110220 and Karlan, D., and J. Zinman. 2011. Microcredit in theory and practice: Using randomized credit scoring for impact evaluation. *Science* 332(6035):1278–1284. doi:10.1126/science.1200138.
2. Centre for Financial Inclusion. https:// centerforfinancialinclusionblog.files. wordpress.com/2011/12/financialinclusion-whats-the-vision.pdf.
3. Demirguc-Kunt, A., L. Klapper, D. Singer, and P. Van Oudheusden. 2015. The Global Findex Database 2014: Measuring financial inclusion around the world (English). Policy Research working paper; no. WPS 7255. Washington, DC: World Bank Group. http://documents.worldbank.org/curated/en/187761468179367706/The-Global-Findex-Database-2014-measuring-financial-inclusion-around-the-world.
4. Alliance for Financial Inclusion. G20 Principles for Innovative Financial Inclusion. http://www.afi-global.org/sites/default/files/afi%20 g20%20principles.pdf.
5. Alliance for Financial Inclusion. 2013. Putting Financial Inclusion on the Global Map. The 2013 Maya Declaration Progress Report. Alliance for Financial Inclusion. https://www.afi-global.org/sites/default/files/publications/afi_2013_maya_declaration_progress_report_final.pdf.
6. 2011. *The Maya Declaration*. Launched at Global Policy Forum 2011 Riviera Maya, Mexico. https://www.afi-global.org/sites/default/files/publications/afi_gpf2014_maya_factsheet_aw_low_res.pdf.
7. Alliance for Financial Inclusion. 2014 Measurable Goals with Optimal Impact. 2014 Maya Declaration Progress Report. https://www.afi-global.org/sites/default/files/publications/2014_maya_declaration_progress_report_final_low_res.pdf.
8. United Nations Capital Development Fund (UNCDF). http://www.uncdf.org/en/pfip.
9. Pacific Financial Inclusion Program. http://www.pfip.org/support/grants/.
10. Some bullet points used directly from Department of Foreign Affairs and Trade. http://dfat. gov.au/geo/pacific/development-assistance/Pages/economic-growth-and-private-sector-development-pacificregional.aspx.
11. May, R. J. 2003. *Harmonizing linguistic diversity in Papua New Guinea*. In Fighting words: Language policy and ethnic relations in Asia, Eds. M. E. Brown and S. Ganguly, pp. 291–317. Cambridge, Massachusetts: MIT Press.
12. Reilly, B. 2008. Ethnic conflict in Papua New Guinea. *Asia Pacific Viewpoint* 49(1):12–22.
13. Some bullet points used directly from Pacific Financial Inclusion Program. http://www.pfip.org/about/16 where-we-work-1/png/.
14. Data provided to the author by the CEFI in PNG.
15. Gates Foundation Financial Services for the Poor. 2013. Fighting poverty, profitably: Transforming the economics of payments to build sustainable, inclusive financial systems. https://docs.gatesfoundation.org/documents/fighting%20poverty%20profitably%20full%20report.pdf.
16. Ernst and Young. 2009. Mobile money: An overview for Global Telecommunications operators. http://www.ey.com/Publication/vwLUAssets/Mobile_Money./%24FILE/Ernst%20%26%20Young%20-%20Mobile%20Money%20-%2015.10.09%20(single%20view).pdf.

17. GSMA. 2010. Mobile Money for the Unbanked. http://www.gsma.com/mobile fordevelopment/wp-content/uploads/2012/06/mobilemoneydefinitionsno marks56.pdf.

18. GSMA. 2014. State of the Industry Mobile Financial Services for the Unbanked. http://www.gsma.com/mobilefordevelopment/new-2014-state-of-the-industry-report-on-mobile-financial-services-for-theunbanked.

19. Karlan, D., A. Ratan, and J. Zinman. 2013. Savings by and for the poor: A research review and agenda. CGD Working Paper 346. Washington, DC: Center for Global Development. https://www.cgdev.org/sites/default/files/savings-poor-research-review-agenda_1.pdf.

20. Kendall, J., P. Machoka, and C. Veniard. 2014. An Emerging Platform: From Money Transfer System to Mobile Money Ecosystem. *Bill Maurer Legal Studies Research Paper Series No. 2011-14.*

21. Suwamaru, J. K. 2014. Impact of mobile phone uage in Papua New Guinea. State, Society and Governance in Melanesia, Australian National University. http://ssgm.bellschool.anu.edu.au/sites/default/files/publications/attachments/2015-12/IB-2014-41-Suwamaru-ONLINE_0.pdf.

22. World Bank. http://www.worldbank.org/en/news/ feature/2012/09/11/papua-new-guinea-connecting-people-in-one-of-themost-isolated-places-on-earth.

23. van der Vlies, M., and A. Watson. 2014. Can mobile phones help reduce teacher absenteeism in Papua New Guinea? Paper submitted for publication in the Proceedings of the Australian and New Zealand Communication Association Annual Conference, Swinburne University, Victoria 9–11 July, 2014.

24. The Economist. 2014. *Government has ambitious plans for telecoms sector.* http://country.eiu.com/article/aspx?articleid=1641419148&Country=Papua%20New%20Guinea&topic=Economy.

25. World Bank. 2014. http://www.worldbank.org/en/news/feature/2012/09/11/papua-new-guinea-connecting-people-in-one-of-themost-isolated-places-on-earth.

26. International Telecommunication Union (ITU). 2014. *The World in 2014: ICT Facts and Figures.* Geneva: ITU.

27. Pacific Financial Inclusion Program. http://www.pfip.org/mediacentre/in-news/2013-1/80-of-png-unbanked-post-courier-png.html.

28. Urkidi, L. 2010. A global environmental movement against gold mining: Pascua-Lama in Chile. *Ecological Economics* 70(2):219–227.

29. See, for example, the body of work sponsored by the International Council of Mining and Metals.

20

Vignette: Local Level Agreement Making in the Extractive Industries—A Viewpoint on Context, Content, and Continuing Evolution

Bruce Harvey

CONTENTS

20.1 Introduction

For 30 years, the emerging political empowerment of land-connected people[1] in both developed and developing economies has presented a challenge to natural resource developers and sovereign nations that wish to simultaneously foster democratic forms of government and subnational economic development (Langton and Longbottom 2012; O'Faircheallaigh 2016). An innovation in social governance that has surfaced in response and continues to evolve is Local Level Agreements (LLAs).[2] LLAs are negotiated between resource developers and local land-connected peoples to secure,

---

[1] Land-connected peoples are defined here as social groups that have demonstrated residency and/or livelihood connections to definable geographic areas that have persisted over more than one generation. This include peoples labeled Indigenous, as well as multi-generational real property owners and others who have an emotional and/or spiritual attachment to the land in question. The land connection may involve exclusive or non-exclusive rights and frequently also include waters within the same geographic domain.

[2] There is a plethora of local agreement nomenclature, much of it value laden, including Community Development Agreements (CDAs), Impact Benefit Agreements (IBAs), and Indigenous Land Use Agreements (ILUAs). The generalised term Local Level Agreement (LLA) is preferred by the author.

under mutually agreed terms, consents and regional development pathways for resource projects and host communities. With many being embryonic, particularly in their implementation, there has been constant innovation in the detail of how LLAs are negotiated and structured to solve particular contextual challenges. LLAs hold much promise as an innovative approach for more direct, collaborative, and transparent interaction between extractive companies and host communities. To fully develop their potential, however, the temptation to impose "top–down" detailed prescription, often a result of involvement by governmental and multilateral policymakers, must be resisted in favor of regulatory oversight restricted to ensuring procedural and distributional fairness.

20.2 A Context for Local Level Agreements: The Challenge of Diffuse Governance

Globalization, among other things, involves a transformation of social geography. In nationalistic political contexts, governance takes a predominantly territorial form, with centralized authority claiming comprehensive, exclusive jurisdiction over a territory and its inhabitants. However, many people now have an opportunity to emphasize identities that supplement and, in some cases, override feelings of national identity. Under these conditions, local, regional, and identity groups have gained important degrees of freedom over their own governance and self-determination. At a substate level, this has involved a devolution of domestic powers from central state governance to more local and provincial forms; for example, the Scottish Parliament and the Welsh Assembly in the United Kingdom, and the Territory of Nunavut in Canada. The resurgence of Indigenous identification around the world, promulgated in sovereign laws that recognize Indigenous rights and interests, is a form of this evolving diffuse governance.

Devolved governance is not a new idea; it is central to the political science concept known as "subsidiarity," a principle of social organization first formally described within the Roman Catholic Church in the nineteenth century. In its most basic formulation, it holds that social and political issues should be dealt with at the most immediate (or local) level that is consistent with their resolution. Successful resource developers have become comfortable with this idea, recognizing and making agreements directly with host communities. The challenge is convincing others, and particularly nationalist governments, that subsidiarity and LLA create optimal conditions for local social and economic development.

LLAs were originally conceived to govern resource development and other aspirations with land-connected Indigenous peoples in Common Law counties such as Canada, Australia, New Zealand, and Papua New Guinea.

LLAs are now evolving globally in a wide range of legal, cultural, and industry sector situations. They have been adopted in some form as a legislated requirement in more than 30 countries (Dupuy 2014) and international financial institutions increasingly recommend them as a condition of finance, often by reference to offered agreement templates or stipulated formats (World Bank 2012).

20.3 Content and Considerations within Local Level Agreements

The purpose and content of LLAs can vary significantly, from simple, non-binding, memoranda to complex, multifaceted contracts that run to hundreds of pages covering many areas of mutual concern and resolution, or hybrids of them (Rio Tinto 2015). Some examples of content are summarized here. Definitional aspects of LLA might include consensus on what constitutes "local" in the context being considered; governance and audit provisions for joint and delegated management bodies; complaints, disputes, and grievance resolution processes; formulation of evaluation criteria to measure agreed commitments; the expectations on relationships with other parties (e.g., governments); and the establishment of dedicated Trusts to receive funds for purposes specified within the agreement. A second content theme relates to shared and/or responsible use of land and resources, considering things like formalized compensation for loss of property and amenities; land and natural resource access and rent provisions (e.g., for water and forests); joint security and land access protocols; environmental co-management arrangements; shared access to major infrastructure (roads, harbors, airports); and consideration of post-closure land use. Methods of fostering closer integration of business and community are also usually an important consideration: for example, training and employment terms and targets for local people; commitments to engaging with local supply chains and business development; potential subsidiary business opportunities; equity provisions within the hosted business; and even partnering for lobbying regional and/or central government and other agencies. Finally, context-specific obligations might arise as being desirable for both parties, such as ensuring contributions to social and civic infrastructure (e.g., health clinics and scholarships) or resources and initiatives aimed at maintaining cultural heritage practices.

While the contents of resource development LLAs can be extremely varied, reflecting the context in which they are negotiated, certain vital principles have proven to be essential for successful implementation. The first of these is that agreements should center on proximal—most often land-connected—communities and their representatives, rather than provincial-level or national governments. It is also important that, regardless of the involvement or non-involvement of governmental authorities, an institutionalized relationship is adopted: that is, a structured relationship between a local community group and a business as institutions is assumed, rather than LLAs being something agreed between certain individuals.

Second, the parameters and underpinning basis of LLA is, understandably, also key to their success. For instance, the concept of a "value exchange" must be established: specifically, there must be clear evidence of the mutual obligation on which local community support for a resource project is based and an agreement that all parties will commit to working toward achieving shared objectives. Relatedly, there should also be clear statements of community support for the resource development, describing the process by which community support for the resource project is given and the terms of that support, and vice versa. Expected outcomes should be clear, with measurable performance indicators and consequences for non-performance, but with scope for non-core implementation elements to be (collaboratively) revised relatively easily to match changing circumstance and learning arising out of implementation. These parameters provide communities and companies with well-defined, mutually agreed terms of who they are dealing with, toward what end, and how the success of the agreement can be monitored. In fact, LLAs frequently include more detail than prescribed in legislative requirements regarding the management of social and environmental interaction.

Experience has shown that the process of reaching agreement is as important as the LLA itself, reflecting people's innate desire for "procedural fairness" as a precursor to trust (also a central facet in a company's "social license to operate)." This has led to continuing innovation. For instance, it can be useful to reach subsidiary agreements along the way. These can include initial non-binding Memoranda of Understanding (MOUs), followed by agreements about process and interim agreements, before proceeding to a comprehensive agreement. Other options include the parties choosing to address specific themes or discrete geographic domains separately, reach agreement about each of them, and treat the amalgam of those agreements as a final composite LLA.

The approach to making LLAs needs to be quite different from the tactics frequently applied in negotiating agreements between commercial enterprises, which can be adversarial and aimed at securing a short-term transaction rather than a long-term relationship. In contrast, LLAs adopt a more reciprocal and incremental approach, reaching agreements along the way, which builds trust between the parties, and confidence in agreement making and their institutional capability. Such interim agreements are also a useful way to prototype longer term, substantive agreements. In most situations, it is useful to start with agreement on principles, process, and the definition of roles and responsibilities. Then the focus can shift to subject-specific areas where early agreement is most likely, before tackling resource-allocation and more onerous obligations. The "ultimate" LLAs can be comprehensive and reflect the joint resolution of all areas of mutual interest between a business and a local community group, as well as transparent implementation commitments and institutional arrangements for the parties' ongoing relationship.

In the longer term, LLAs need to be flexible and allow for the reality of intergenerational social flux and changes in the business over time. Flexibility needs to be simultaneously balanced against a business's need for certainty for the life of its operations. One way to achieve this balance is to structure a comprehensive LLA in components. The overarching, "umbrella" component includes only those core elements that will remain constant. This component can be supported by a suite of annexures that represent stand-alone, binding management plans, each capable of being independently updated by agreement of the parties as required. The advantage of this innovation is that operational components of the LLA can be varied, while leaving the core financial, legal, and governance elements immutable.

While many LLAs have been successful in creating governance conditions that support stable and predictable societal conditions at a local level, thus enabling extractive businesses to operate through generational transition without social conflict, some have unfortunately failed to achieve this balance. LLAs generally fail when agreement making has been approached expediently as a commercial transaction to be completed as quickly as possible without much thought to implementation and governance. As such, it would appear that the ethos and implementation of LLAs are inextricably linked.

A refined understanding is required of the governance imperatives that need to be in place to ensure long term stability and success. This requires a spectrum of measures, the most important of which include, first, balanced representation, where there is an insistence that representatives from various credible community sub-groups, such as women, youth, elders, and church, are all present on governance committees. Second, the agreement should emphasize cultural fit and endowment, in which governance arrangements have enough traditional resonance in place to ensure customary self-regulating behavior. Third, the agreement should be "broadspectrum," being comprehensive in its scope and not overly focused on cash benefactions even when these are present. Finally, a series of checks and balances being in place, which might include various sub-bodies set up with responsibilities for different aspects of an LLA, with specific powers in that regard and a mutual ability to keep each other in check alongside formal custodial and independent provisions, such as custodial trustees and fiscal fiduciary arrangements. Collectively these measures provide for a form of power separation like company governance, where executive functions are separate to and constrained by "wise heads" situated at board/policy level.

Such arrangements provide a good governance mesh that prevents intemperate decision making, dishonesty, and factional capture. While they can also be bureaucratic and prevent fast action, to a large extent this is the point.

That is, it is better to stifle the potential for corruption of public office and goods, and leave innovation to flourish in the private realm of individual families, people, and enterprises. Such are the lessons of thousands of years of history, by trial and error, gradually evolving political order for increasingly complex social contexts. The lessons are available for selective innovation to fit local circumstance by adjusting scale, scope, and "contextual fit."

20.4 The Future: Protecting a New Approach from Old Habits

The frequency of community conflict with resource projects and the negative consequences for resource developers and affected communities has increased in the past two decades (Davis and Franks 2014), demonstrating that resource developers can no longer rely on central government decree for ongoing consent and the operational certainty they need. In response, LLA-making is spreading and evolving rapidly as developers and governments realize it addresses local aspiration and mobilizes the power of self-destiny without threatening national sovereignty (O'Faircheallaigh 2013). In short, demonstrating an earned social license is increasingly a necessary pre-condition for the issuance of a legal license. The innovation of LLAs does this and meets the diffuse governance needs of a globalizing world while promoting the ability of local peoples to govern their own lives. Indeed, the idea is catching on beyond the resource sector; other industries, service groups, NGOs, and governments are adopting LLA approaches.

The direction of LLA innovation will, pragmatically, be governed by experiment, and some LLA processes will be more successful than others. Experience suggests that placing the principle of procedural and distributional fairness, as determined by all parties, ahead of prescribed content is necessary to form sustainable LLAs. Currently, however, much of the LLA experimentation is unfortunately going in the other direction. A recurring trend among many governments in Africa, Asia, and South America (Chilenye 2016), and even in international agencies, is to mistakenly work from the idea that their regulatory role is to dictate terms and content by prescription and offered templates. This approach effectively re-appropriates the local empowerment that LLAs seek to promote.

Regulation and policymakers should instead focus on ensuring procedural equity, allowing the LLA parties to reach their own accommodation relevant to circumstance, not some deemed universality. Procedural equity means governments and financial institute umpires should focus on ensuring all parties have equal access to advice, representation, and resources to counterbalance each other. Achieving power balance is the key to successful and appropriate LLA outcomes. Attempting to prescribe the content of LLA is antithetical to innovation. Conversely, focusing on creating an open-ended and fair playing field creates conditions for infinite innovation. LLAs and diverse processes for making them will continue to open opportunity for resource developers, other enterprises, and host communities, but only where the latter conducive conditions prevail.

References

Chilenye, N. 2017. Legal and institutional frameworks for community development agreements in the mining sector in Africa. *Extractive Industries and Society*, 4, 202–215.

Davis, R. and D. Franks. 2014. Cost of company-community conflict in the extractive sector. Corporate Social Responsibility Initiative Report No. 66, Harvard Kennedy School: Cambridge, MA.

Dupuy, K. E. 2014. Community development requirements in mining laws, 1993–2012. *Extractive Industries and Society*, 1(2), 200–215.

Langton, M. and J. Longbottom (Eds.). 2012. *Foundations for Indigenous Peoples in the Global Mining Boom*. Routledge: Oxford, UK.

O'Faircheallaigh, C. 2013. Community development agreements in the mining industry: An emerging global phenomenon. *Community Development Journal*, 44(2), 222–238.

O'Faircheallaigh, C. 2016. *Negotiations in the Indigenous World: Aboriginal Peoples and the Extractive Industry in Australia and Canada*. Routledge: New York.

Rio, T. 2015. *Why Agreements Matter: A Resource Guide for Integrating Agreement Making into Communities and Social Performance Work at Rio Tinto*. Rio Tinto and Centre for Socially Responsible Mining, University of Queensland: Brisbane, Australia.

World Bank. 2012. *Mining Community Development Agreements Source Book*. The World Bank: Washington, DC.

21

Social Incident Investigation in Mining: Thinking Outside the Fence

Deanna Kemp, John R. Owen, and Jill Harris

CONTENTS

21.1 Introduction

This chapter is about social incident investigation in the global mining industry. We are writing this chapter as academic practitioners who are actively engaged either by industry or by other stakeholders to investigate the causes, dynamics, and effects of social issues and incidents that involve local communities. Many social issues and incidents have common features; for example, they occur in and around the industrial complex of the mine, in remote locations, in less developed locations with weak systems of governance, and often with a history of attempts by the mining company to resolve at least part of the problem we are presented with. The circumstances surrounding our engagement with social incident investigations in mining also have common features: a sense of crisis on the part of the company or the government, a lack of confidence in existing approaches to understanding and managing issues related to the incident, internal disagreement as to the causes and contributing factors of the situation, and a breakdown in trust between the primary stakeholders embroiled in the event.

Social science, we believe, has a great deal to contribute to the mining industry. Because social scientists work with a diversity of stakeholders, across cultures, locations, languages, and at different points in a project's life cycle, it is not possible to develop a standardized, one-size-fits-all approach to examining and responding to social issues and incidents in mining. This situation puts the social sciences at a manifest disadvantage, given that in most of the other functional disciplines, such as geology, mining, minerals processing, logistics, safety and security, and human resources, companies have established standardized instruments and methodologies for identifying, analyzing, predicting, controlling, and optimizing risk and opportunity events. We believe that the mining industry is at a crossroads in its relationship with the social sciences. We observe that a critical tension exists between increasing levels of demand for expertise by lenders and international agencies for skills that apply "outside the fence" and an entrenched institutional culture that has grave difficulty in enabling social scientists to operate effectively "inside the fence."

In this chapter, we provide an outline of a social incident investigation model that has strongly influenced our own investigative practice. This model, known as the "flashpoints" model originated in the 1980s following a series of public disorder events that were based in mine-community conflicts (Waddington et al., 1991). The model takes the flashpoint—or the point at which an issue is explicitly and often dramatically brought to the forefront— as but one part of a much larger picture of events and circumstances. Since the late 1980s this model has become an established approach for analyzing public disorder events such as demonstrations, strikes, and riots that occur globally. The model is not formally used by mining companies, and we have not cited research where social scientists have explicitly used the model to investigate social incidents in and around mining complexes. We have drawn on the approach in our own work—noting the importance of understanding interactions, situations, and context in the events prior to the incident or issue arising (Kemp et al., 2008, 2013; Kemp and Owen, 2015b; Martin et al., 2016).

As a basis for comparison and learning we also review the "bowtie method," commonly used by the mining industry (and other high risk industries) to manage risks associated with workplace health and safety hazards. While we are not active practitioners or advocates of the bowtie method for the management of social interactions, we recognize that the safety discipline in understanding unwanted accidents and incidents prioritizes a range of human factors that contribute to the prehistory and ultimately to the event itself. In this brief chapter we engage these two models and propose an approach aimed at generating a deeper understanding of social incidents in the global mining industry. Thus, we are explicitly seeking to improve the social capability of the mining industry so that it can meet an expanding set of local, national, and international social performance obligations in a more optimum manner (Owen and Kemp, 2017).

21.2 The Flashpoints Model

David Waddington et al. (1989) developed the flashpoints model in response to a series of public disorder events in Britain during the early 1980s. In their preliminary approach to the model they began with a flashpoint incident and then sought to use expanding levels of social analysis to contextualize it. Figure 21.1 depicts the approach diagrammatically with the flashpoint positioned at the center supported by a series of concentric circles indicating the different levels of knowledge required to understand the precursors and development of the flashpoint incident itself.

Table 21.1 describes the six levels of structuration in the preliminary model (Waddington et al., 1989, 20–22).

The purpose of the approach was to progressively move through each level as a means for developing a deeper understanding of the factors surrounding the flashpoint. As the authors (Waddington et al., 1989) note in their preliminary presentation of the model, they were at that time:

> "attempting to represent different levels of influence in a way which avoided two ways of reducing the complexity of flashpoints. The first, most typical of the psychological perspectives… confines itself to the dynamics of the interaction in its immediate situation, so that only the wider context is that of the mentality of the crowd. The second, most prevalent amongst sociologists, sees flashpoints as the inevitable outcome of racial, political, or industrial conflict, with little attention to what actually happens on the ground" (pp. 22–23).

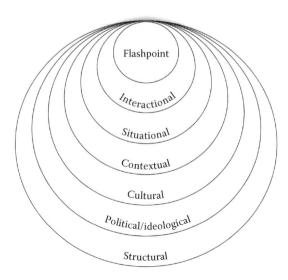

FIGURE 21.1
Levels of structuration in public order situations.

TABLE 21.1

Levels of Structuration in the Flashpoints Model

Level	Description
Interactional	Where the event represents dramatic shift or change in the usual pattern of interactions
Situational	The precise locality and timing of the event
Contextual	The prehistory of the event, and the perceptions and preconceptions relating to interactions/situation in which the event is based
Cultural	Incorporates definitions of the situation by the various parties and their understanding of their own and other's interests associated with the event
Political/Ideological	The arena in which elites discuss, judge, and debate the various "rights and wrongs of the issue and its manifestations"
Structural	The underlying interests, organizations, conflicts, and dynamics of the society as a whole

What the authors were seeking to achieve was a working pattern of factors, which taken "cumulatively and in interrelation" could assist in explaining the emergence and character of disorderly events. Following decades of development, critique, and reflection, Moran and Waddington (2016, 37–38) revised the flashpoints model (shown in Figure 21.2), which focuses on riots as the flashpoint event and includes an additional level of analysis (organizations and institutions), a schematic that arranges the levels to assist in explaining causal effects, time dimensions (or what they call "lull"), the potential for flashpoint events *not* to occur, and finally a recognition that once flashpoint events do occur these events themselves unfold in a patterned way. The implication of this last point is that institutions (such as mining companies) can strongly influence what the flashpoint looks like based on how they first engage it.

From a methodological standpoint, the flashpoint analysis begins with "obtaining as many accounts of what had happened as could be collected in the time available" (Waddington et al., 1989, 17). The approach requires intensive levels of fieldwork with interviews with as many of the direct participants from each interest group, as well as with observers and commentators connected with the event but slightly removed in terms of physical location or time. In the analysis of six public disorder events from August 1981 to October 1984 by Waddington et al. (1989), their primary sources of data included interviews and statements from crowd members and police officers, field observations, local and national media reports, and legal documents relating to the events. In half of the cases, they had been able to rapidly deploy a fieldworker to record extensive notes that were then used to cross-check and validate subsequent sources of data gathered after the events through interviews, media reports, and legal sources.

In the context of social investigations in mining-related incidents, the flashpoints model has some limitations. We note four in particular. The first of these limitations is that the model implies that the users will have intermediate to

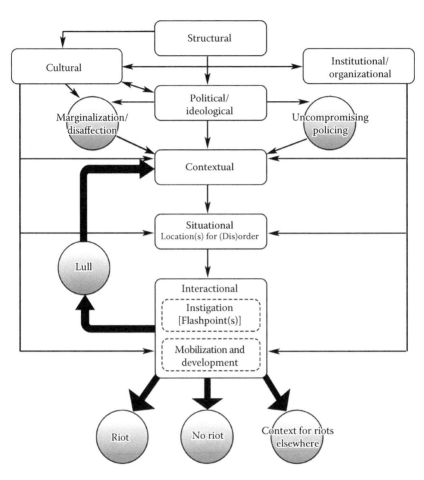

FIGURE 21.2
Revised flashpoints model of public disorder.

advanced levels of social science training. As we have noted elsewhere, the investment in and application of social scientific knowledge by the mining sector is almost universally low (Owen and Kemp, 2013; Kemp and Owen, 2015a). The prospect of a mining company rapidly and voluntarily deploying a team of social scientists to gather data on an incident that occurs wholly or even partially in the public domain is unlikely.

The second limitation is that while mining operations are generally prepared to commit resources to support a comprehensive investigation of serious safety incidents that occur within the industrial environment, the flashpoints model introduces levels of analysis and complexity that go beyond the industry's current practice of incident investigation. While social incidents may have been triggered or exacerbated by the industry, they usually occur "outside the fence." Moreover, there is often a lack of willingness within mining companies to investigate social incidents that have no obvious or immediate impact on the business and thus no perceived need for investigation and analysis (Owen, 2016).

A third limitation may relate to the limited jurisdiction mining companies have in commissioning investigations in the broader social environment outside of its legally mandated area of operation. Companies are often reluctant to engage in comprehensive investigations involving sensitive events or people outside the parameters of the business. This reluctance can lead to subversive or informal intelligence gathering about social incidents where mining is considered complicit in adverse impacts and outcomes. Our experience suggests that few mining companies are willing to initiate or participate in a more open and rigorous form of social inquiry unless external expectations or circumstances compel them to do so (Kemp and Owen, 2013).

The fourth limitation is that the flashpoints model has not been developed to serve an operational-level purpose. This is curious given that the model was first developed in response to mine-community conflicts. The model has largely been used and extended by academics in explaining public disorder events. However, since neither the mining industry nor social scientists have formally incorporated this model as an operational-level instrument, "control" or "management" components are not featured in the model's development over the past three decades.

21.3 The Bowtie Method

The bowtie method is a risk mapping framework developed for application in industrial settings. The focal point of a bowtie diagram (Figure 21.3) is the unwanted event or the "knot" in what looks like a man's bowtie. Although its exact origins are not known, the bowtie method has been linked to a public inquiry report by Lord Cullen following the catastrophic Piper Alpha oil platform incident in the North Sea in 1998. Established to investigate the cause of the disaster, the inquiry concluded that the operator, Occidental Petroleum, did not adequately manage the relationship between hazards and risks (Cullen, 1990). The oil industry subsequently became acutely aware of the need to develop systematic approaches to controlling those hazards that may lead to catastrophic events.

The bowtie method was adopted by the Royal Dutch/Shell Group in the early 1990s as part of a new approach to proactive risk management (Alizadeh and Moshashaei, 2015) and has since been applied throughout the extractive industries including within the mining sector and particularly in the occupational health and safety discipline. Commercial software packages and specialist consultants are available to help the industry map risks, draw bowtie diagrams, and manage institutional knowledge about actual and potential unwanted events.

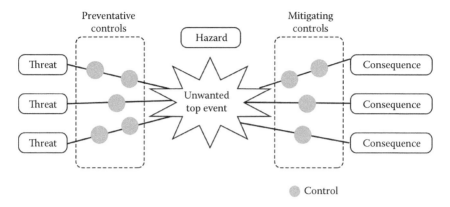

FIGURE 21.3
Basic elements of the bowtie framework.

The framework itself is a hazard-centric and control or barrier-based approach to managing risks (Hollnagel, 2008). The starting point in the bowtie framework is identification of the "hazard" which may be a source, a situation, or an act that has the potential to cause harm. Loss of control over the hazard is known as the "top event." "Proactive" interventions are positioned to the left of the top event and "reactive" interventions are positioned to the right.

The main element to the far left of the event is "threats," which are those factors that could cause the top event. There may be multiple parallel threats to the event occurring. This side of the bowtie also enables analysts to list barriers that could prevent the threats from causing the top event. The main element to the far right of the top event are "consequences" that are the result of outcome of the top event occurring. There can be multiple consequences to the top event. Barriers that may lessen the impact or prevent the realization of consequences are also captured on this side of the framework. A further level of detail can include the addition of "escalation factors" which define how preventative barriers on either side of the bowtie may themselves fail.

In the safety discipline, the bowtie method is best conducted with the involvement of controllers and operators—those people who have the deepest knowledge of the preventative and mitigating barriers used to control risks (Dodshon and Hassall, 2017). The ideal output is a bowtie map that is accessible to those people involved in everyday risk prevention and mitigation, and, in addition, to those with oversight and responsibility of the industrial system or subsystem. It is a commonly held view that a bowtie map should avoid being either too detailed or too generic. Detailed information about specific elements, such as a deep analysis of a specific hazard, should be linked to the bowtie map and stored elsewhere. By applying bowties to different issues and events over time, a repository of knowledge can form leading to the development of risk maps for a range of processes and projects. Standard lists of consolidated hazards, threats, and consequences progressively improve the consistency and precision of the overall process.

As a tool for understanding and managing actual and potential social incidents in a mining context, the bowtie has two primary limitations. First, since the method has largely been applied within an industrial system, there is a latent assumption that the operating environment can be "controlled" by the industrial entity, its employees, or its agents. However, outside of a project or processing environment, this is not the case. Many social factors do not sit within the immediate control of the company. Events that transcend the boundary of mining leases and company boundaries, and which can involve a multitude of stakeholders, necessarily requires a comprehensive understanding of the wider social and political context. While the bowtie encourages practitioners to identify a diverse set of causal factors, the model was not developed for the purposes of examining complex social issues and incidents where the locus of control sits predominantly outside of the industrial setting.

The second limitation of the bowtie model is its linear presentation of information. Presentation of event and incident data is both linear and compartmentalized and does not draw attention to the possible influence of feedback loops. For example, causes are depicted as leading to a single and central event that has consequences, and barriers are tested for their preventative or mitigating effects, depending on which side of the central event they are located in the framework. How consequences might feedback into the framework and moderate the effect of causal or control factors is not explicitly included. As the ultimate objective of social inquiry is to "discover" what happened and how it happened, the bowtie's compartmentalization of information is likely to hamper its effectiveness in surfacing and explaining complex and dynamic social events. For incidents that occur outside of the industrial domain, the process of discovery must be exploratory and inclusive of nonindustrial actors and contextual or other factors.

21.4 Utilizing Flashpoint Analysis Inside and Outside the Industrial Complex

The bowtie method is an instrument that is used to view the control systems that are in place (or not) to manage the risk associated with a (often technical) hazard in industrial settings. In Figure 21.4 we offer a diagram incorporating the relative attributes of the flashpoints and bowtie models. This diagram brings into focus areas of complementarity, as well as points where the explicit use of the flashpoints model could be improved to more directly link to operational-level processes and systems to the external domain. We refer to this model as the "flashpoint incident analysis and response" model.

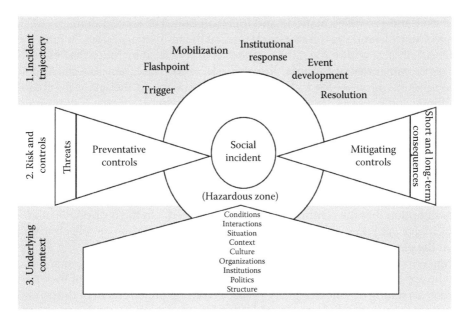

FIGURE 21.4
Flashpoint incident analysis and response model.

The bowtie model has been successful for at least two reasons. First, the model was profiled following the much-publicized Piper Alpha disaster. The oil and gas industry found immediate value in the methods and messages that emanated from the public inquiry, and the mining sector, among others, quickly adapted the model. Secondly, the model is both systematic and accessible. Users are brought into the investigative and analytical stages through an inclusive process of discovery. This second set of reasons offers particularly helpful insights into terms of enhancing the industry's understanding and future use of social scientific and sociological approaches such as the flashpoints model.

The proposed flashpoint incident analysis and response model operates from three connected domains (listed on the left and shaded for emphasis). Starting at the top of the diagram, the first domain—Incident Trajectory—tracks the trajectory of the incident, assuming that the incident has occurred in relation to, or as the result of, the mine's presence or activities. An underlying assumption carried in the flashpoints model is that social incidents are dynamic with layers of interaction occurring across an event timeline. The timeline includes six temporal phases: (1) a trigger (surface cause), (2) a flashpoint (the point at which an issue is dramatically brought to the forefront), (3) mobilization (where additional actors are brought into the incident), (4) institutional responses (formal and informal authority that can influence the trajectory of the incident), (5) event development (subevents that result in escalation or de-escalation), and (6) resolution. The second domain—Risk and Control—reflects the structural logic of the industrial system "within the fence"; that is, within the mine's sphere of influence. It includes the "classic" bowtie categories of hazards, threats, risks, and consequences—short- and long-term. To allow for a comprehensive approach to the analysis of conditions and controls, the third domain—Underlying Context—is concerned with the underlying social conditions of the incident and brings into focus the "levels of structuration" contained in the flashpoints model.

At the center of our combined model is the social incident. This provides a broader focus than the classic bowtie model's tightly scoped "top event," which the combined model expresses within the trajectory as the "flash-point." A second variation is that the combined model does not begin with the assumption that the central incident is "unwanted." In the flashpoints model incidents are taken as arising as an expression of individual or collective will on the part of the aggrieved with the expressed purpose of addressing a deep concern. To describe incidents as "unwanted" introduces a bias against those who are actively trying, in their own way, to correct something that they perceive as wrong or fundamentally unjust. Avoiding prejudgment of the central incident as inherently "unacceptable" is a key feature of the combined model.

A "hazardous zone" nested around the incident is also included, which is characterized by either a heightened likelihood of the event or serious consequence occurring. From the mine's perspective, a well-scoped and developed management system would include preventative or proactive controls that minimize the likelihood of the incident occurring. If an incident does occur, reactive controls are in place to reduce the severity of the consequences. Added to our combined model are factors that could potentially add "noise" and disrupt anticipated outcomes—like the "escalation factors" in the bowtie model. These factors include unique external conditions or levels of structuration (see Table 21.1).

A persistent limitation of the effectiveness of social science methods in mining is the perceived incongruence between exploratory approaches to knowledge and discovery, and the need for precision and distilled controls in a risk-sensitive environment. Conversely, an overly constrained, linear frame of analysis is unlikely to yield sufficient knowledge to assist mining companies in navigating the social context in which their operations are based. Our approach explicitly connects (1) the exploratory method of social incident investigation, (2) through the operation's interaction with the incident, (3) to the system of risk management and controls.

Previously we noted that the bowtie method succeeded in developing an operational-level framework that connects users to incident and risk analysis through systems in a fashion that is legible and accessible. We have stated here, and elsewhere, the structural challenges associated with grounding social analysis within the operational knowledge and decision-making systems. One of the objectives of the "flashpoint incident analysis and response" model is, therefore, to make comprehensive social data and analysis similarly useable to nonsocial scientists for improving the overall responsiveness of the business.

International governance frameworks, such as the United Nations Guiding Principles on Business and Human Rights, have established in no uncertain terms the absolute necessity of corporations fully understanding the human rights risk landscape in which they find themselves (Ruggie, 2011). It is no longer sufficient for companies to confine their analysis to interests and events within the operational fence line, or that reflect narrow self-interest. Developments in the international system require business to engage a multitude of actors and institutions, and to understand the situational, cultural, and political context of rights and protections. The approach we have taken in presenting this model, therefore, has not been to simply "bolt-on" a mainstream industry investigative instrument (i.e., the bowtie method), but to connect the bowties to an established sociological model to support increasing demands for enhanced social capabilities by mining companies.

21.5 Conclusion

Our purpose in this chapter has been to profile a distinctly sociological approach to social incident investigation in the mining sector. We have presented this approach on the basis that the industry is at a crossroads in terms of its social capabilities. There is increasingly powerful external demand for companies to demonstrate technically rigorous social due diligence across a range of complex topics areas, while at the same time, there is a living legacy within the industry that appears to prevent the investment, application, and mainstreaming of these capabilities.

We have reviewed two models of incident investigation: (1) the flashpoints model and (2) the bowtie method. This latter method has succeeded in entering the mainstream logic of the global mining industry. In promoting a modified version of the flashpoints model, we have drawn on key attributes from the bowtie method, namely "accessibility" and "operability." Our incorporating of these features could be interpreted by social science colleagues as a form of industrial "capture"; however, this is far from what we are hoping to achieve. In its current state, our modified flashpoints model would require further development before it could be incorporated as an integrated industrial system. The larger play from our perspective is that making technically robust social approaches accessible to the mining industry is a critical step in improving its overall social literacy and, through that, providing a better awareness of what is required to respond to an ever-increasing set of complex expectations and challenges.

References

Alizadeh, S., and Moshashaei, P. (2015) The bowtie method in safety management system: A literature review. *Scientific Journal of Review*, 4(9), 133–138.

Cullen, W.D. (1990) *The Public Inquiry into the Piper Alpha Disasters Vol I and II*. HMSO: London.

Dodshon, P., and Hassall, M. (2017) Practitioners' perspectives on incident investigations. *Safety Science*, 93, 187–198.

Hollnagel, E. (2008) Risk + barriers = safety. *Safety Science*, 46(2), 221–229.

Kemp, D., Evans, R., Plavina, J., and Sharp, B. (2008) Newmont's Global Community Relationships Review, 'Organisational Learnings from the Minahasa Case Study'. Centre for Social Responsibility in Mining (CSRM).

Kemp, D., and Owen, J.R. (2013) Community relations and mining: Core to business but not "core business." *Resources Policy*, 38, 523–531.

Kemp, D., and Owen, J.R. (2015a) Social science and the mining sector: Contemporary roles and dilemmas for engagement. In Price, S. and Robinson, K.M. (Eds.), *Making a Difference? Social Assessment Policy and Praxis and its Emergence in China*. Berghahn Books: New York.

Kemp, D., and Owen, J.R. (2015b) A third party review of the Barrick/Porgera Joint Venture off-lease resettlement pilot: Operating context and opinion on suitability. Centre for Social Responsibility in Mining (CSRM), The University of Queensland: Brisbane, Australia.

Kemp, D., Owen, J., Cervantes, D., Arbelaez-Ruiz, D., and Benavides Rueda, J. (2013) Listening to the city of Cajamarca. Research Paper, CSRM, Sustainable Minerals Institute, University of Queensland: Brisbane, Australia.

Martin, T., Cervantes, M., Mendes, M., and Kemp, D. (2016) Tragadero Grande: Land, human rights and international standards in the conflict between the Chaupe family and Minera Yanacocha: Report of the Independent Fact Finding Mission. RESOLVE: Washington DC. Retrieved from http://www.resolv.org/site-yiffm/files/2015/08/YIFFM-report_280916-Final.pdf (accessed 1 February 2018).

Moran, M., and Waddington, D. (2016) *Riots: An International Comparison*. Palgrave MacMillan: London.

Owen, J.R. (2016) Social license and the fear of mineras interruptus. *Geoforum*, 77, 102–105.

Owen, J.R., and Kemp, D. (2013) Mining and community relations: Mapping the internal dimensions of practice. *The Extractive Industries and Society*, 1, 12–19.

Owen, J.R., and Kemp, D. (2017) *Extractive Relations: Countervailing Power and the Global Mining Industry*. Greenleaf Publishing: Sheffield.

Ruggie, J. (2011). Guiding principles on business and human rights: Implementing the United Nations Protect, Respect and Remedy Framework. Report of the Special Representative of the Secretary-General on the Issue of Human Rights and Transnational Corporations and Other Business Enterprises. Seventeenth Session, Agenda Item 3, A/HRC/17/31. Human Rights Council. Retrieved from http://www.ohchr.org/Documents/Issues/Business/A-HRC-17-31_AEV.pdf.

Waddington, D., Jones, K., and Critcher, C. (1989) *Flashpoint: Studies in Public Disorder*. Routledge: London.

Waddington, D., Wykes, M., and Critcher, C. (1991) *Split at the Seams: Community, Continuity and Change after the 1984–1985 Coal Dispute*. Open University Press: Philadelphia, PA.

22

Social Media and Community Relations: Five Key Challenges and Opportunities for Future Practice

Colette Einfeld, Sara Bice, and Chen Li

CONTENTS

22.1 Introduction: What Does Social Media Mean for Community Relations Practices?

New and enhanced technologies are transforming the mining and extractives industries. As this book attests, technological advancements involving automation, capacity to efficiently mine declining ore grades, a focus on late production chain activities, and means of reducing energy consumption are but a few of the innovations shaping the future of the sector (Deloitte, 2016). Other chapters in this volume demonstrate the breadth of these innovations and the depths to which they may revolutionize the industry's core tasks. But mining and extractives (M&E) is not about engineering in isolation. Indeed, many of the industry's greatest challenges relate more to communities than commodities.

In the past two decades a community relations function to support the global mining industry has become institutionalized within M&E companies, providing them with "a mechanism through which to engage and manage their relationship with key stakeholder groups and protect their business interests" (Bice & Moffat, 2014). Support for community relations roles is generally driven by broader concerns about corporate social performance, increasing expectations to develop partnerships with local communities (especially in developing countries), and the industry's growing rhetorical focus on companies' earning and maintaining a "social license to operate" (SLO) (Bice & Moffat, 2014). Community relations, therefore, encompasses a variety of activities and purposes, including but not limited to media and communications, public relations, stakeholder dialog and relationship building, and community development (Kemp, 2010).

Consequently, community relations are now, arguably, just as critical to the success of a M&E project as good engineering. Recent research concerning the costs of community conflict to M&E project delivery supports this assertion. Franks et al. (2014), for example, found that corporations could endure financial costs averaging US$20 million per week because of closure or project delays from such conflicts. Community opposition to projects also bears a more intangible cost through its effects on company legitimacy (Jenkins, 2004). Consequently, M&E companies often now approach issues of negative social, environmental, or financial impacts on communities with more care than they have historically (Bice, Brueckner, & Pforr, 2017). But conflicts with communities and, by extension, governments, continue, seriously impacting the viability and profitability of certain M&E operations (Jenkins, 2004).

The effects of community opposition on M&E projects are especially visible where projects are more controversial in nature. These cases might involve community resettlements (Hilson, 2002), indigenous land access or use or related threats to cultural heritage (Ali, 2009), or the use of new or controversial mining or extraction techniques, as in the cases of deep sea mining (Filer & Gabriel, 2017) or hydraulic fracturing in coal seam gas (CSG) (Dodge, 2014; Lacey & Lamont, 2013). The latter type of controversial cases is particularly relevant for consideration in this volume, as they occur at the intersection of "new" mining technologies,[1] technical advancement, and communities.

This chapter turns the book's focus on technological innovations to the ways in which social media is (re)shaping M&E companies' community relations spheres and practices. To suggest that social media holds the potential to redefine, restructure, and reconstitute what is meant by "best" community relations practice may seem hyperbolic. But, we argue, it is not. Indeed, the term "community relations 2.0" was coined in *Harvard Business Review* to reflect the expansive and abiding implications of online technologies for the practice, especially social media including Facebook and Twitter (Kane, Fichman, Gallaugher, & Glaser, 2009). As Gerald Kane and his coauthors explain, the advent of internet and social media is transforming community relations primarily through space-time compression and the dispersal of greater power to a greater number of interested stakeholders who are simultaneously better networked than ever before. Or, in the authors' words:

> "Before the internet, firms had far more time to methodically monitor and respond to community activity. With the rise of social media, that luxury has vanished, leaving a community-management vacuum in dire need of fresh skills, adaptive tactics, and a coherent strategy. In fact, in today's hyperconnected world, a company's community has few geographical barriers; it comprises all customers and interested parties, not just local neighbors." (Kane et al., 2009, p. 45)

For the M&E industry, these findings are both a bellwether of challenges to come and a confirmation of what many companies are already beginning to experience. The immediacy and reach of social media poses especially important considerations for M&E companies' SLO. Generally understood as the "the ongoing acceptance and approval of a [project] by local community members and other stakeholders that can affect its profitability" (Moffat & Zhang, 2014), SLO remains a debated but widely used concept to explain the state of relationship between an M&E company or project and a local community (Bice, 2014). Social media presents a real-time, influential platform on which SLO could be theoretically won and lost.

Within the M&E sector, the battle to attain SLO is typically waged by community relations professionals whose companies commit them to developing and maintaining the license. Practices to support the SLO-earning/maintenance process can include philanthropic efforts such as the donation of sports equipment, contributions to local facilities and sponsorship

---

[1] New in relation to certain of these technologies is a relative term. Deep seabed mining, for instance, has been possible in theory since the 1970s but such mining has yet to be undertaken. Similarly, hydraulic fracturing dates to at least the 1940s in the United States, but its contemporary application remains new to certain places, including Australia.

of local teams, and ensuring the communities are consulted and informed through newsletters and community meetings (Harvey & Bice, 2014). Companies also often—and in some areas, are legislatively required to—establish "community consultative committees" (CCCs), which represent the company, local community, and local businesses, partly as a means of monitoring SLO and ensuring that local representatives are consulted and informed about proximal operations. While such endeavours related to SLO theoretically offer many positives, including reduction or mitigation of conflict, online technologies pose many challenges to how SLO is understood, its boundaries, and measurement. Traditional community relations practices are similarly being challenged as social media is increasingly used to influence and shape community and political debate (Jeffares, 2014).

This chapter draws upon two years of social media data related to Australia's burgeoning CSG industry to provide insights and recommendations about how, exactly, social media may be changing the community relations landscape for the M&E sector. In doing so, it contributes what we believe to be the first major study into social media use in relation to M&E projects. While we recognize that findings from this case are limited in their generalizability, Australia does offer a strong leading indicator of changes in the M&E industry. The nation's economy is based heavily on its resources extraction. Although Australia's mining boom appears to be over, in 2012–2013 exports of natural resources still amounted to just over AUD$200 billion (Australian Government Department of Industry, Office of the Chief Economist, 2014), 81% of Australia's total goods exports (valued at AUD$246.9b) for the same period (ABS, 2014b). Many of the world's major M&E companies, including BHP Billiton, Shell, and Rio Tinto, were either born in Australia or run major operations on the island continent. Australian-based M&E companies, therefore, have substantial operational histories, including community relations that make them prime case examples for the global sector.

Our discussion of the Australian case is guided by three central questions. First, how is social media being used by communities, companies, and governments in relation to controversial M&E projects? Second, to what extent might social media affect a project, company, or even an industry's social license to operate? Third, how might community relations practice need to change or adapt to accommodate social media's key influences?

The chapter proceeds by introducing the reader to Australia's controversial CSG industry, the source of our social media case data. We then explain Twitter and highlight its usage and the development of an Australian "Twittersphere" before proceeding to explain the findings of capturing, coding, and analyzing a database of almost one million Tweets, recorded over a 2-year period, 2014–2016. Findings from this illustrative case are used to distill five critical challenges and opportunities that social media poses for traditional community relations practices, including:

1. Geographic boundaries that have traditionally provided focus for community engagement practices may no longer apply.
2. Stakeholders are better informed, and better connected, than ever before.
3. New forms of activism are developing through social media.
4. Community voices are louder and more impactful than ever before, meaning companies' communications will need to respond.
5. Where SLO was previously applicable at the project- or company-level, social media signals the emergence of an industry-level SLO.

22.2 Coal Seam Gas in Australia

Australia's CSG industry is the latest "boom" in a country which has a long history in resources extraction (Bec, Moyle, & Char-lee, 2015). CSG is an "unconventional gas," mainly composed of methane, and released from coal seams through drilling. This drilling can include a process of hydraulic fracturing, known commonly as "fracking," which involves widening and creating new fractures in existing coal seams to release gas, which is then extracted from the ground (Commonwealth of Australia, 2014). Advocates for CSG argue that it is a "transitional" resource, allowing economies to move from a reliance on coal and petroleum toward energy resources that produce fewer carbon emissions (Porter, Gee, & Pope, 2015).

Despite these promises, however, the unconventional gas industry has instead been the subject of great concern and anxiety in communities. Much of this concern seems to have been generated from media reports about environmental and social harm, contrasted with an industry that does not seem to be, at least publicly, changing its business practices (House, 2013). The very public nature of concerns about CSG in Australia has contributed to a substantial civil society protest movement and government policy action, and policies on CSG have become heavily politicized. In New South Wales (NSW) and Victoria, for example, state governments have placed moratoria on CSG licenses.

During this controversy, governments attempted to gather evidence on which to base future policy decisions while companies argued for their rights to explore and extract, pointing to future gas shortages and lost economic potential. In NSW, the Chief Scientist and Engineers Office undertook a review of the state's CSG activity, largely in response to community "unease" about the industry. The report, released after a thorough, 13-month investigation, prompted a raft of regulatory measures covering landholder access and areas able to be released for CSG exploration (Chief Scientist and Engineer NSW, 2014). These regulatory measures now require companies to negotiate access with landholders, and restrict CSG exploration and operation in certain areas. Meanwhile, in Victoria, the Auditor General's Office undertook investigations, concluding in its 2015 report that the state's current regulatory environment was not properly equipped to manage CSG risks (Victorian Auditor General's Office, 2015). The Victorian Government ultimately proclaimed a "permanent ban" on onshore CSG in 2016.

As governments and companies wrangled over the prospects for CSG along Australia's east coast, communities became exceptionally well organized to protest the industry. In NSW, for example, landowners in the Hunter Valley formed the Hunter Valley Protection Alliance. Many of these landowners were "tree changers" who had retired from corporate life in Sydney to the vigernon lifestyle prominent in the Hunter Valley's verdant wine region, an area also known for its thoroughbred horse-breeding, but also equally well known for coal mining. Consequently, these protest movements were well organized, well-connected, and well-funded in comparison to other grassroots protests. A major component of this strategic organization involved substantial online campaigns, involving websites, Facebook pages, and Twitter accounts. The virtual presence of these groups appears only to have amplified their protest messages and success, drawing critical masses of people to both online and in-person protests and engaging support from areas far beyond those directly affected by CSG exploration or drilling in Australia and overseas. Before continuing to examine just how the social media component of these movements affected the development of CSG along Australia's east coast, it is helpful to take a moment to explore what, exactly, social media is, and how it has emerged as a tool for social movements and debate.

22.3 The Emerging Role of Social Media in Protest and Policy

Social media has the power to shape debates, and the potential to be an influential policy platform (Jeffares, 2014). Twitter, a social media tool, offers an interactive, accessible opportunity to have multiple stakeholders interact around issues of interest. As Stephen Jeffares writes:

> "The widespread adoption of mobile communications and social media channels offers a new environment for policy-making. With social media, never has it been cheaper, easier or quicker to coin and disseminate an idea. Similarly, never has it been easier to expedite the demise of a policy idea, or to mobilise an alternative viewpoint (*Interpreting Hashtag Politics* [2014:2])."

For the discussion on CSG in Australia, nowhere was this more visible than on Twitter.

22.3.1 Twitter

Twitter is a micro-blogging site, launched in 2006. It now has 313 million active monthly users (as of June 2016). Tweets are no more than 140 characters, and often include hashtags used to signal discussion of topics or trends. For example, the birth of Princess Charlotte to the Duke and Duchess of Cambridge was discussed under the hashtags #Royalbaby and #princess-charlotte. Every Twitter user has a unique "handle," denoted by the "@" symbol before the handle name. Twitter maintains a cache of metadata about its Tweets, meaning that it is possible to access a wide range of information, including who is Tweeting about what, locations from where Tweets are sent, users' demographic profiles, replies to others' comments, and URL sharing, among other options. The depth and variety of data is considerable and its newness as a data source requires careful attention, not only from researchers, but from policy makers, corporate representatives, and civil society (Bruns & Burgess, 2012).

There are about 2.8 million Australian Twitter accounts. A recent project to map the Australian Twittersphere demonstrates that these users employ the technology for a variety of reasons, and that usage tends to cluster around specific issues (Bruns, Burgess, Kirchhoff, & Nicolai, 2012). For example, and most relevant to this project, certain users employ Twitter for political activism and community organising or to discuss political concerns.

Twitter can be used for political and activist purposes, and has fewer barriers to entry than having to write policy submissions. The shape of debate on Twitter, and the stakeholders involved, may therefore be different from that seen through more formal mechanisms. Indeed, at the time of writing, emerging behaviors from the current President of the United States through his personal Twitter account signal a historical shift from more traditional policy debate platforms, highlighting the power and reach of Twitter. It is also an inherently accessible, open, timely, and reactive form that has the potential to connect disparate groups around issues. In relation to the CSG debate in Australia, it seems that the Twittersphere is capturing in real-time the changing the nature of the discussion.

22.4 Project Methodology

The illustrative cases in this chapter are based on a multiyear study investigating social license to operate in the CSG industry. One aspect of this research was to understand how CSG debates are shaped online, for which we extensively mined Twitter. Before exploring the findings, we briefly detail the research method.

We began our Twitter investigation by undertaking an extensive document analysis focused on government documents, civil society reports, public submissions to parliamentary inquiries, media, and related websites. Through this analysis, we aimed to construct a concourse of debate—the multilayered, diverse, and competing perspectives publicly voiced about a specific issue (Stephenson, 1980). Importantly, given that a concourse is potentially infinite (McKeown & Thomas, 2013), we bounded our inquiry using time and saturation point. To do this, we searched a variety of publicly available documents to establish the breadth and depth of discussion concerning SLO and CSG in Australia. During the document data collection period (February–April 2015), we retrieved 89 sources before reaching saturation point. We sampled documents until the views presented "approximate[d] the total commentary" (McKeown & Thomas, 2013, p. 22) on CSG in Australian communities. News articles represented the most common source of SLO mentions, comprising 44% of all sources retrieved (excluding public submissions to parliamentary inquiries). Submissions to inquiries on CSG were the second most common source of SLO mentions, accounting for 22% of all sources retrieved. Finally, SLO was mentioned in reports produced by various stakeholders related to the CSG sector, including advocacy groups, government organizations (at federal, state, and local levels), M&E companies, and peak bodies. These reports included discussion papers, reports following parliamentary inquiries, corporate social responsibility reports, and corporate annual reports (12% of all sources). Less frequently retrieved sources were newsletters (5%), presentations (3%), policy documents (2%), and speeches (2%). A limited number of agenda papers, blog entries, letters, media releases, and petitions were also retrieved (Figure 22.1). A full analysis of these documents is beyond the scope of this paper, but it is instructive to be aware of the main outlets in which SLO tends to appear beyond social media. The findings of this analysis allowed us to identify key search terms describing the concourse of debate concerning SLO and CSG in Australia.

Following establishment of a search register in 2014, we began regular, fortnightly Twitter captures on March 26, 2015. The fortnightly schedule is aligned with Twitter's proprietary usage of Tweets, through which they allow access to two weeks of historical Tweets. These captures were completed in March 2016, thus allowing us to mine two years of data. In total we captured 930,699 Tweets (including Retweets) and performed preliminary

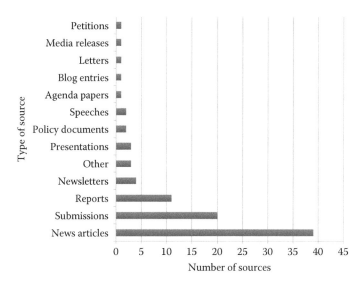

FIGURE 22.1
Number of document sources retrieved by type of resource, February to April 2015.

analysis. The hashtags and handles captured relate to Australia's CSG industry and the major companies and issues involved at the time of the Twitter
capture. For example, a small sample of hashtags and handles searched for
included: #fracking, #CSG @AGLEnergy, @SantosLtd. @Shell. For purposes
of this chapter, we focus on two specific cases that emerged organically
from the broader data captured. These incidents offer a timely and interesting opportunity to explore the relationship of social media to community
relations because each captured real-time reactions to controversies as they
emerged. Before we detail the cases and findings related to Santos' Narrabri
Gas operation and AGL's total divestment from the CSG industry, however,
it is helpful to explain how we treated the data, once captured.

To analyse this data, we first cleaned it using R Statistical Software to
remove all the duplicate Tweets that were captured due to users' use of multiple hashtags. For example, a Tweet that included the hashtags #csg and
#nswpol would be captured twice in each separate hashtag search. We
then used the R "text mining package" (tm package) to uncover trends in
the text and to determine word frequencies. This also allowed us to identify Tweets that contained synonyms for issues of interest. For example, by
programming R as index1 = grep ("social license," Tweet), we were able to
find all Tweets that contained words of interest and their synonyms, one by
one. This search included "social license," "social licence," "social license to
operate," and "social licence to operate," as shown in Figure 22.2 as index1,
index2, index3, and index4. We then merged all these Tweets through index =
unique(c(index1, index2, index3, index4). This process was applied to explore
a range of topics of interest in the Twitter data.

```
# package "tm" installed
library (tm)
library (snowballc)

setwd("R")
# list.files()
df = read.csv("Final.csv"), header = TRUE)
tweet = df[, "Tweet"]

# find "social license"
index1 = grep("social license", tweet)

# find "social license"
index2 = grep("social license", tweet)

# find "social license to operate"
index3 = grep("social license to operate", tweet)

# find "social license to operate"
index4 = grep("social license to operate", tweet)

# merge all index into one
index = unique(c(index1, index2, index3, index4))

# save all columns
write.csv (df[index,], "Final SLO.csv", row.names = FALSE)
```

FIGURE 22.2
R text manager: Program search example.

Following the automated analysis in R, we also performed certain manual analyses. We developed a separate database of Tweets directly relating to SLO which we draw upon throughout this chapter. We then performed primary and secondary coding of this smaller database of 428 Tweets or Retweets. To do this, we examined each handle's previous Tweets and information included in the "about" sections of their Twitter accounts. Handles were then coded into groups, depending on whether they were an individual, affiliated with or representing an NGO, a community group, government, private sector, or other actor. We also coded the content of each handle's Tweets. This included whether most of their Tweets were original and composed by them, or whether their Tweeting behavior consisted primarily of Retweeting others' comments, opinions, or information.

While a full discussion of our method is beyond the scope of this chapter, the above review provides some insight into the methods employed to explore the issues in which we were most interested. As our engagement with the data deepened, we performed further secondary analysis to understand better:

- Where Tweets were coming from and what types of Tweets were coming from which places.

- The different, common styles of Tweeting and how these linked to emergent types of activism (discussed in detail later in this chapter).
- The most common uses for Twitter, in relation to CSG.
- The types of information linked to within Tweets (e.g., other community webpages, news media, reports, or videos).

In the sections that follow, we present the Santos Narrabri CSG operation and AGL divestment cases to distill the research findings most pertinent to the contemporary practice of community relations in the M&E sector. We also used other cases from our SLO database.

22.5 Santos' Narrabri Gas Operation: Pilliga State Forest, New South Wales

Santos is a majority Australian-owned gas producer, with several resources in Queensland and NSW. In March 2014, during the period when we were conducting our initial Twitter capture, Santos faced accusations of polluting aquifers at a site in the Pilliga state forest, NSW. Uranium levels 20 times higher than safe drinking water guidelines were recorded (ABC Illawarra, 2014). Following an Environmental Protection Authority (EPA) investigation that took several months, Santos was ultimately fined AUD $1,500, with further reviews by the EPA identifying two other incidents in 2013 and 2015. Although no further fines were issued, the environmental regulators ordered Santos "to improve the operations and transparency" of their Narrabri CSG project (Hannam, 2015). This incident, and the subsequent media attention it received, coincided with the NSW State elections. The incident inflamed pre-existing controversy and protest, making NSW's CSG sector the focus of political ire.

Twitter snapshots taken shortly after initial media reports concerning the aquifer incident surfaced (March 18 and 25, 2014) and included search terms and hashtags, such as CSG, Santos, water, fracking, NSW, gas, Pilliga, #nswpol, and #auspol (hashtags for "NSW politics" and "Australian politics").

22.6 AGL's Investment and Divestment from CSG

AGL is one of Australia's biggest energy companies. AGL formed its Upstream Gas Business as the investor and operator of its CSG businesses. AGL had hoped to build internal capability to operate services and develop technological expertise in CSG operations, and consequently made many investments, becoming owner/operator on four main projects: Camden (producing CSG), Galilee, Hunter Valley, and Gloucester (all at pilot stage).

During the period of AGL's investment in CSG, the NSW Government developed its gas plan in response to the NSW Chief Scientist and Engineer's independent review of CSG activities in NSW. AGL's submission to the plan provides an interesting insight into its position on CSG. AGL argued that CSG should be supported for the following reasons: CSG has lower carbon emissions than coal, and is in line with AGL's commitment to lower its emissions; CSG is the only source of energy other than coal that can meet the base load in NSW and can provide a secure, reliable source of energy; it is the bridge between coal and renewable sources of energy; it has a smaller physical footprint than coal; CSG can coexist with agricultural land use, and the income provided for farmers may be their only reliable source of income.

Like Santos, AGL's operations were also a focus of protest for communities and stakeholders. It is worthwhile here noting two incidents that occurred during our Twitter capture, as they became central to "building the evidence base" against AGL's CSG projects. In January 2015, elevated levels of BTEX (benzene, toluene, ethylbenzene, and zylenes) were found in samples drawn from the Waukivory pilot project, a small operation which was established to determine the viability of the Gloucester project. BTEX chemicals are naturally occurring in the environment, and therefore also occur in water produced from CSG activities, but sometimes are also added to fracking fluids to improve the extraction process. In high concentrations, BTEX can be harmful to the population. The NSW Government banned the use of BTEX chemicals in fracking in 2012 (Division of Resources and Energy, 2015). The EPA was notified about the detection of BTEX, but the notification was 12 days after the detection. The EPA then launched a full investigation into the incident to investigate the contamination and the delay in notification. Ultimately the EPA found that AGL had not breached its license, but recommended that AGL improve its monitoring and management of BTEX. Meanwhile, community groups expressed concerns about the impact of the contamination on local agriculture.

In October 2014, information about a series of incidents at the Camden site was released following Freedom of Information (FOI) requests from local activist groups. The report outlined a series of methane leaks at the site that appeared to contradict figures and reports released by AGL. Community activist groups highlighted the contradiction between the EPA findings and a self-funded report from AGL which found no leaks. AGL argued that their report covered a different period of time; however, community groups questioned the transparency of AGL and effectiveness of self-regulation, given the reports were only released after an FOI request.

Despite publicly supporting CSG activity, AGL announced a review of the Upstream Gas business in 2015, which led to AGL abandoning attempts to be a self-sufficient gas producer. AGL cited the uncertainty around gas prices and its unwillingness to commit further investment to gas production as the key reasons for the decision. It also pointed to the political landscape in NSW as having changed the viability of CSG exploration. In February 2016, AGL announced it was divesting from all CSG production and moved to immediately close its pilot project at Gloucester. Camden, its only site in full production, will close 12 years earlier than anticipated, and it is currently expected that AGL's investment in CSG will cease in 2023. Despite assertions that it was the market that determined AGL's decision decision to divest, there are suggestions that stakeholder pressure and community unease also played a part. During the period we tracked Twitter, we captured all Tweets that contained AGL, #AGL, @aglenergy and @YourSayAGL and that referred to the CSG industry.

22.7 Results: What Does Social Media Mean for Community Relations?

The findings of these two cases and investigation of our SLO database indicate that social media is transforming key aspects of community relations. Our discussion of these findings responds directly to our central research questions, detailing how social media is used by key players in CSG and demonstrating how such usage appears to have affected an SLO related to the studied cases. We distill five key challenges and opportunities that social media poses for contemporary community relations practice from these findings. We then suggest how community relations practice may need to change or adapt to accommodate the key influences of social media identified, at the very least for the cases we studied.

22.7.1 Traditional Geographic Boundaries No Longer Apply

The empirical cases we chose for this chapter relate to very specific geographic areas located along the east coast of Australia. Santos' case related primarily to a specific aquifer in the Pilliga State Forest, while AGL's operations were based in the town of Camden and in the NSW regions of the Galilee Basin, Hunter Valley, and Gloucester Basin. The very "localness" of these cases reflects a common approach in which industry, government, and researchers emphasize the importance of local engagement and local community, especially when defining audiences for community relations (Kemp, 2010) and when talking about developing a SLO (Thomson & Boutilier, 2011).

The historical importance of the "local" in M&E operations speaks not only to the geographically defined nature of the extractive processes, but also to an important means by which companies may define the boundaries of their social and environmental performance obligations. Local approaches have been emphasized by leading bodies, including the International Energy Agency (IEA) in its 2012 "Golden Rules" for unconventional gas. The agency argues that environmental and social concerns about unconventional gas must be addressed for the industry to gain an SLO, which is required for the development of the industry:

> "The Golden Rules underline that full transparency, measuring and monitoring of environmental impacts and engagement with *local* [emphasis added] communities are critical to addressing public concerns." (IEA, 2012)

Within the Australian CSG sector, concerns for the local are also center stage. For example, the Council of Australian Governments (CoAG) 2013 study, *National Harmonised Regulatory Framework for Natural Gas for Coal Seams*, which aimed to provide a best practice, consistent approach to CSG regulation throughout Australia's diverse states and territories, highlights the importance of community approval:

> "Community engagement should include upfront and honest conversations and negotiations - providing information on activities and operations in the short and long term and the impacts that those activities may have on *local* [emphasis added] communities."

Attention to the local is also traditionally closely linked to the concept of a SLO. Leading academics on SLO, Moffat and Zhang (2014), for example, discuss how companies can avoid costs associated with conflicts by developing and maintaining a SLO with local communities. Thomson and Boutilier (2011), upon whose work much consultancy on social license is based, discuss how their idea of a SLO is one that must be developed with local stakeholders and the local community.

Social media presents an unprecedented challenge to the conceptualization of "local" and to the already difficult task of placing boundaries around affected communities. Our research suggests that through its capacity to connect disparate and distant groups around concerns, social media is reshaping the boundaries of stakeholder influence. AGL, for example, developed four investor, operator projects. Of the projects, Camden, Hunter Valley, and Gloucester were in the states of NSW and Galilee in Queensland. Despite the location of the operations in NSW and Queensland, Tweets came from as far away as Canada and the United Kingdom. In the United States alone, we saw Tweets about the NSW government buy-back scheme, sharing of information about AGLs divestment in CSG, and Tweets about farmers' excitement over the divestment decision. Within Australia, but as far away as Western Australia, many of these Tweets expressed their concern about CSG operations and the development of the CSG industry.

Our findings suggest that, at least in the case of CSG in Australia, community engagement staff must expand their conceptions of community beyond the local level as it has been historically defined. While it remains true that many of the immediate impacts of operations, such as increased trucks and noise during construction, or the potential to have wells located on nearby properties, will continue to be felt in a physically local way, social media allows non-proximal groups and individuals to take a stake. Such stakes can be articulated in many ways that demonstrate strongly held perceptions that a project's negative impacts, however distant, are meaningful to that stakeholder.

At this juncture, one might argue that, even if an individual or organization expresses concern about an issue which is geographically distant to them, this would not necessarily qualify that individual or organization as a stakeholder. But given that stakeholders are commonly defined as those affected by a project or entity or those with the capacity to affect a project or entity, a strong argument can be made that social media participants should qualify as stakeholders in the eyes of community relations staff. We will return to this assertion in more detail in a moment but for now it is worthwhile considering that, through social media, these stakeholders do have the capacity to run effective campaigns and build coalitions that can materially affect an operation.

In the case of CSG in Australia, many distal individuals and organizations are claiming a stake based on their concerns about the relationship between CSG and fossil fuels or CSG and water security. Our findings show that these individuals are actively committed to expressing their concerns about Australia's CSG industry as a means of protesting activities that they view as contributing to negative environmental impacts. This leads us to our second major finding about what social media means for contemporary community relations practice.

22.7.2 More Informed and Better Connected Stakeholders

Social media not only connects disparate stakeholders; it allows for information to be spread through stakeholder networks in real-time and with ease. Additionally, social media has the potential for exponential sharing of positions and ideas. It also offers a platform for local communities to share their experiences with a wider audience, thus building broader stakeholder coalitions.

For example, findings from our main CSG dataset, that contained over 900,000 Tweets and Retweets, found "http" to be among the most frequently used words. "Http" indicates a weblink is being shared by the

Tweeter. Findings reveal that weblinks commonly directed readers to media stories or NGO websites, which typically discussed dangers of the CSG industry, contaminations or spills, or similar environmental concerns. In this way, community members are using Twitter to share information of interest, while also building an evidence base that supports their perspectives.

Within our broader dataset where SLO was mentioned specifically, two-thirds of those Tweets included a weblink. These links led to a variety of content, including newspaper articles, petitions, photos of protests, and YouTube videos. This sharing of information is not only visible through the large number of weblinks shared, but also through the large proportion of those who are "Retweeters." On average, in our large CSG database, every Tweet was Retweeted twice, spreading this information through exponentially growing online networks. Throughout our dataset, we saw the relatively effortless Retweet spreading information swiftly and widely, reinforcing messages to those already interested in the CSG debate (e.g., followers of the Tweeter and Retweeters), or providing new information to a Tweeter's broader networks.

Information sharing through Twitter was also closely linked to political debates related to this study's focus on CSG. For example, the incident in the Pilliga occurred during the NSW State election. Of the 3,000 Tweets relating to the Pilliga incident, "http" was the most frequently appearing term, with the #nswvotes appearing third-most often. This linking of CSG information and concerns with the broader election further emphasizes the ability of social media to allow stakeholders to put forward their perspectives and shape debates, thus connecting local concerns to broader environmental and political issues.

22.7.3 New Forms of Activism Are Emerging

Our findings suggest that Twitter is used primarily as a means of sharing information and influencing debates. But we also found that new forms of activism are emerging which are specific to social media. Community relations professionals usually devote considerable time to engaging with stakeholder activists, advocacy organizations, and protest and grassroots community action groups. The findings of this study indicate that, not only are those potential activist nets widening, but also the roles and activities adopted are specific to a social media context. Through manual, thematic coding of the smaller, SLO database, we identified four key types of activist activity: endorsers, activists, clicktivists, and watchdogs (Table 22.1).

"Endorsers" comprised the most common and perhaps most effective type of activists in our SLO database. These individuals typically refrained from penning their own original Tweets. Instead, they actively used Twitter as a means of spreading others' information and ideas in support of their

TABLE 22.1

Table of Twitter Activity Types

Twitter Activity Type	Description of Behavior/Activity
Activist	Individuals or Groups who actively organize petitions or protests on or offline
Clicktivist	Individuals or groups who may express strong activist sentiments, and typically post original tweets and link to videos, articles, and so on, but may not be organizers of offline events
Endorser	Primarily retweeters, occasionally own content
Watchdog	Tweets photos/videos of CSG impact, alerts others to corporate behavior

own values, positions, and beliefs. Of the 186 individual handles in the SLO database, 110 tended primarily to Retweet rather than create their own content. In doing so, they effectively created an exponential spread of information through their networks while also enhancing the visibility of the original; the more "likes" and Retweets a Tweet receives, the more likely it is to appear on trendlists and in individuals' "notifications" pages as "most mentioned."

Endorsers were followed closely by more traditional "activists": individuals who use Twitter as an online means of organizing their more traditional activist work. "Activists" tended to announce NGO or protest meetings and also used Twitter to share photos and outcomes of protest events, and to advertise petitions or other "real world," offline activities. There were 41 activists in our dataset, with many Activist handles representing community groups and NGOs (24/41) and the remaining 17 operating as Activists on an individual basis.

The third most common type of online activist to emerge from our data set was the "Clicktivist" (25/186). These individuals regularly created their own original content to express strong activist sentiments but stopped short of organizing offline protests or activities.

Social media also appears to be giving rise to a fourth type of novel activist, the "Watchdog." While smaller in number (10/186), Watchdogs also appear to be influential. These Tweeters assumed the role of an informal watchdog, monitoring corporate behaviors and even taking photos and videos that were then Tweeted and Retweeted. Watchdogs appeared to mine government sites for media releases and reports, monitoring information about exploration licenses, environmental assessments, and other data that, while technically publicly available, often requires some digging to locate. Certain Watchdogs also shared pictures that allegedly showed toxic waste or pictures of rivers and waterways with bubbles, which communities have argued indicates methane leakages from gas wells. Other posts used real-time photographs to alert locals to the activities of companies near them, including movement of trucks.

We also found evidence of Watchdogs in the Pilliga case. Watchdogs on Twitter monitored government approval processes, local sites, and publicly available reports to alert others to potential CSG activities. For example, one community action group regularly posted videos of local CSG operations to YouTube and Tweeted links to them. One Watchdog Tweet linked to a 44 second video showing a gas well cap sparking and emitting water droplets. This one Tweet was then Retweeted 25 times, all by individuals in the same community.

These emergent categories of online activists show the distinct and varied ways that Twitter is used. Our findings also show that social media helps to link up diverse types of activists, with Clicktivist, Activist, and Watchdog Tweets Retweeted by all different categories, creating a varied and wide web of organizations and individuals acting on an issue of concern in distinct but influential ways.

Most users who Tweeted or Retweeted about the CSG industry were sharing and spreading information, but a notable number of others were using Twitter for activism to encourage and organize around CSG protests and conflicts. Still others are creating content and using Twitter to share perceived evidence against the CSG industry. While watchdogs have long been in action prior to the advent of social media, Twitter provides a vehicle through which these messages can be spread quickly and easily by the Endorsers, while simultaneously raising the profile of the Watchdogs themselves, as their Tweets are shared.

22.7.4 The Rise of Community Voice

Community relations in the time of social media must also respond to a rising community voice. Throughout the analysis, we were struck by the largely absent voices of corporate actors in these Twitter conversations. While our findings show Twitter participation by individuals, civil society organizations, media, government, and all the major CSG companies operating in Australia, corporate and government voices were relatively mute in comparison to others. In other words, although all central parties to Australia's CSG debate are participating via social media, our findings reveal that corporate actors' social media use is parsimonious and has a very difference pattern than non-corporate actors.

In our SLO database, for example, of those who were Tweeting and Retweeting about SLO and CSG, 141 were individuals, 31 were from NGOs or community groups, 7 were government representatives, and 3 represented the private sector. Of the private sector Tweeters, two were from media outlets, and only one was from a mining company. This finding was particularly surprising, given that SLO as a term or concept has traditionally been used by business and industry. Yet on social media it is individuals and groups using the term, most often to deny or withdraw a social license from the companies who would traditionally lay claim to it.

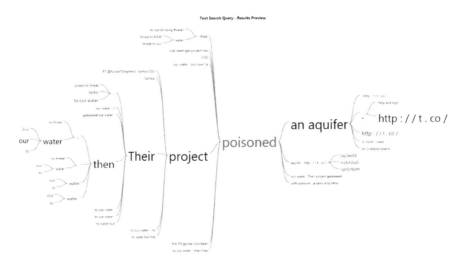

FIGURE 22.3
"Poison" or synonyms in tweets, March 18–25, 2014. (From ARC Discovery project, Fear and Licensing in Australia: Investigating the social licence to operate of an emerging coal seam gas industry.)

The flavor of the discourse and debates between these groups is also diverse. Many individuals Tweeting about CSG in Australia usually adopt an anti-CSG perspective. Corporate accounts tend to be reactive, responding to select issues with facts (or alleged facts) or public relations-like statements. For instance, a subset of 709 Tweets/Retweets captured in relation to the Pilliga case shows non-corporate actors Tweeting highly emotive ideas and statements, with links to news stories about the aquifer contamination, which many Tweeters described as "poisoning" (Figure 22.3). The Tweets also clearly conveyed that Santos did not have and would never have a social license for the project, according to those individuals and groups. The official Santos account, meanwhile, Tweeted only 13 times during this period in relation to the Pilliga aquifer contamination issue. Many of these Tweets aimed to share information or to address certain other Tweeters' concerns. Many Tweets linked to a Santos-authored report that sought to correct information about the incident which Santos saw as false (Figure 22.4). Our thematic analysis of words used by community and by the corporate Santos account (@SantosLtd) showed a divergent discourse between these actors, suggesting that Santos was not seeking to engage in debates, or address the issues raised by other stakeholders, but to start an entirely different conversation.

Before social media, corporations held positions of considerable power in relation to messaging and agenda setting, as they influenced or controlled the key means of mass dissemination. Corporates may have sought, for example, to influence conversations and debates through large advertising budgets and lobbying. Social media presents a different terrain. Our data suggests that social media offers a more level playing field.

FIGURE 22.4
@SantosLtd official Twitter, Tweet about Pilliga incident. (From Twitter, March 12, 2014.)

Platforms like Twitter create an accessible and influential space in which non-corporate and non-governmental actors have the capacity to set agendas, not only for public debate but also in relation to influencing policy directions (Jeffares, 2014).

22.7.5 Emergence of an Industry Level SLO

Social media may also have implications for how community relations professionals and corporations attend to a social license to operate. Today's community relations professionals have a necessary familiarity with the concept of SLO. It is regularly invoked by companies to lay claim to a right to operate specific projects and also as a means of demonstrating corporate concern about broadly defined social issues (Bice, 2014; Parsons & Moffat, 2014). But what happens when the level at which SLO is applied is elevated beyond individual projects or corporations to an entire industry?

Findings from this study suggest that social media is contributing to the already complex challenges of defining and laying claim to SLO. As discussed, social license has traditionally been defined in scholarship and in practice as a project-level license that involves local communities and depends upon local acceptance, as particular to each project (Thomson & Boutilier, 2011; Owen & Kemp, 2013). The implications of social media for the defining of local communities and stakes were discussed earlier, and these changes relate to SLO, too. In relation to CSG, social media not only allows groups and individuals located beyond local boundaries to lay claim to a stake in a project's social license, it also facilitates questioning of the legitimacy of the entire industry. Throughout our data Tweeters denied a social license for CSG at an industry level, using the concept as a means of rejecting an industry and technology they view as dangerous and harmful.

In the Pilliga case, for example, despite the contamination of the aquifer relating to only one project at one closely defined location, and only one company, conclusions drawn by Tweeters questioned the safety and legitimacy of the entire CSG industry. The extension of concern from individual projects to the broader industry is partly influenced by hashtags, which Tweeters use to signal items of interest to other Twitter users. In the Pilliga Tweets, #CSG was the most commonly used term (Figure 22.5), extending the boundaries of concern from one project to the entire industry.

Throughout the findings, Tweets were regularly sparked by one specific project, media story, or incident, such as the Pilliga example, protests against AGL, a BTEX contamination, and AGL's ultimate divestment, but conclusions drawn tended to condemn the entire CSG industry. While such generalizations may be critiqued as unfounded, they appear no less influential, especially in the social media space where such observations are both common and accepted. But given that social media demonstrates considerable and progressive influence, even despite these flaws, what does this mean for community relations, especially in relation to SLO?

FIGURE 22.5
Tweets concerning Pilliga National Forest aquifer contamination, March 25, 2014. (From authors' data.)

Our findings suggest that a rethink of the ways in which SLO is defined and understood may be required, regardless of whether social media is reflecting or influencing this trend. If social media facilitates a public extension of expectations for procedural fairness beyond traditionally defined, geographically based stakeholders, and if social media encourages the generalization of project-level concerns from one project or company to an entire industry, then notions of procedural fairness and the building of social capital must also be extended. It is possible that, in future, companies seeking to earn and maintain a SLO may be forced to work in greater concert with their industry colleagues to address cumulative impacts of their industry and attend to an industry level SLO. This is particularly the case in relation to CSG in Australia, where our findings indicate that communities seem to no longer discriminate between the individual corporate actors and the industry itself. The spread of SLO from project/corporate to industry level signals an important challenge for the future of community relations and operational legitimacy.

22.8 Considerations for Future Research

While the discussion in this chapter draws on a major study of Twitter to distill five key challenges and opportunities for community relations in the digital age, it remains limited in its focus on one specific mining and extractives sector within Australia. The concentration of community actors and civil society organizations within the Twittersphere also raises the question of whether social media is simply an echo chamber? Or are Tweets truly spreading their influence to wider audiences beyond those included in the direct social media networks of Tweeters, Retweeters, and Followers.

Our research also raises the question of whether this phenomenon is being seen outside of the CSG industry, in other, potentially more established industries such as coal mining? This study offers a foundation for other researchers to consider the usage, influence, and meaning of social media for other M&E sectors and for other impactful industries beyond M&E.

The study is also limited by its focus only on Twitter. While a focus on Twitter allowed us to collect a substantial and readily analyzable database for the study, other social media, including Facebook, LinkedIn, Vine, and YouTube is also likely playing a role in influencing stakeholder sentiment and on- and off-line debates. Future studies might explore these other media and it would be interesting to compare the different means of usage on various platforms. Do similar types of activists emerge on Facebook, for example, or is there something unique in the Twitter platform's set-up that encourages certain behaviors?

22.9 Conclusion and Recommendations

This chapter has demonstrated that social media is a powerful new platform that is better informed and better connected stakeholders. It is expanding the population of organizations and individuals who might be considered legitimate project stakeholders. The accessibility and rapidity of social media is also giving communities swifter and greater access to information and allows them to connect readily to one another. New activist identities are emerging through this access to information and networks, with social media being used to accelerate traditional activist practices, spread information and ideas, spout opinion, or monitor corporate and government behaviors. Online connections are supporting a powerful community voice that is influencing debates and setting agendas for corporate and policy concerns. These changes are further evidenced by the emergence of a SLO at an industry level, expanding stakeholder interests and expectations beyond traditional project- or corporate-level concerns. Combined, these five developments suggest that social media is transforming the terms of engagement and expectations for community relations concerning mining and extractives projects.

Although the insights detailed in this chapter could be posed as challenges, we suggest that they are better perceived as opportunities. Community relations professionals are operating in an increasingly dynamic environment which circumscribes actual and virtual communities in progressively active and meaningful ways. The profession, and the corporations that depend upon it for their SLOs, now has an opportunity to develop and improve community engagement practices. While this will necessarily require new, more, and different ways of working, it may also assist smoother or more effective engagement. For example, social media offers companies unprecedented access into stakeholders' networks and thinking. For communities, this presents a two-edged sword. While the community relations practitioners with whom they engage may be better able to understand them, social media comes at the cost of lost privacy and an openness to activist discussion and organizing that is likely to transform civil society activity as much as corporate practice.

Should mining and extractives companies choose to become more active on social media, or at least more active than in the cases studied here, they would have an opportunity to engage immediately and directly with stakeholder concerns. This could have many advantages, including ensuring that the companies remain abreast of emerging concerns or complaints, allowing corporations to respond directly to early allegations before they result in the spread of rumour, or allowing early redress of impacts to reduce negative ramifications for communities and the company.

The emergence of industry-level SLO in social media discussions also suggests that the platform holds the potential to spark improved corporate consideration of cumulative impacts. If social media campaigns have the capacity to bring the SLO of an entire M&E sector into question, then companies must unite to present a sector-wide response. Such a response will require collaboration and careful consideration of how the activities and impacts of one project or corporate actor can amplify or sully the activities of another. Through its generalizations—whether one agrees with them or not—social media has the potential to force collective corporate attention rather than losing SLO altogether.

Our research demonstrates that M&E companies do need to consider their social media presence and (re)evaluate their use of platforms like Twitter. Many community members have adopted social media and the community relations profession must be prepared to engage. This chapter has identified five key areas in which that work could commence.

Acknowledgments

The authors thank Mr Martin Bortz for his research assistance. Martin's work capturing Twitter data and analyzing documents for the 'concourse of debate' was critical to the ideas presented here. This project was funded by the Australian Research Council, Discovery Projects program, DP140102779.

References

ABC Illawarra. (2014). CSG Pilliga contamination sounds alarm bells for anti-mining advocates, 10 March. *ABC News Online*. Retrieved January 24, 2018, from http://www.abc.net.au/local/stories/2014/03/10/3960324.htm.

Ali, S. H. (2009). *Mining, the environment, and indigenous development conflicts*. Tucson, AZ: University of Arizona Press.

Australian Government Department of Industry, Office of the Chief Economist. (2014). Resources and Energy Statistics 2014. Canberra, Australia: Australian Government.

Bec, A., Moyle, B. D., & Char-lee, J. M. (2015). Drilling into community perceptions of coal seam gas in Roma, Australia. *The Extractive Industries and Society, 3*(3), 716–726.

Bice, S. (2014). What gives you a social licence? An exploration of the social licence to operate in the Australian mining industry. *Resources, 3*(1), 62–80. doi:10.3390/resources3010062.

Bice, S., Brueckner, M., & Pforr, C. (2017). Putting social license to operate on the map: A social, actuarial and political risk and licensing model (SAP Model). *Resources Policy, 53*, 46–55.

Birol, F., & Besson, C. (2012). *Golden rules for a golden age of gas*, World Energy Outlook Special Report on Unconventional Gas. International Energy Agency, 12.

Bice, S., & Moffat, K. (2014). Social licence to operate and impact assessment. *Impact Assessment and Project Appraisal, 32*(4), 257–262.

Bruns, A., & Burgess, J. (2012). Notes towards the scientific study of public communication on Twitter. In A. Tokar, M. Beurskens, S. Keuneke, M. Mahrt, I. Peters, & C. Puschmann (Eds.), *Science and the Internet*. Düsseldorf, Germany: Düsseldorf University Press.

Bruns, A., Burgess, J., Kirchhoff, L., & Nicolai, T. (2012). *Mapping the Australian Twittersphere*.

Chief Scientist and Engineer NSW. (2014). Final report of the independent review of coal seam gas activities in NSW. Sydney, Australia. Retrieved from http://www.chiefscientist.nsw.gov.au/reports/coal-seam-gas-review.

CoAG Energy Council. (2013). National Harmonised Regulatory Framework for Natural Gas from Coal Seams. Council of Australian Governments.

Commonwealth of Australia. (2014). *Hydraulic fracturing ("fracking") techniques, including reporting requirements and governance arrangements*. Department of the Environment, Canberra.

Deloitte. (2016). *Tracking the Trends*. Retrieved from https://www2.deloitte.com/content/dam/Deloitte/ca/Documents/international-business/ca-en-ib-tracking-the-trends-2016.pdf

Department of Resources and Energy. (2015). BTEX and Coal Seam Gas. In Division of Resources and Energy, State of New South Wales, Australia Department of Trade and Investment (Ed.).

Dodge, J. (2014). Civil society organizations and deliberative policy making: Interpreting environmental controversies in the deliberative system. *Policy Sciences, 47*(2), 161–185.

Filer, C., & Gabriel, J. (2017). How could nautilus minerals get a social license to operate the world's first deep sea mine? *Marine Policy*, 1–7.

Franks, D. M., Davis, R., Bebbington, A. J., Ali, S. H., Kemp, D., & Scurrah, M. (2014). Conflict translates environmental and social risk into business costs. *Proceedings of the National Academy of Sciences, 111*(21), 7576–7581.

Hannam, P. (2015). Santos receives warning over CSG waste water leaks, *The Age*, May 15, 2015.

Harvey, B., & Bice, S. (2014). Social impact assessment, social development programmes and social licence to operate: Tensions and contradictions in intent and practice in the extractive sector. *Impact Assessment and Project Appraisal, 32*(4), 327–335.

Hilson, G. (2002). An overview of land use conflicts in mining communities. *Land Use Policy, 19*(1), 65–73. doi:10.1016/S0264-8377(01)00043-6.

House, E. J. (2013). Fractured fairytales: The failed social license for unconventional oil and gas development. *Wyoming Law Review, 13*, 5.

Jeffares, S. (2014). *Interpreting hashtag politics: Policy ideas in an era of social media*. Basingstoke, UK: Palgrave Macmillan.

Jenkins, H. (2004). Corporate social responsibility and the mining industry: Conflicts and constructs. *Corporate Social Responsibility and Environmental Management, 11*(1), 23–34.

Kane, G. C., Fichman, R. G., Gallaugher, J., & Glaser, J. (2009). Community relations 2.0. *Harvard Business Review, 87*(11), 45–50.

Kemp, D. (2010). Community relations in the global mining industry: Exploring the internal dimensions of externally orientated work. *Corporate Social Responsibility and Environmental Management, 17*(1), 1–14.

Kemp, D., & Owen, J. R. (2013). Community relations and mining: Core to business but not "core business." *Resources Policy, 38*(4), 523–531. doi:10.1016/j.resourpol.2013.08.003.

Lacey, J., & Lamont, J. (2013). Using social contract to inform social licence to operate: An application in the Australian coal seam gas industry. *Journal of Cleaner Production*. doi:10.1016/j.jclepro.2013.11.047.

McKeown, B., & Thomas, D. (2013). *Q Methodology* (2nd ed.). Thousand Oaks, CA: Sage.

Moffat, K., & Zhang, A. (2014). The paths to social license to operate: An integrative model explaining community acceptance of mining. *Resources Policy, 39*, 61–70. doi:10.1016/j.resourpol.2013.11.003

Owen, J. R., & Kemp, D. (2013). Social licence and mining: A critical perspective. *Resources Policy, 38*(1), 29–35.

Parsons, R., & Moffat, K. (2014). Constructing the meaning of social licence. *Social Epistemology, 28*(3–4), 340–363. doi:10.1080/02691728.2014.922645

Porter, M. E., Gee, D. S., & Pope, G. J. (2015). *America's unconventional energy opportunity: A win-win plan for the economy, the environment, and a lower-carbon, cleaner-energy future.* Cambridge, MA: Harvard Business School and Boston Consulting Group.

Stephenson, W. (1980). Consciring: A general theory for subjective communicability. In D. Nimmo (Ed.), *Communication Yearbook 4.* Columbia, MO: University of Missouri.

Thomson, I., & Boutilier, R. (2011). The social license to operate. In P. Darling (Ed.), *SME Mining Engineering Handbook* (pp. 1779–1796). Littleton, CO: Society of Mining Metallurgy and Exploration.

Victorian Auditor General's Office. (2015). *Unconventional Gas: Managing Risks and Impacts.* Melbourne, Australia. Retrieved September 2015, from http://www.audit.vic.gov.au/publications/20150819-Unconventional-gas/20150819-Unconventional-gas.pdf.

23

Phytomining: Using Plants to Extract Valuable Metals from Mineralized Wastes and Uneconomic Resources

Philip N. Nkrumah, Guillaume Echevarria, Peter D. Erskine, and Anthony van der Ent

CONTENTS

23.1 Introduction

Hyperaccumulators are plants that share the ability to grow on metalliferous soils and to accumulate exceptional concentrations of specific metallic and metalloid elements in their shoots (Reeves 2003; van der Ent et al. 2013a). These plants can be utilized as "metal crops" and grown on unconventional resources to recover strategic metals in phytomining (also called "agromining") operations (Chaney et al. 1998, 2007; Baker et al. 2010; van der Ent et al. 2013a, 2015a). Although there are currently over 700 plant species identified as metal hyperaccumulators, including arsenic, cadmium, cobalt, copper, manganese, zinc, nickel, thallium, and selenium, over 70% are nickel hyperaccumulators. Furthermore, subeconomic nickel contaminated and mineralized soils cover extensive land areas; hence most research has focused on the development of nickel phytomining. Large-scale demonstration of nickel phytomining with *Alyssum murale* (yellowtuft) has been undertaken in United States and Albania (Li et al. 2003; Bani et al. 2015a), and substantial potential exists in tropical regions using native "metal crops" (van der Ent et al. 2013b; Nkrumah et al. 2016). Here we discuss the status of phytomining operations with a particular focus on nickel, and highlight the progress our research team has made in providing "real-life" evidence of tropical phytomining.

23.2 Phytomining Technology

Phytomining relies on hyperaccumulators to extract metals in biomass for economic gain rather than pollution remediation (Chaney 1983; Brooks et al. 1998). In this approach hyperaccumulator plants are grown over (spatially large) subeconomic ore bodies or ultramafic soils followed by harvesting and incineration of the biomass to produce a commercial high-grade bio-ore. In the case of nickel, the bio-ore may be turned into a range of different nickel products that may include nickel metal, nickel-based catalysts, and pure nickel salts (Chaney et al. 2007; Barbaroux et al. 2012; van der Ent et al. 2015a).

The nickel hyperaccumulator, *Streptanthus polygaloides* (Brassicaceae), recovered 100 kg ha^{-1} nickel from ultramafic substrates in initial experiments (Nicks and Chambers 1995, 1998). Similar success was achieved with other hyperaccumulators: for nickel with *Alyssum bertolonii* (Brassicaceae) in Italy (72 kg ha^{-1} nickel) by Robinson et al. (1997) and *Berkheya coddii* in South Africa (100 kg ha^{-1} nickel); with a thallium hyperaccumulator, *Biscutella laevigata* (Brassicaceae) in France (8 kg ha^{-1} thallium); and with gold by induced hyperaccumulation using ammonium thiocyanate in *Brassica juncea* (Brassicaceae) (up to 57 mg kg^{-1} gold per plant) (Anderson et al. 1998, 1999, 2005). Large-scale field trials with *A. murale* (Figure 23.1) suggest that >100 kg ha^{-1} nickel could be achieved (Li et al. 2003; Bani et al. 2015a).

Societal pressure to reduce the environmental impacts of conventional mining, technical difficulties in economic recovery of metals from low-grade ores, and high metal prices have contributed to an increased interest in phytomining research (Harris et al. 2009). Compared to strip-mining operations, phytomining has an environmental impact like agriculture or agroforestry and it does not require mine site rehabilitation at the end of life (Harris et al. 2009). Phytomining might also generate renewable energy from the incineration of plant biomass before smelting or hydrometallurgical refining (Anderson et al. 1999). For a detailed life cycle assessment of phytomining supply chain, see Rodrigues et al. (2016).

FIGURE 23.1
Large-scale demonstration of phytomining of nickel with *Alyssum murale* growing on ultra-mafic substrate in the Balkans (Albania).

23.3 Suitable Sites for Phytomining

The cultivation of "metal crops" could be undertaken on large metal-rich surface areas. For nickel, cultivation is feasible on ultramafic areas with suitable topography, where soils are of poor normal agricultural utility, or degraded nickel-rich land which includes nickel laterite mine sites, smelter-contaminated areas, and ore beneficiation tailings (van der Ent et al. 2015a). Criteria to consider for use of land for phytomining include the ownership arrangements, location outside any protected areas/nature reserves, road accessibility, slope aspect, availability of (gravity) irrigation, and local soil properties.

23.4 Selection of "Metal Crops" for Nickel Phytomining Operations

Only hyperaccumulator plant species that accumulate reasonably high concentrations of nickel (>1 wt. %, but preferably >2 wt. %) in their biomass are suitable as a "metal crop." Other desirable traits include a high growth-rate and high biomass of the shoot, the ability to thrive in exposed conditions, low irrigation requirements, ease of mass propagation, resistance to disease, and so forth. Table 23.1 lists the nickel species that have been identified as having especially high potential as "metal crops."

TABLE 23.1

Nickel Metal Crop Species with High Potential for Application as Metal Crops in Nickel Phytomining Operations

Plant Species	Potential Application Area	Native Distribution	Height (m)	Cropping System	Shoot Nickel (wt. %)	References
Alyssum spp.	Mediterranean and Eurasian Region	S & SE Europe, Turkey, Armenia, Iraq, Syria	0.5–1	Perennial herb	1–2.5	(Brooks 1998)
Buxus spp.	Tropical Central America	Cuba	0.3–12	Ligneous shrub	1–2.5	(Reeves et al. 1996)
Phyllanthus spp.	Tropical Asia-Pacific Region	Southeast Asia and Central America	1–6	Ligneous shrub	2–6	(Baker et al. 1992; van der Ent et al. 2015b)
Rinorea bengalensis	Tropical Asia-Pacific Region	Southeast Asia	5–20	Ligneous shrub	1–2.7	(Brooks and Wither 1977)
Berkheya coddii	Southern Africa	South Africa, Zimbabwe	1–2	Perennial herb	1.1	(Morrey et al. 1989)
Pearsonia metallifera	Southern Africa	Zimbabwe	0.35–1.5	Perennial herb	1.4	(Wild 1974)

Source: From Nkrumah, P. N. et al., *Plant Soil*, 406, 55–69, 2016.

Although local plant species are recommended because of their adaptation to local climatic and edaphic conditions (Baker 1999; Bani 2007), there are two species, *A. murale* (Brassicaceae), originating from the Balkans, and *Berkheya coddii* (Asteraceae), originating from South Africa that may be regarded as universal "metal crops" which could be widely used in Mediterranean and Steppe climates, respectively. However, careful climatic matching remains important, as an experiment with *Alyssum* spp. in Indonesia did not yield any useful outcome (van der Ent et al. 2013b). Therefore, the potential exists for *Alyssum* spp. to find application in phytomining operations in nickel-enriched soils of Australia (Queensland), China, Balkans, Iran, Greece, Russia, Turkey, and the Unites States, while *B. coddii* might be utilised in Brazil, South Africa, the United States, and Zimbabwe. We stress that the possible introduction of these species in these locations must comply with applicable national biosecurity legislation and appropriate crop management. For example, poor management subsequent to the scientific trials with *A. murale* in the United States resulted in the species becoming invasive and eventually being listed as a noxious weed in Oregon (USDA 2015).

23.5 Insights from Laboratory and Field Tests to Maximise Nickel Yields

High biomass production and shoot nickel content of "metal crops" are very important considerations in nickel phytomining. Appropriate agronomic systems have been proposed to maximize the yields of the selected "metal crop" (Li et al. 2003; Bani 2007; Bani et al. 2015a; Nkrumah et al. 2016). Inorganic fertilization plays a significant role in maximizing the growth and metal yield of "metal crops" (Li et al. 2003; Bani 2007; Bani et al. 2015a; Álvarez-López et al. 2016). Phosphorus appeared to have a strong effect on the biomass yield and nickel uptake by hyperaccumulator species growing on soil not previously fertilized, while previously fertilized soils show a lesser response to phosphorus fertilization (Robinson et al. 1997). In Albania, Bani et al. (2015a) found *A. murale* biomass yield was increased 10-fold with 120 kg NPK fertilizer and 77 kg calcium ha^{-1} plus monocot herbicide to control grasses. Furthermore, these agronomic practices increased nickel phytoextraction yield from 1.7 to 105 kg ha^{-1}.

Broadhurst and Chaney (2016) investigated the effect of organic matter amendments on growth and metal yield of *A. murale*. The authors observed negligible effect on the biomass and yield of the "metal crop." However, the extreme conditions of some substrates (e.g., industrial waste material or mine spoil) may require the use of organic amendments to improve soil fertility (Séré et al. 2008; Chaney and Mahoney 2014).

Furthermore, the effect of soil pH on nickel accumulation in *Alyssum* spp. is unusual. Whereas increasing soil pH reduces the solubility of nickel, and hence reduces nickel concentration in "normal" crop plant species (Kukier and Chaney 2001), the nickel concentration in the biomass of *Alyssum* increased as soil pH was raised depending on soil properties (Nkrumah et al. 2016).

Beyond fertilizer treatment and pH adjustment, many plant management practices need to be employed to enhance metal yields in phytomining. First, plant density is important to optimize biomass production per unit area, and evidence suggests intermediate density results in optimum nickel yield (Angle et al. 2001; Bani et al. 2015b). Second, weed control minimizes the competition between the "metal crop" and weeds for essential nutrients and water (Chaney et al. 2007; Bani et al. 2015a). Third, Plant Growth Promoting Rhizobacteria (PGPR) might be an interesting option as some PGPR isolated from the native rhizosphere of hyperaccumulators were shown to significantly improve the phytoextraction yield of hyperaccumulator plants grown in inoculated soils in pot experiments (Durand et al. 2016). The management of propagation and harvest will necessarily be dependent upon the species being used for phytomining (Nkrumah et al. 2016).

23.6 Processing of Nickel Biomass and Bio-Ore

Early nickel phytomining trials employed an arc furnace to smelt nickel metal from the bio-ore (Chaney et al. 2007). Recent studies suggest other methods could further capitalize on the "biopurity" of the bio-ore to increase the profitability of nickel phytomining. For instance, the hydrometallurgical processing method could be a suitable alternative to derive higher value products from the bio-ore: (1) nickel catalysts for the organic chemistry industry (Losfeld et al. 2012) and (2) nickel chemicals for the electroplating industry (Barbaroux et al. 2012). Research is still needed to explore more efficient methods to synthesize nickel products from the biomass.

23.7 Tropical Phytomining

Tropical regions (e.g., in the Asia-Pacific region, Indonesia, Malaysia, Philippines, Papua New Guinea, and New Caledonia) have the greatest potential for phytomining as large expanses of ultramafic soils exist (van der Ent et al. 2013b, 2016). Phytomining operations in this region may be a complementary process to existing mining operations, as part of the progressive rehabilitation process after conventional resource extraction. Agromining could also replace existing marginal agriculture on poor ultramafic soils. The application of agromining is envisaged to provide opportunities for an income source for communities in Malaysia, Indonesia, and the Philippines as an alternative type of agriculture or agroforestry pursuit ("farming for nickel").

Substantial progress is currently being made in developing nickel phytomining in the Asia-Pacific Region by our research team. We have recently discovered over 20 new hyperaccumulator plant species in Sabah (Malaysia) and Halmahera (Indonesia), which is indicative of the very high potential of this untapped resource in the Asia–Pacific Region (van der Ent et al. unpublished data). From the different hyperaccumulator plant species that have been discovered, suitable "metal crops" are now being selected for agronomic trials to assess growth performance, fertilizer requirements, and sustained nickel yields of the crops when successively harvested. Suitable "metal crops" are selected based on their relative growth rates, nickel accumulation, and effective propagation methods. Currently, experimental studies (Figure 23.2) are undertaken to establish optimal agronomic systems to stimulate biomass production and nickel yield in two prospective species: *Phyllanthus securinegoides* (Phyllanthaceae) and *Rinorea bengalensis* (Violaceae).

FIGURE 23.2
Pot trial undertaken over a period of 12 months in Sabah, Malaysia, with *Phyllanthus securinegoides* (small leaf blades) and *Rinorea bengalensis* (large leaves) to determine optimal agronomic systems for tropical nickel phytomining.

23.8 Potential Lifespan and Economics of Nickel Phytomining

As with all methods for resource extraction, phytomining will be finite due to the diminishing concentrations of nickel in the zone accessible by plant roots (Chaney et al. 2014a, b). Nevertheless, considering soil materials with 0.2 wt. % nickel and "metal crops" with a yield of 100 kg nickel ha^{-1} the phytomining venture may be sustainable for decades (van der Ent et al. 2015a). Table 23.2 presents the economic potential of nickel phytomining.

We summarize the economic analysis under two main production systems: (1) an intensive system such as demonstrated in the United States (Li et al. 2003), and (2) an extensive system as demonstrated in Albania (Bani et al. 2015a). Here we define an intensive system as a fully mechanized production system where the cost of operation includes costs for seed stock, fertilizers, labor, and equipment, whereas an extensive system mainly employs manual labor in its operation, and the production cost involves the use of fertilizers, herbicides, and complementary agricultural management practices. The production costs in the extensive system are relatively low, and this system is recommended for places with readily available and relatively low-cost manual labor. It is evident that nickel phytomining is a highly profitable agricultural technology for the respective systems. The profitability could increase when recovery of energy of combustion and sale of carbon credits are considered. We stress that nickel metal product is profitable; however, other higher value nickel products such as pure nickel salts may further increase the profitability of nickel agromining in the near future.

TABLE 23.2

Economic Analysis of an Annual Nickel Phytomining Crop

Expense and Income Category	Intensive System (ha^{-1} yr^{-1})	Extensive System (ha^{-1} yr^{-1})
Cost of production in 2016	$1074	$600
Cost of metal recovery	$720	$396
Gross value	$3600	$1980
Net value	$1806	$984

Economic analysis of an annual nickel phytomining crop per ha under two main production systems: (1) an intensive system such as demonstrated in the USA (Li et al. 2003a) and (2) an extensive system as demonstrated in Albania (Bani et al. 2015a). The cost of production in the intensive system is high, including costs for seed stock, fertilisers, labour and equipment, whereas the production costs in the extensive system are relatively low because it mainly involves the use of fertilisers, herbicides and complementary agricultural management practices. The annual crop nickel yield for an intensive system and an extensive system are 200 and 110 kg ha^{-1}, respectively. The commercial value of nickel of $18 per kg was estimated as an average value of nickel over a period of 5 years (2010–2015) at the London Metal Exchange. The cost of metal recovery was estimated at 20% of nickel value.

23.9 Conclusions

Phytomining technology has been successfully demonstrated in Mediterranean and temperate climates for nickel using *Alyssum* spp., and our ongoing research in Southeast Asia using *P. securinegoides* and *R. bengalensis* will be critical to provide "real-life" evidence of tropical phytomining. The nickel mining industry needs to test phytomining as a supplement to traditional mining as it uses only a small portion of sub-economic ultramafic soil deposits and could be highly profitable. Nickel phytomining will also improve soil fertility and reduce toxicity due to soil nickel; this is a significant service rendered through phytomining which we then defined as agromining. As such it will make the land suitable for other future usage, including forestry and some types of traditional agriculture. It is envisaged that agromining could also support local livelihoods with income opportunities as an alternative type of agriculture: to farm nickel. The demonstration of phytomining other strategic elements (e.g., cobalt, manganese, rare earths) is underway and should use the same general approach as nickel phytomining.

Acknowledgments

P. N. Nkrumah is the recipient of an Australian Government Research Training Program Scholarship at The University of Queensland. A. van der Ent is the recipient of an Australian Research Council Post Doctoral Fellowship (DE160100429).

References

Álvarez-López, V., Á. Prieto-Fernández, M. I. Cabello-Conejo, and P. S. Kidd. 2016. Organic amendments for improving biomass production and metal yield of Ni-hyperaccumulating plants. *Science of Total Environment* 548–549:370–379.

Anderson, C. W. N., R. R. Brooks, A. Chiarucci et al. 1999. Phytomining for nickel, thallium and gold. *Journal of Geochemical Exploration* 67(1–3):407–415.

Anderson, C. W. N., R. R. Brooks, R. B. Stewart, and R. Simcock. 1998. Harvesting a crop of gold in plants. *Nature* 395(6702):553–554.

Anderson, C., F. Moreno, and J. Meech. 2005. A field demonstration of gold phytoextraction technology. *Minerals Engineering* 18(4):385–392.

Angle, J. S., R. L. Chaney, A. J. M. Baker et al. 2001. Developing commercial phytoextraction technologies: Practical considerations. *South African Journal of Science* 97(11–12):619–623.

Baker, A. J. M. 1999. *Revegetation of asbestos mine wastes.* Princeton Architectural Press, New York.

Baker, A. J. M., W. H. O. Ernst, A. van der Ent, F. Malaisse, and R. Ginocchio. 2010. Metallophytes: The unique biological resource, its ecology and conservational status in Europe, central Africa and Latin America. In *Ecology of Industrial Pollution,* L.C. Batty, K.B. Hallberg (Eds.). Cambridge University Press, Cambridge, UK, pp. 7–40.

Baker, A. J. M., J. Proctor, M. M. J. Van Balgooy, and R. D. Reeves. 1992. Hyperaccumulation of nickel by the flora of the ultramafics of Palawan, Republic of the Philippines. In *The Vegetation of Ultramafic (Serpentine) Soils,* A. J. M. Baker, J. Proctor, R. D. Reeves (Eds.). Intercept, Andover, UK, pp. 291–304.

Bani, A. 2007. In-situ phytoextraction of Ni by a native population of *Alyssum murale* on an ultramafic site (Albania). *Plant and Soil* 293:79–89.

Bani, A., G. Echevarria, S. Sulçe, and J. L. Morel. 2015a. Improving the agronomy of *Alyssum murale* for extensive phytomining: A five-year field study. *International Journal of Phytoremediation* 17(1–6):117–127.

Bani, A., G. Echevarria, X. Zhang et al. 2015b. The effect of plant density in nickel-phytomining field experiments with *Alyssum murale* in Albania. *Australian Journal of Botany* 63:72–77.

Barbaroux, R., E. Plasari, G. Mercier, M. O. Simonnot, J. L. Morel, and J. F. Blais. 2012. A new process for nickel ammonium disulfate production from ash of the hyper-accumulating plant *Alyssum murale.Science of Total Environment* 423:111–119.

Broadhurst, C. L., and R. L. Chaney. 2016. Growth and metal accumulation of an *Alyssum murale* nickel hyperaccumulator ecotype co-cropped with *Alyssum montanum* and perennial ryegrass in serpentine soil. *Frontiers in Plant Science* 7:451.

Brooks, R. R. 1998. *Plants that Hyperaccumulate Heavy Metals: their Role in Phytoremediation, Microbiology, Archaeology, Mineral Exploration, and Phytomining.* CAB International, Wallingford, UK.

Brooks, R. R., and E. D. Wither. 1977. Nickel accumulation by *Rinorea bengalensis* (Wall.) O.K. *Journal of Geochemical Exploration* 7:295–300.

Brooks, R. R., M. Chambers, L. Nicks, and B. H. Robinson. 1998. Phytomining. *Trends Plant Science* 3:359–362.

Chaney, R. L. 1983. Plant uptake of inorganic waste constituents. In *Land Treatment of Hazardous Wastes,* J.F. Parr, P.B. Marsh, Kla J.M (Eds.). Noyes Data Corp, Park Ridge, New Jersey, pp. 50–76.

Chaney, R. L., and M. Mahoney. 2014a. Phytostabilization and phytomining: Principles and successes. Paper 104, *Proceedings of the Life of Mines Conference,* July 15–17. Australasian Institute of Mining and Metallurgy, Brisbane, Australia.

Chaney, R. L., I. Baklanov, T. Centofanti et al. 2014b. Phytoremediation and phytomining: Using plants to remediate contaminated or mineralized environments. In *Plant Ecology and Evolution in Harsh Environments,* N. Rajakaruna, R.S. Boyd, and T. Harris (Eds.). Nova Science Publishers, New York, pp. 365–391.

Chaney, R. L., J. S. Angle, A. J. M. Baker, and Y-M. Li. (1998). Method for phytomining of nickel, cobalt and other metals from soil. U.S. Patent 1998, 5, 711,784.

Chaney, R. L., J. S. Angle, C. L. Broadhurst, C. A. Peters, R. V. Tappero, and D. L. Sparks. 2007. Improved understanding of hyperaccumulation yields commercial phytoextraction and phytomining technologies. *Journal of Environmental Quality* 36(5):1429–1433.

Durand, A., S. Piutti, M. Rue, J. L. Morel, G. Echevarria, and E. Benizri. 2016. Improving nickel phytoextraction by co-cropping hyperaccumulator plants inoculated by plant growth promoting rhizobacteria. *Plant and Soil* 399:179–192.

Harris, A. T., K. Naidoo, J. Nokes, T. Walker, and F. Orton. 2009. Indicative assessment of the feasibility of Ni and Au phytomining in Australia. *Journal of Cleaner Production* 17(2):194–200.

Kukier, U., and R. L. Chaney. 2001. Amelioration of nickel phytotoxicity in muck and mineral soils. *Journal of Environmental Quality* 30(6):1949–1960.

Li, Y. M., R. L. Chaney, E. Brewer et al. 2003. Development of a technology for commercial phytoextraction of nickel: economic and technical considerations. *Plant and Soil* 249:107–115

Losfeld, G., V. Escande, T. Jaffré, L. L'Huillier, C. Grison. 2012. The chemical exploitation of nickel phytoextraction: An environmental, ecologic and economic opportunity for New Caledonia. *Chemosphere* 89(7):907–910.

Morrey, D. R., K. Balkwill, and M. J. Balkwill. 1989. Studies on serpentine flora—Preliminary analyses of soils and vegetation associated with serpentinite rock formations in the Southeastern Transvaal. *South African Journal of Botany* 55:171–177.

Nicks, L., and M. Chambers. 1995. Farming for metals. *Mining and Environmental Management* 3:15–18.

Nicks, L., and M. Chambers. 1998. Pioneering study of the potential of phytomining for nickel. In *Plants that Hyperaccumulate Heavy Metals: Their Role in Phytoremediation, Microbiology, Archaeology, Mineral Exploration and Phytomining*, R. R. Brooks (Ed.). CAB International, Wallingford, UK.

Nkrumah, P. N., A, J. M. Baker, R. L. Chaney et al. 2016. Current status and challenges in developing nickel phytomining: An agronomic perspective. *Plant and Soil* 406(1):55–69.

Reeves, R. D. 2003. Tropical hyperaccumulators of metals and their potential for phytoextraction. *Plant and Soil* 249:57–65.

Reeves, R. D., A. J. M. Baker, A. Borhidi, and R. Berazain. 1996. Nickel-accumulating plants from the ancient serpentine soils of Cuba. *New Phytologist* 133:217–224.

Robinson, B. H., R. R. Brooks, A. W. Howes, J. H. Kirkman, and P. E. H. Gregg. 1997. The potential of the high-biomass nickel hyperaccumulator Berkheya coddii for phytoremediation and phytomining. *Journal of Geochemical Exploration* 60(2):115–126.

Rodrigues, J., V. Houzelot, F. Ferrari et al. 2016. Life cycle assessment of agromining chain highlights role of erosion control and bioenergy. *Journal of Cleaner Production* 139:770–778.

Séré, G., C. Schwartz, S. Ouvrard, C. Sauvage, J. C. Renat, and J. L. Morel. 2008. Soil construction: A step for ecological reclamation of derelict lands. *Journal of Soils and Sediments* 8:130–136.

USDA. 2015. United States Department of Agriculture, Natural Resources Conservation Service, Plants Profile. http://plants.usda.gov/core/profile?symbol=ALMU (accessed December 24, 2015).

van der Ent, A., A. J. M. Baker, M. M. J. Van Balgooy, and A. Tjoa. 2013b. Ultramafic nickel laterites in Indonesia: Mining, plant diversity, conservation and nickel phytomining. *Journal of Geochemical Exploration* 128:72–79.

van der Ent, A., A. J. M. Baker, R. D. Reeves, et al. 2015a. "Agromining": Farming for metals in the future? *Environmental Science and Technology* 49(8):4773–4780.

van der Ent, A., D. R. Mulligan, R. Repin, and P. D. Erskine. 2017. Foliar elemental profiles in the ultramafic flora of Kinabalu Park (Sabah, Malaysia). *Ecological Research* [In Press].

van der Ent, A., G. Echevarria, and M. Tibbett. 2016. Delimiting soil chemistry thresholds for nickel hyperaccumulator plants in Sabah (Malaysia). *Chemoecology* 26(2):67–82.

van der Ent, A., R. D. Reeves, A. J. M. Baker, J. Pollard, and H. Schat. 2013a. Hyperaccumulators of metal and metalloid trace elements: facts and fiction. *Plant and Soil* 362(1–2):319–334.

Wild, H. 1974. Indigenous plants and chromium in Rhodesia. *Kirkia* 9(2):233–241.

24

Vignette: The Eden Project—Innovative Restoration of a Mine Site for Tourism

Saleem H. Ali

Mined for metal and industrial minerals like clay, the Cornish landscape has been altered by humanity for centuries, and not until the turn of the twenty-first century has there been a concerted effort to rehabilitate the damage. Pits of china clay have dominated the mid-Cornwall region and the wastes generated from the extraction have been immense to the extent that artificial hills have been created over vast expanses of the region and are called the "Cornish Alps."[1] The goal was to reclaim the mining heritage of this area by not just building a museum as a tribute to what had occurred in the past, but rather to celebrate the prospects for a sustainable future. Establishing a productive economic enterprise, albeit as a charity, was still very much part of the equation for Tim Smit, the founder of the project, which coincided with Britain's "millennium" development efforts in 2000. Unlike the billion-dollar dome that was funded by the Millennium Commission and faltered into insolvency in 2003, the Eden project flourished and became an icon restoration success. The organization continues to be largely self-reliant to this day despite occasional rumblings of discontent about excessive traffic and congestion from the influx of tourists to this remote rural area.

A 12-acre gray pit that was a blight to any aesthetic eye was the property with which the designers had to work. Clay pits do not usually have to contend with heavy metal pollution concerns, nor do they have the impending threat of acidic drainage that is a foremost challenge to remediation efforts for metal and coal mines. The easiest reclamation idea would have been to vegetate the area and create a sort of sunken garden with perhaps a lake in the deepest portions of the cavity. One of the world's finest gardens, The Butchart in Victoria, Canada, had the pedigree of a former quarry as well. An earlier mine reclamation in Cornwall had followed a similar path by creating "The Lost Garden of Heligan" a few years earlier. The Bodiva clay pit would have made for an interesting garden excursion, but generating large-scale tourist revenues from such a reclamation needed more than just

---

[1] Details of the Cornwall Rehabilitation efforts taken from an editorial article in Mining Environmental Management Magazine (January 2004).

a garden. However, the theme of renewability was vitally important to the designers of the project and plants were central to any process of terrestrial biotic renewal.

Botany and engineering have a natural convergence in greenhouses, and so the Eden project designers set out to construct the world's largest greenhouse in the Bodiva pit. Shaped as a series of huge domes within the pit, the structure might appear to some as symbolic of eggs being hatched out of the mine's womb. The spherical shape with pentagonal and hexagonal structural geometries was also reminiscent of the geodesic structures envisaged by Buckminster Fuller (1895–1983). Much of Fuller's erratic but productive life had been devoted to understanding chances of human survival with constrained resources and how to engineer our way out of intractable environmental challenges. He is perhaps best known for coining the term "Spaceship Earth," which became an evocative metaphor for the environmental movement.[2] The United States pavilion at the 1967 Montreal exposition was designed by Fuller and now stands as a kind of greenhouse museum operated by Environment Canada.

Using the greenhouse to cultivate exotic plants can also have an important role in restoration research itself. Often, selective plants, fungi, and their symbiotic bacteria can play a cleansing role in ecosystems depending on the kind of pollutant in play. For sites contaminated with metals, a range of plants such as those of the genus *Cnidoscolus* from Brazil have been known to be "attracted" to metals and can metabolize elements like nickel.[3] Growing such plants for experiments in restoration efforts in a facility like the Eden project with a clear mandate for post-mining research is a productive enterprise on its own.

Keeping the pit stable with all this engineering being undertaken required a major investment in slope stabilization on the walls of the cavity, which took 2 years of planning. The designers also wanted to show how the 85,000 tons of topsoil needed for the greenhouse plants could be locally produced. So they set out to partner with soil scientists to come up with a suitable recipe using the waste materials within the pit. This ecologically responsible approach also saved the builders thousands of dollars in production and transportation cost for the topsoil. Architect Nicholas Grimshaw was quite

---

[2] A good exposition of such plants can be found in an anonymous editorial story titled "Mining, metallophytes and land reclamation," *Mining Environmental Management Magazine* (March 2002, pp. 11–16).

[3] A competitive ecological challenge prize in Fuller's honor, worth $100,000 was inaugurated in 2008 and the first recipient was applied biologist John Todd, whose proposal for remediating coal mining lands was commended by the jury for "weaving together a set of processes—from restoration of land to geosequestration of carbon, to community involvement, to long-term economic vitality—to create a blueprint for a future for Appalachia that envisions a harmonious self-sustaining community." Details at: http://www.bfi.org. For an account of the life and times of Buckminster Fuller see Lloyd Steven Sieden, *Buckminster Fuller's Universe: His Life and Work* (New York: Basic Books, 2000).

intent on developing the structures of the project to mimic nature in various ways. For example, the latest addition to the project is a copper-roofed building called "The Core," which has been designed geometrically with the botanic principles of phyllotaxis. The biomimicry of the design follows a mathematical structure that most plants use for their growth and is manifest in the opposing spiral structures in sunflower heads, pineapples, and pine cones. The copper in the room has also come from traceable mine sources that were mined according to criteria set forth by various environmental and social advisors to the project.

25

The Frugal Rehabilitation Methodology for Artisanal and Small-Scale Mining in Mongolia: An Innovative Approach to Formalization and Environmental Governance with Potential for International Adaptation through the BEST-ASM Initiative

Jonathan Stacey, Yolande Kyngdon-McKay,
Estelle Levin-Nally, and Andrew Cooke

CONTENTS

25.1 Introduction

Innovation in the environmental management of artisanal and small-scale mining (ASM) has gained notable traction in the past few years. The sector's growing recognition as a legitimate economic contributor has encouraged select governments, civil society organizations, development agencies, and social ventures to pioneer new methodologies for bringing the sector's environmental practices more in line with the United Nation's Sustainable Development Goals, including Responsible Consumption and Production, Life on Land, Sustainable Cities and Communities, and Climate Action. However, despite this recent positive trend, ASM typically remains the lowest priority in any conservation situation vis-à-vis logging, forest clearance for plantations, agriculture, charcoal making, and bushmeat hunting.

The Frugal Rehabilitation Methodology (FRM) and the Biodiversity and Ecosystem Services Transformative ASM ("BEST-ASM") initiatives are ASM environmental management programs that can help to address this oversight by pragmatically institutionalizing sustainable environmental management in the ASM sector. This chapter discusses the design of the FRM based on the successful development and implementation of Frugal Rehabilitation Demonstration (FRD) on ASM-degraded lands in Mongolia, and how the BEST-ASM initiative is seeking to broaden the application of the FRM, along with additional methodologies shown to improve the ASM sector's environmental management, to additional active ASM sites around the world.

25.2 Background: History and Development of the Artisanal and Small-Scale Mining Sector in Mongolia

Following the withdrawal of Soviet economic support from Mongolia during the early 1990s, a sudden transition to renewed economic independence forced the country to look to the value of its natural resources and forge its place as a free market economy. Its mineral reserves—of coal, copper, and gold in particular—were to play a key role in its economic positioning, not only with respect to its Russian and Chinese neighbors but also to the global mining community (World Bank 2009). During the 1990s and early 2000s, over 3,500 mining exploration licenses and 1,000 extractive licenses were issued across the country. This licensing was done partly to initiate economic growth through the establishment of large-scale mining activities in Mongolia but also to generate the financial resources required to support the establishment of a new democratically elected government (BirdLife Asia 2009). The withdrawal of Soviet economic support resulted in increased unemployment in both urban and rural environments. During this transition period, with Mongolia's rural economy still very much dependent on traditional livestock herding, the country experienced significant economic hardships, exacerbated by *dzud* (winter storm) events that decimated livestock herds (Villegas 2013). These events were key drivers behind the emergence of informal ASM[1] across the country, as unemployed communities in the countryside desperately sought out alternative and accessible options for livelihood (Swiss Agency for Development and Cooperation).

The ASM gold "rushes" that took place between 1999 and 2002 were particularly triggered by *dzud* events, during which time a total of 11 million livestock were lost. ASM had suddenly become the only employment opportunity for herders who had lost their livelihoods. The ASM sector grew further when the global gold price trended upwards throughout the early 2000s, and spiked following the Global Financial Crisis of 2008–2009 (Carlson 2008). This series of events saw the establishment of a widespread informal ASM sector in Mongolia, largely focused on alluvial gold, but also extending locally to hard-rock gold, coal, and fluorspar mining.

While the Mongolian government made progress in developing and implementing mining legislation with the 2006 Minerals Law, there was no legal provision recognizing ASM, or requiring its formalization. ASM had nonetheless increased in geographic scope and activity across the country, leading to a range of problems impacting communities and other stakeholders, such as land degradation and broader environmental impacts (an analysis of which will be the focus of this paper).

The Sustainable Artisanal Mining (SAM) Project, introduced by the Swiss Agency for Development and Cooperation (SDC) in 2005, sought to encourage ASM formalization in Mongolia (Swiss Agency for Development and Cooperation (SDC) 2011). Through a number of phased initiatives, SAM aimed to assist in the organization, formalization and institutionalization of the sector, as well as the building of appropriate technological and health and safety capacity (Swiss Agency for Development and Cooperation (SDC) 2015a, 2016a, 2016b). This included an improvement in job security, and an appreciation of human rights for both artisanal miners and impacted stakeholders. Furthermore, SAM supported the development of a regulatory framework for the ASM sector (in collaboration with key stakeholders), to enable informal ASM to enter the legal sphere. As a result of SAM's efforts, ASM was incorporated into Mongolia's revised Minerals Law of 2010, (Swiss Agency for Development and Cooperation (SDC) 2015b) cementing its role in the country's mining industry.

Official statistics suggest that 38,000 artisanal and small-scale miners are currently active in Mongolia, whereas unofficial estimates suggest a number closer to 100,000, with an additional 400,000 people indirectly dependent on the sector (Swiss Agency for Development and Cooperation (SDC)). Despite the abovementioned legislation there was a perceived need to improve it to address environmental performance of the ASM sector.

---

[1] ASM is a major source of mineral resources production in the world. It is largely informal, and is associated with low levels of safety measures, health care or environmental protection. http://mneguidelines.oecd.org/artisanal-small-scale-miner-hub.htm.

25.3 Rationale: Targeting Environmental Performance for an Improved Artisanal and Small-Scale Mining Sector

Poor environmental performance is a primary obstacle to the formalization of ASM in Mongolia. This obstacle underpinned the rationale for targeting environmental performance as a gateway to enhanced integration and formalization of ASM in the country. By 2013, the recognition of Mongolia's ASM sector as a valid economic contributor had also highlighted the need to determine why it was plagued by poor environmental practices and negative social impacts, and what effects these were having on the sector's "license to operate".[2] ASM, which is typically a subsistence-level economic activity carried out by atomistic actors, has few incentives in place to make environmental management rational (from a social, economic, or legal perspective). As a result, informal ASM within Mongolia (and around the world) operates with little regard for its environmental impacts. Voluntary rehabilitation, or any other kind of environmental management, is rare. An analysis of ASM in Mongolia through the lenses of formalization and better environmental management was needed to ultimately inform the identification of a range of incentives and disincentives that could encourage further ASM organization and formalization, and the sector's integration into local and national economies.

During SAM's third phase of implementation, a separate but complementary project—the Engaging Stakeholders in Environmental Conservation Project (Phase II) (ESEC II)—was initiated to explore these issues in Mongolia's ASM sector. It was primarily funded by SDC and codesigned and implemented by The Asia Foundation (TAF).[3] Prior to ESEC II launching in 2013, a joint SDC/TAF Project Preparation Mission met with representatives of national, provincial, and local government as well as the large-scale and ASM sectors, to assess and identify the key issues limiting ASM's contribution to sustainable development in Mongolia. Three key causes limiting positive progress in the sector were identified:

---

[2] It is important to note that ASM's negative environmental impacts do not make the sector an outlier in the mining industry; mining has long been associated with poor environmental performance. Until the late 20th century, environmental rehabilitation was the exception rather than the rule, and was seen by all producer groups (ASM to large-scale) as an unnecessary economic burden.

[3] TAF had already had some experience in Mongolia in developing methodologies for measuring and improving environmental health through its ESEC I project. It was well placed to partner with the Mongolian government (Ministry of Mining) and with SDC to develop and implement a project that would complement the SAM project by focusing on the development and implementation of environmental best practices in terms of both an ASM-specific rehabilitation methodology and an environmental governance tool to help ASM integrate into impacted stakeholder communities by demonstrating tangible commitments to environmental responsibility.

- The lack of access to methodologies and green technologies for environmental rehabilitation and associated "best practice" environmental management.
- The lack of local fora for discussing and addressing stakeholder concerns regarding ASM and the environment.
- Limited demonstration of environmental mitigation and rehabilitation at ASM sites.

As a result, the project's mandate was designed to address these issues, aiming to enhance the contribution of Mongolia's ASM sector to sustainable human rights-based local development, including respect for the right to decent work and the right to a healthy environment. The project's emphasis was on developing and demonstrating environmental best practices, through the development of

- A nationally endorsed environmental rehabilitation methodology.
- Human rights-based approaches (HRBA).
- Local stakeholder-inclusive negotiation platforms.
- ASM-inclusive environmental management planning tools.

It was during the project development phase that the concept of Frugal Rehabilitation (FR) was conceived as an accessible approach to environmental rehabilitation of ASM-degraded and abandoned lands. FR was to be characterized by three key attributes. It needed to be:

1. Economically affordable, so that it could be undertaken by artisanal miners with limited access to resources.
2. Socially acceptable, so that the results of rehabilitation addressed the concerns, requirements, and standards of local and national stakeholders.
3. Ecologically viable, so that degraded lands would be left in a condition that was technically stabilized and on the path to an ecological recovery appropriate to locality and ecozone.

ESEC II first needed to understand what technologies, if any, were being utilized for rehabilitation in-country, and which best rehabilitation practices were in use within or across the ASM sector internationally. To do this, the project commissioned a mining specialist consultancy to undertake an international review of environmental rehabilitation approaches for ASM (Butler et al. 2014). The consultant profiled 18 case studies from ASM-producer countries, including Liberia, Central African Republic, Mongolia, Brazil, Mozambique, Sierra Leone, the United States, and Ecuador. As of early 2014, a key finding of the research was that, while some examples of "FR" in an

ASM context existed around the world, they were limited in number and poorly documented. The review indicated that this lack of information was a product of the rarity of environmental rehabilitation in ASM. While the research identified a range of both theoretical and practical approaches and short-term, small-scale pilot projects, it was clear that internationally there was a relative lack of experience in successfully implementing long-term, self-sustaining rehabilitation projects at the ASM scale.

As a result of these findings, ESEC II recognized that in Mongolia it would need to do two things: (1) pioneer environmental rehabilitation in ASM and (2) bring together, for the first time, an array of disparate approaches to land rehabilitation relevant to the local ASM context. By doing so, it could lead this practice and bring innovation and value to the international community by documenting and communicating its approaches, progress, demonstration successes, and lessons learned.

25.4 Project Implementation: The Development of a Frugal Rehabilitation Methodology

The FRM for the ASM sector in Mongolia was designed to enable its outputs and proposed outcomes to be assessed and evaluated against a monitoring and evaluation log framework. The project needed to be aware of various stakeholder perceptions of the ASM sector's performance, as well as have an appreciation of the incentives and disincentives that could be used to facilitate enhanced formalization and environmental responsibility within the ASM sector. It was important that a baseline perception survey be undertaken at an early stage so that the project's impacts and influences could be evaluated when the project was completed in 2016 (in an endline perception survey). The results of this assessment will be discussed later in this chapter.

A key objective was to develop an FRM that would be endorsed by relevant government ministries (Ministry of Mining, associated executive agencies, and the Ministry of Nature, Environment, and Tourism), thus ultimately enabling the FRM to be incorporated into legislation. To ensure that the FRM was based on evidence-based experience relevant to Mongolia's diverse ecosystems, a Frugal Rehabilitation Demonstration (FRD) program was initiated and launched, which would provide the case studies needed to inform the proper development of the FRM.

The scope of the ESEC II project implementation was to cover 45 soums (counties) across nine aimags (provinces). A Ministry of Mining-chaired site selection committee was established, comprising representatives from government ministries and agencies, including the Generalized Agency for Specialized Inspection (GASI), the body responsible for environmental inspectors posted countrywide. Early in the field season, as winter conditions receded and sites became accessible to assessment, field excursions were undertaken to visit a range of sites that were scoped beforehand, in consultation with local government, the Mineral Resources Agency of Mongolia (MRAM), and the Ministry of Environment, as well as locally based formalized ASM nongovernmental organizations (NGOs).

The selection of demonstration sites needed to meet specific criteria to ensure that the rehabilitation site was:

- Logistically accessible for demonstration purposes.
- Representative of ASM product type and approach.
- To be rehabilitated by an ASM NGO that had adequate capacity with organized and formalized status.
- Able to provide value in terms of a range of environmental parameters such as biodiversity and land-use value.
- An ASM degraded site that was abandoned and would not be vulnerable to further informal ASM activity. This latter point was critically important—the rehabilitation investment needed to be secure and serve as a demonstration for FR into the future.

All potential sites were assessed as to their suitability for FRD through the site selection committee. Ten sites were selected for FRD in 2014 and seven sites in 2015. These 17 FRD project sites were in 11 of Mongolia's 16 ecological zones, thus ensuring the project's representativeness in terms of biogeographic context.

In 2014, 10 FRDs were undertaken through program-funded partnerships with 10 ASM NGOs in five aimags: Selenge, Khentii, Dornogobi, Bayankhongor, and Govi-Altai, representing forest steppe, mountain steppe, gobi-steppe, and Khangai mountain steppe environments. Accumulatively, 73 hectares of degraded lands resulting from alluvial and hard-rock artisanal mining of gold and fluorspar were successfully rehabilitated, some of which was in protected areas. These 10 case studies were the basis for the development of the FRM, which was submitted to consultation through a Project Advisory Committee comprised of relevant government ministries, associated agencies, sectoral associations, and civil society NGOs representing human rights, biodiversity conservation, and, of course, the interests of the ASM sector.

Through 2015 a similar process saw a further seven FRD projects selected and completed in the southern and western aimags of Dundgovi, Umnogovi, Khovd, and Uvs, located in a variety of Gobi and Altai ecosystems, again covering portions of protected areas. The findings of the 2015 rehabilitation effort also contributed to minor refinements of the draft FRM, as it proceeded through a process of internal and external consultation and ministerial evaluation. The accumulative area rehabilitated in 2015 was 70 ha, bringing the FRD programme total during 2014–2015 to over 140 ha, at a total cost of U.S. $390,000. FR costs averaged U.S. $2750/ha.

In April 2016, the *Frugal Rehabilitation Demonstration Case Studies Handbook* (Ministry of Mining, Mongolia 2016a) was published, documenting the 17 FRD projects. This publication was accompanied by the launch of the *Frugal Rehabilitation Methodology Field Handbook*, (Ministry of Mining, Mongolia, Swiss Agency for Development and Cooperation (SDC) 2016b) which detailed the stakeholder-approved FRM. The FRM was intended as an attachment to the revised regulation on Small-scale Mining in Mongolia, and at the time of writing this process remains on going.

The FRM is structured around three main components:

a. Technical Rehabilitation

b. Topsoil Management

c. Biological Rehabilitation

The FRM is designed to be implemented either as a response of ASM communities to environmental degradation resulting from their activities, or to assist other stakeholders in rehabilitating abandoned ASM alluvial or hardrock deposits. However, some elements of the methodology could be applied at other mineral or energy deposits, or to rehabilitation efforts outside of the extractives sector.

FR is a series of comprehensive activities designed to improve value and productivity of degraded lands, to recreate acceptable living conditions for local residents and their livestock, and re-establish such lands on a route to ecological recovery and productivity. The methodology is comprised of the following six steps:

1. *Preparation and planning*: Land degradation and boundary assessments; labor costs, volume estimates, and equipment assessments; hydrological assessment; waste management; Occupational Health and Safety (OHS) standards.
2. *Technical rehabilitation*: Infilling of pits, shafts, tunnels; regrading and reprofiling to appropriate topography; appropriate use of limited mechanized approaches.

3. *Topsoils*: Identification of topsoils; targeted winning of available soils on and around site; conservation/storage and re-distribution.

4. *Biological rehabilitation*: Topsoil enrichment techniques (where necessary, using natural organic fertilizers); natural regeneration assessments; identification of native vegetation communities and key species (dominant, codominant, and successional species); seed collection of target species; distribution and mixing of target seeds and natural fertilizers into topsoils; tree and shrub and grass/herb plantings (where necessary and feasible).

5. *Mitigation hierarchy/whole mine cycle approach*: Integrating rehabilitation planning into active ASM design and operations to reduce primary environmental impacts (footprint), and so reduce unnecessary rehabilitation effort.

6. *Handover of completed rehabilitation site to relevant government administrations for approval/sign-off*: Local environmental inspectors will be engaged in monitoring and evaluating the rehabilitation process, using the FRM as an assessment tool. Recommendations will be made to the local governor who will formally approve the rehabilitation effort and receive the land back under government responsibility.

The technical aspects of the FRM are carefully assessed and planned, given a presumption of labor-intensive manual approaches, occasionally assisted by appropriate mechanization. Biological rehabilitation is focused on the identification of target rehabilitation and successional species that are native to the contextual environment, and does not depend on cultivated or imported species (which is commonplace in the rehabilitation of large-scale mining sites). The botanical and vegetation ecology aspects are based on a field survey and floristic evidence, identifying and utilizing vegetation community classification systems where they occur, or an adaptation of such, where they are not available. An appreciation of ecological successional species is useful, given that certain plants rapidly recolonize rehabilitated surfaces, advancing a rapid recovery of native species.

25.5 Critical Factors in the Success and Sustainability of the FRM in Mongolia

The strength and value of the FRM lay in its foundation in the FRD, and the extensive stakeholder consultation that contributed to its development. The FRD meant that the methodology was seen to be evidence-based, evincing its core attributes of economic affordability, social acceptability, and ecological viability, and its usefulness to organized and formalized artisanal miners. Economic affordability, social acceptability, and ecological viability are linked defining attributes of the methodology, in that demonstrating one attribute needs to be balanced with the demands of the other two. In terms of affordability, rehabilitation budgets were determined on a case-by-case basis by fair but rigorously negotiated labor costs. In any one year, such labour costs were generally consistent across the program.[4]

Following finalization of the FRM, a formal training programme on the methodology and its Whole Mine Cycle Approach (WMCA) was deployed across project soums, delivered by the Responsible Mining Institute of Mongolia. The WMCA introduced the concept of the Mitigation Hierarchy into the ASM operational cycle, seeking to encourage miners to reduce their environmental impacts by planning for impact avoidance, mitigation of impacts caused through mining operations (such as loss and conservation of topsoil), and maximizing the conditions for effective frugal rehabilitation. Such training was to be followed up in the final year of the project in the mentoring and development of ASM NGO Rehabilitation Action Plans (RAPs) by ESEC II's partner NGOs, including the National ASM Federation, the industry association for ASM in Mongolia.

ESEC II could not focus only on developing the FRM. Other tools needed to be developed and applied as the project progressed, such as a *Multi-stakeholder Engagement Training (MET) Programme* (and a published manual) (The Asia Foundation 2015) that introduced broad environmental education to local communities. The MET program detailed issues such as climate change, sustainable land-use, water and waste management, the value of natural resources as ecosystem services, and biodiversity conservation, and was delivered across 45 soums (counties) nationally, where ASM was a key development issue. The MET training also advocated and facilitated the establishment of Local Multi-stakeholder Councils (LMCs) in the soums. LMCs were established through invited representation of environmental stakeholders, including ASM, LSM, livestock herders, and tourism operators.[5] Government officials also were able to advise and participate in the process; LMCs were typically chaired by the elected head of the Citizens

---

[4] as indicated in Section 3 - at a total cost of USD390,000 - Frugal Rehabilitation Demonstration costs averaged USD 2,750/ha through 2014-2015.

[5] The establishment of Local Multi-stakeholder Councils had already been initiated in The Asia Foundation's ESEC I project, and the model had been well received.

Rural Khural (or council). By the end of 2015, 39 of the 45 soums that participated in the MET program had an LMC established.

It was recognized by ESEC II that local democratic platforms alone would
not effectively deliver better planning and decision-making if participating
stakeholders were not aware of their rights and duties as responsible citizens. Human rights transgressions frequently characterized relationships
between informal ASMs and local communities, in both directions. The
SDC's SAM project had previously delivered trainings on HRBA in the ASM
sector (Swiss Agency for Development and Cooperation (SDC) 2013, 2015c).
However, the ESEC II project also needed to advocate for HRBA in enabling
the development and establishment of ASM-inclusive LMCs at soums across
the project area.

A program of trainings in HRBA was delivered to government officials and
local communities through selected NGO partners across a majority of the
project's 45 soums, which complemented the teachings of the MET program.
Such capacity-building efforts helped stakeholder involvement in LMCs to
become more participatory, confident, and effective.

Derived from the FRM was the Rehabilitation Action Plan, (Ministry of
Mining, Mongolia, Swiss Agency for Development and Cooperation (SDC)
2016) a planning tool that ASM NGOs could use to document past achievements and commitments to FR (for example, through their FRD efforts), or
present their current and future activities with regards to FRM implementation. A completed RAP (either for documenting previous approved rehabilitation or for rehabilitation alongside proposed future mining) would
support ASM NGOs when applying to local government to access new land.[6]
A RAP would represent an ASM NGO's formal contribution to the soum's
Environmental Management Plan (EMP), a plan which enabled LMCs to
identify, better understand, and strategically manage environmental problems between local stakeholders.

The process of project implementation, environmental tool development
and capacity-building is represented in the following table.

---

[6] A local governor would then submit the proposal to the Minerals Resources Agency (MRAM)
for final approval.

Improved ASM environmental capacity and inclusive environmental governance	
Environmental toolkits	Capacity-building: educational; local democratic representation and process
Frugal rehabilitation methodology (FRM) through co-financed/funded FRD programs	Multi-stakeholder engagement training (MET)
FRM/WMCA/RAP training	Human rights-based approach (HRBA) training
ASM Rehabilitation action plan (RAP) development	Local multi-stakeholder council (LMC) establishment

LMC convened with local government
Environmental management plan training and development
Formalized and ASM-inclusive environmental management plan (EMP)

In late 2015, ESEC II commissioned consultants to undertake pilot trainings and workshops to explore in what format EMPs would best work. This was further refined in early 2016 and the resulting template adopted by a team of Mongolian civil-society NGOs to facilitate the development of soum EMPs with 22 LMCs and local government. The EMPs were designed for a 3–5-year planning period. ASM-inclusive EMPs provided a forum for relevant environmental stakeholders to voice their concerns, discuss shared environmental issues, and seek consensus for problem-solving. A simultaneous training and development program for ASM NGO-led RAPs ensured that ASM NGOs were well placed to submit and integrate their completed and approved RAPs into the EMP development process in a timely way.

An additional aim of the ESEC II project was to encourage local government to use appropriate public finances to support the replication and scaling-up of the FR agenda. While active formalized ASM practitioners are now expected to fund and incorporate the FRM into their ongoing and future operations, it falls to government to fund the rehabilitation of abandoned ASM land. During 2015, ESEC II commissioned a Public Finance Mechanism assessment that identified the legal frameworks whereby taxes from a variety of sources could be re-invested in publicly funded works, including environmental rehabilitation, through Local Development Funds (LDFs) (Enkhbat 2015). It became clear that the PFMs were not working in a way that would readily see local government taking the lead in funding rehabilitation efforts. This was in part due to the severe economic downturn Mongolia was experiencing from 2014 to the time of writing this paper (Riley 2016). With the agreement of the SDC, ESEC II was able to design and implement cofinancing packages with several aimag and soum governments for FR at 13 sites in 2016, resulting in completed rehabilitation of over 60 hectares. This demonstrated a willingness of soum and aimag governments to financially support initiatives for frugal rehabilitation, drawing from LDFs and other sources, and undertaken by skilled ASM NGOs.

By the ESEC II project's end, not only was there a demonstrated successful rehabilitation methodology, there was an agreed consultative process for RAPs to feed into LMC-led soum EMPs. Such plans were also being approved and utilized at local and national government levels. Furthermore, by 2016 "frugal rehabilitation" had developed a profile as an alternative livelihood option for ASM NGOs and a source of employment for others in the community who were able to appreciate the value of the activity.

Considering all the above, will the FRM become mainstreamed into formalized ASM operations? Given that the FRM has been annexed to the Revised Regulation on Small-scale Mining (151), the situation is very positive (Swiss Agency for Development and Cooperation (SDC) 2017). Importantly, the Ministry for Nature, Environment, and Tourism (MNET) was closely involved in the FRM's development. Early in the project, MNET was skeptical about the project's aims and objectives, and held the view that empowering and building capacity within the ASM sector could only contribute to further environmental damage. The project's engagement with MNET sought to ensure that any methodology, despite being frugal, would be authentic, practical, and ecologically viable. The FRD program and its results were convincing in this regard.

Furthermore, from the outset the ESEC II project highlighted the importance of biodiversity values with respect to the potential impacts of ASM activities. It scoped a national profiling of geographic relationship between soums with ASM activity and biodiversity importance, for species, sites, and habitats (Batbayar and Purev-Ochir 2015). Illegal ASM within Protected Areas was especially an ongoing problem for the Protected Areas Administrations within MNET. The project's selection of rehabilitation demonstration projects within Protected Areas highlighted a common commitment to conserving these sites.[7] It was also imperative for MNET to have the opportunity to contribute to the development of the FRM through Project Advisory Group consultation and through its involvement in the ministerial working group established to progress the regulation on small-scale mining. This process helped secure a more positive regard for the FRM and its deployment, and resulted in both the Ministry of Mining and MNET signing a joint decree with respect to the FRM and its function.

Having two major ministries supporting the FRM is a strong factor in its resilience and value, both in terms of political support and potential impact. However, in a country where enforcement of environmental laws is often inconsistent and constrained by various factors, including a lack of institutional capacity in a countryside characterized by great distances, remoteness, and low population density, adherence to the law cannot be taken for granted.

ESEC II also recognized a need for transparency and consistency in land access decision-making responsibilities of the Ministry of Mining, MRAM, and the GASI (responsible for environmental monitoring and inspection). A consistent call from ASM communities keen to adopt the FRM as a formal undertaking was that access to land was key. ASM's legal access to viable mining reserves is critical in determining whether the sector takes a responsible formalized track or operates illegally—if no viable mining reserves are made available to ASM communities then this can perpetuate informality. Formalized ASM NGOs can be the best protectors of areas that are vulnerable to illegal "ninja" mining. In response to the success of the FRD, local and national government authorities are now recognizing that ASM can be undertaken responsibly, and are facilitating the sector's access to appropriate mining lands.

Despite the positive progress enabled by ESEC II, the mainstream performance and adoption of the FRM by the wider informal ASM community remains a significant challenge. A key threat to this is the sector's widespread transition to a greater dependency on heavy machinery. Such operators go beyond the legal threshold for mechanized ASM and enter a grey area of mining that is now an ongoing and significant problem for government. While the FRM can be adapted to increased mechanization within certain parameters, the rehabilitation of land mined by heavy machinery demands a departure from the three key principles of the FRM: economic affordability, social acceptability, and ecological viability. Loss of topsoils, a lack of organic soil enrichment, and particularly soil compaction are frequent problems with this type of rehabilitation. Where companies undertake rehabilitation dominated by heavy machinery, there is widespread evidence suggesting that ecological recovery

---

[7] Some of the highest profile frugal rehabilitation demonstrations occurred within Special Protected Areas and National Parks in the south and west of the country.

is compromised, and this may well impact the interests of other stakeholders (especially those in the agricultural sector). Examples of highly mechanized rehabilitation efforts associated with small- or medium-scale mining were encountered in many locations across Mongolia, such as in Jargalant, Bayan Ovoo, and Galuut soums of Bayankhongor province, and Yusunbulag soum in Govi-Altai. Such sites were often in close juxtaposition to project FRD sites, and useful comparisons could be made between contrasting methodologies. Some examples were documented in the FRD case Studies Handbook.

Implementing the ESEC II project over the 3.5 years was an intensive process, communicating across many levels, from national government ministries to provincial and local government, and from international and national consultants, NGOs, and other organisations, to grass-roots NGOs and Local Multi-stakeholder Councils, and to the artisanal miners themselves. Project implementation and communication at all levels was dependent on a variety of digital tools, such as PCs, laptops, the various Microsoft Office tools, as well as digital cameras, projectors, and, of course, cell phones. The widespread availability of digital hardware across the country, even in the most remote locations, enabled the project to deliver its various outputs and trainings within a constrained timescale. Widespread stakeholder capacity to use digital hardware and software was a key, if understated, value. Implementing the project within such a timescale would have been impossible but for the enabling capacity provided by such digital technologies. In this sense, the project was very much an initiative of the digital age, as indeed are many during such times.

25.6 The Success of Developing the FRM in Mongolia and Potential to Export and Adapt the Methodology

The success of the FRM in both design and execution is evidenced by its:

- Endorsement by relevant (key) line ministries in national government.
- Attachment to a statutory regulation (151).
- Uptake and demonstration by the ASM sector.
- Successful rehabilitation of significant portions of land in Mongolia (which will become available for alternative economic activities after a period of regeneration).
- Acceptance by wider communities and stakeholders.

It is believed that the FRM's conceptual structure and approach can be applied to a broad array of ASM contexts. Adapting such techniques and prescriptions would need to take careful account of the following:

- Labor costs within the ASM environment, both manual and technical.
- The degree of dependence on heavy machinery.
- General or specialized technical approaches to ASM.
- Climatic and weather conditions, including length of growing season, degrees of aridity and vulnerability to drought and flood.
- Hydrological environments, including mining in alluvial floodplains, the seasonality of flooding.
- The presence of mercury and other key pollutants in the mining process and in the wider environment.
- Soil types and management constraints.
- Biomes and vegetation community context. Is there a National Vegetation Classification (NVC) standard in place? If not available, then standard phytosociological approaches should be accessed and applied through national experts.
- Successional processes.
- Grazing regimes.
- Contextual land-use competition and pressures.
- Cultural, educational and/or political perspectives on environmental protection.

As to whether the FRM approach can stand alone as a tool-based introduction to developing ASM best environmental practice needs to be evaluated against wider socio-economic and political criteria within the country in question. In Mongolia, although the FRM was prominent as a stand-alone approach, it was conceived as a critical component in an integrated, layered, and holistic approach to capacity building for sectoral behavior change. In other contexts, it may be that existing or new development initiatives are underway in-country to address ASM issues, and the FRM could bring added value to such efforts. Alternatively, it may be that there are no other similar initiatives underway and an FRM project must start from scratch.

The consideration and analysis of assumptions and risks would also be crucial to project success, including the following:

- *Political*: Stability; national and local government capacities; election cycles; budget management; ministerial policy alignments and practices; transparency; potential for corruption; etc.
- *Legislative*: Progressive/conservative; sufficiently developed to accommodate ASM formalization; mercury issues; effectiveness of law enforcement; legal disincentives for informal ASM; potential for corruption.

- *Economic*: Public finance mechanisms in place/status; national and global trajectory assessments; incentives (or lack thereof) to support ASM formality/informality; large-scale mining sector influence/ role; interactions between LSM and ASM.
- *Social*: Stakeholder opposition; presence and effectiveness of local community-based negotiation platforms; context of human rights abuses; capacity for inclusion of formal/informal ASM sectors, including interactions between local and migratory ASM; stakeholders' educational capacities including literacy.
- *Environmental*: Climate change context; mercury pollution; Protected Areas/biodiversity impact issues influencing government policy; wider ASM-related land-use conflicts; environmental education capacities; environmental planning and governance capacities.

25.7 The BEST-ASM Initiative and Its Planned Expansion of the FRM into New ASM Contexts

The authors feel confident that adapting and applying the FRM to new ASM contexts would be a feasible method for building stakeholder desire and capacity to pursue environmentally responsible ASM in a variety of countries. The impact proposition appears high and the FRM merits promotion to others to adapt and adopt this approach to their own situations.

Among the many development initiatives financed with millions of dollars of public money, there are a number whose success is felt by few and carried forward by even fewer. In the authors' experience, it is rare that development organizations allocate resources to promoting their successes, particularly in the years following implementation and project completion. Yet environmental management is typically the lowest priority in any ASM community, and ASM is typically the lowest priority in any strategic conservation response situation vis-à-vis ecosystem degradation due to forest-loss, agriculture, charcoal making, illegal wildlife trade and poaching, and bushmeat hunting. The international policy arena often prioritizes attention to human rights issues associated with the ASM arena whose impacts are more immediate. That ASM is frequently illegal or informal compounds the problem of institutional policy-based engagement, for conservation and social NGOs as well as for government, and the issues are often not easily or readily confronted. This perpetual side-lining of ASM's impact on the environment due to a relative perception of low urgency means that its environmental impacts are not fully acknowledged and attempts to address them underfunded. Yet environment and mining ministry officials from developing and emerging countries often lament their inability to get a handle on environmental management and rehabilitation in ASM,[8] and governments continue to evict artisanal miners from protected areas and critical ecosystems (PACE) with force, only to find them returning again (Salo et al. 2016).

Inspired by the successes of ESEC II's FRD, the social venture, Levin Sources, formerly Estelle Levin Ltd. (ELL), also the authors of this chapter, decided to attempt to raise international awareness of the FRM, its associated FRD, and other useful tools and successful approaches to environmental management in ASM. The Biodiversity and Ecosystem Services Transformative ASM ("BEST-ASM") initiative aims to enable ASM actors to reduce and mitigate their impacts on the environment and affected stakeholder communities, ultimately enhancing the contribution of ASM to economic development through an enhanced license to operate and environmental and social resilience.

The BEST-ASM initiative seeks to build on existing toolkits already piloted, demonstrated, and implemented in the sector, including the FRM. It is also developing new tools to fill the gaps where necessary, and incorporate tools from third parties into its "toolbox" over time. The initial existing toolkits to be used in this approach are informed by three leading ASM development programs:

1. The ESEC II Project.
2. The ASM in Protected Areas and Critical Ecosystems (ASM–PACE) program.
3. The Gold and Illicit Financial Flows (GIFF) Project.

The ASM-PACE programme has been active since 2010, building on the research and development program planning experience of Levin Sources, the World Wildlife Fund for Nature, and other partners into the sustainability and management of ASM in PACE around the world. It focuses on addressing the impacts of ASM in PACE by finding workable, sustainable, and win-win solutions that balance environmental concerns with the human rights and economic development potential of ASM. The program's Global Solutions Study catalogued incidents of ASM in PACE in 36 countries and approaches conservation organizations and authorities were taking to mitigate its impacts, thereby establishing the extent of the issue and lessons from existing efforts to address it. ASM–PACE's Methodological Toolkit presents a methodology for determining where ASM is taking place in PACE, its economic contributions, the incentives that drive engagement in the sector, the stakeholders involved in and affected by the industry, the successes and failures of governance measures, and its potential and actual social, economic, and environmental harms. The toolkit can also provide guidance on how to

---

[8] AMDC Artisanal and Small-Scale Mining Study Expert Group Meeting, 7-8th April 2016, Addis Ababa, Ethiopia.

engage with miners and their stakeholders, map the mining activities, and assess the nature of the problem to develop informed, feasible solutions that are more likely to produce sustainable, long-term outcomes. ASM–PACE and Levin Sources have applied the Methodological Toolkit to ASM impacts on globally important PAs in many "megadiversity" countries, including Ecuador, DRC, Indonesia, Madagascar, Sierra Leone, Liberia, and Gabon.[9]

The GIFF Project was cofounded by Levin Sources and the Global Initiative against Transnational and Organized Crime in 2015 to investigate the role of illicit financial flows (IFFs) in inhibiting the creation of more equitable and sustainable modes of ASM gold production and trade. This issue had previously been overlooked by initiatives seeking to increase rates of formalization in ASM. The BEST-ASM initiative will utilize the GIFF project's Supply Chain Mapping Tool to understand the commercial and political economy interests at play in the ASM sector, particularly unpacking the ways in which organized crime groups and corrupt officials instigate or perpetuate informality (intentionally or otherwise), make stakeholders vulnerable to human rights abuses, and undermine the ability of miners and traders to improve their practices and thus enhance their legitimacy as economic contributors. The GIFF Project published a handbook for mapping financial flows in the ASM sector in 2017 (Hunter et al.2017). This handbook has specific relevance for analyzing and strategizing on ASM in protected areas, where corruption, collusion, and criminality are typical characteristics of the protection economies surrounding ASM in such illegal situations.

The BEST-ASM initiative will mitigate environmental and social impacts by doing the following:

1. Broadening the scope of environmental impact mitigation in ASM operations.
    a. Addressing environmental impacts at different points of the mining life cycle and through applying the mitigation hierarchy in a way that is appropriate to the capacity of the ASM sector.
    b. Informing ecosystem and landscape approaches to biodiversity conservation in the ASM sector, across all producer groups (and throughout the mining life cycle).
2. Improving the availability of and access to tools and expertise, tailoring them to the specific needs of the ASM sector and of impacted stakeholder communities.
    a. Leveraging existing successes for mitigating environmental impact in ASM.
    b. Complementing and providing resources to other ASM development and sourcing initiatives.
    c. Tailoring tools for different users.

---

[9] Some of these reports are publicly available at www.asm-pace.org.

3. Focusing on measurement, scalability, and knowledge mobilization in program design and implementation.

The BEST-ASM initiative utilizes a diverse range of tools that are applicable to ASM occurring generally, and also for ASM that is taking place within PACE. When applied generally, the initiative aims to assist ASM to develop, implement, and demonstrate environmental best practices, enable best practice rehabilitation of degraded habitats, and engage more positively with affected stakeholders. In PACE, it aims to go further to contain ASM activities, enable best practice rehabilitation of degraded habitats, and facilitate a managed and sustained exit of ASM from within such areas, through a Containment and Exit Strategy (CES). It is also recognised that different stakeholders will value different components and tools within the proposed BEST-ASM toolkit, depending on specific interest, perspective, and context of application. The toolkit is thus flexible and adaptable to a wide range of contexts.

The FRM, combined as required with additional tools, can create a suite of tools that can be used by practitioners to, at a minimum, do the following:

1. In cases where ASM is not occurring in PACE, productively identify and engage with the miners and associated stakeholders, and use the ESEC II model to build their capacity to implement the FRM on an ongoing basis to limit environmental impacts (based on the mitigation hierarchy) and ultimately restore the site's biodiversity and ecosystem services to the acceptability of local stakeholders (once the deposit has been exhausted) or to an alternative productive use.

2. In cases where ASM is occurring in PACE, identify, research, and engage with the miners and associated stakeholders, to contain mining activities and facilitate the miners' responsible exit out of that area (CES), and using the ESEC II environmental governance model, ensure collaboration with and training of local stakeholders to use the FRM framework to restore the area's biodiversity and ecosystem services, while potentially realizing alternative livelihoods that support such valued areas.

The GIFF Project's Follow the Money Tool can be used in both scenarios to bring clarity to the protection economies that may entrench illegal mining in protected areas and/or inhibit ASM's progression toward more responsible mining methods and the protection of human rights. This information is vital to determining which "key logs" may need to be shifted, with cooperation from government and other stakeholders, to allow the ASM sector to embark on a process of steady change and improvement, and how that can be facilitated.

Through a strategic distribution of the BEST-ASM concept note, Levin Sources is currently consulting a range of selected organizations that have an interest in developing, piloting, and implementing the BEST-ASM toolkit in their respective geographic and institutional areas of interest. Levin Sources is seeking constructive criticism and positive feedback on how such stakeholders may work together with the company to implement the desired tools in the BEST-ASM initiative.[10]

The BEST-ASM Initiative has thus far identified at least 25 initiatives that could benefit from an application of the proposed toolkit and/or could more effectively share lessons learned and their own tools through BEST-ASM. Potential partners and implementing organizations come from national and local government, international and national civil society organizations, ASM development programs and entities, Protected Area management agencies, and responsible sourcing initiatives. ASM communities suitable for the application of FRM can either be operating without any kind of external support/intervention, or be part of an existing development program. The fact that both the ESEC II model and ASM-PACE Methodological Toolkit provide guidance on facilitating ASM formalization as a key component of biodiversity conservation does not preclude informal ASM communities from benefiting from such piloting, subject to government approval.

Business, government, and civil society can operate at different paces and with different expectations. Additionally, one of the major barriers to sustainable transformation of mineral sectors is the short-termism often found in interventions. While acknowledging institutional differences in operational practice, such collaborations for BEST-ASM should seek to find a balance in managing expectations and facilitating progressive change. The indicator for continuous improvement as a measure of success is derived from the "OECD's Due Diligence Guidance and the OECD's Guidelines for Multinational Enterprises."[11] BEST-ASM also aspires to take the long-term view, as far as possible, to maximize the conditions for sustainability and success.

Following the completion of demonstrations, Levin Sources anticipates the following activities:

1. Replication and scaling-up of BEST-ASM in ASM regions around the world.
2. Piloting in additional ASM communities.
3. Standardizing the FRM in industry responsible sourcing standards, ASM development programs, and so on.
4. Developing a global map and baseline impact assessment of ASM on BES within PACE.

---

[10] To participate in this consultation, please contact Andrew Cooke: andrew@estellelevin.com.
[11] For further information, see: http://www.oecd.org/corporate/mne/45534720.pdf.

The BEST-ASM initiative is open to multiple models of engaging with an opportunity. Levin Sources' aim is to ensure BEST-ASM becomes a standardized approach in the ASM sector, and to allow others to become aware of and harness its demonstrated benefits in ways that are fitting to context.

25.8 Conclusions

The ASM sector is going to remain an important part of the world's mining industry for the foreseeable future. Institutionalizing the effective long-term management of its environmental impacts is thus vital for not only ensuring ASM-producer countries are on the right track with regards to conservation and meeting the United Nations Sustainable Development Goals, but also for enabling artisanal and small-scale miners to enjoy a legitimate license to operate in the communities they inhabit. The development and demonstration of a FRM approach in Mongolia, and the additional methodologies utilized in the Biodiversity and Ecosystem Services Transformative ASM initiative have demonstrated efficacy in the realization of these goals, in Mongolia and in a suite of other countries. Applying these innovative methodologies more broadly across the world's ASM-degraded habitats thus has strong potential to help normalize sustainable environmental management in the sector, in turn reinforcing its status as a valid economic contributor in developing and emerging economies.

Acknowledgments

The authors wish to acknowledge the considerable efforts and support of those within the SDC, as well as the SAM Project, in Mongolia. Their long-standing commitment to building capacity for responsible ASM in Mongolia provided the basis for the ESEC II Project that followed. Thanks go to The Asia Foundation and its representatives in Mongolia, and to those members of the ESEC II Team who worked so tirelessly to implement this comprehensive project within a constrained timescale. Thanks also to the many NGOs and other organizations who played a contributory role in the project's realization across the country.

We also wish to acknowledge the considerable efforts of all those within Levin Sources' international team in the convergent effort of developing the BEST-ASM initiative, drawing on both the many ASM-PACE case studies and the GIFF project.

References

Aldama, Z. 2016. Winners and losers in Mongolia's mining gold rush. *Post Magazine*, June 13, 2016.

Batbayar, N. and P. Gankhuyag. 2015. *Biodiversity and Artisanal and Small-scale Mining in Mongolia: Scoping High Biodiversity Values in soums with active ASM*. Wildlife Science and Conservation Center of Mongolia (WSCC) for The Asia Foundation, Mongolia.

BirdLife Asia. 2009. Safeguarding important areas of natural habitat alongside economic development. *Mongolia Discussion Papers*. East Asia and Pacific Region Sustainable Development Department, World Bank, Washington, DC.

Butler, L., P. Mitchell, and E. Levin. 2014. *International Review of Environmental Rehabilitation Approaches for Artisanal and Small-Scale Mining: A Review of Best Practices for Frugal Rehabilitation of ASM in Mongolia* Estelle Levin for The Asia Foundation, Mongolia.

Carlson, D. 2014. 2008 financial crisis set stage for gold rally. *Kitco News*, October 13, 2014.

Enkhbat, A., and G. Bolormaa. 2015. Public finance mechanisms for resourcing environmental rehabilitation. Commissioned report to The Asia Foundation, Mongolia.

Hunter, M., A. Smith, and E. Levin-Nally. 2017. Follow the money. Financial flows linked to artisanal and small-scale gold mining. A tool for intervention. *The GIFF Project (The Global Initiative Against Transnational Organised Crime and Estelle Levin)*. At http://globalinitiative.net/wp-content/uploads/2017/03/illicit-financial-flows-linked-to-artisanal_06.03.17.compressed.pdf.

Lamb, D. 2016. No to Rehab? The mining downturn risks making mine clean-ups even more of an afterthought. *The Conversation*. April 29, 2016.

Ministry of Mining (MoM), Mongolia, Swiss Agency for Development and Cooperation (SDC), and The Asia Foundation (TAF). 2016a. *Frugal Rehabilitation Demonstration in Mongolia – A Case Studies Handbook*. The Asia Foundation, Mongolia.

Ministry of Mining (MoM), Mongolia, Swiss Agency for Development and Cooperation (SDC), and The Asia Foundation (TAF). 2016b. *The Frugal Rehabilitation Methodology (FRM) – A Field Handbook*. The Asia Foundation, Mongolia.

New Internationalist. 2013. World Development book case study: Mining in Mongolia. Mega-mining in Mongolia—A development bonus or resource curse? At https://newint.org/books/reference/world-development/case-studies/mining-mongolia-development-resource-curse/.

Reichardt, M. 2013. The Wasted Years: A history of mine waste rehabilitation methodology in the South African mining industry from its origins to 1991. PhD thesis, University of the Witwatersrand, South Africa.

Riley, C. 2016. This country went from boom to economic nightmare in 5 years. *CNN Money*. August 23, 2016.

Salo, M. et al., 2016. Local perspectives on the formalization of artisanal and small-scale mining in the Madre de Dios gold fields, Peru. *The Extractives Industries and Society*. 3(4), 1058–1066.

Swiss Agency for Development and Cooperation (SDC). n.d. Sustainable artisanal mining project. At http://sam.mn/sustainable-artisanal-mining-project/.

Swiss Agency for Development and Cooperation (SDC). 2011. SDC experiences with ASM formalization and responsible environmental practices in Latin America and Asia (Mongolia). SDC, Mongolia.

Swiss Agency for Development and Cooperation (SDC). 2013. The human rights in small-scale mining in Mongolia. *Sustainable Artisanal Mining Project and National Human Rights Commission of Mongolia*. SDC, Mongolia.

Swiss Agency for Development and Cooperation (SDC). 2015a. Small-scale miners' occupational safety and health—A simplified handbook. *Sustainable Artisanal Mining Project*. SDC, Mongolia.

Swiss Agency for Development and Cooperation (SDC). 2015b. Legislation related to artisanal and small-scale mining. *Sustainable Artisanal Mining Project*. SDC, Mongolia. pp 13–23.

Swiss Agency for Development and Cooperation (SDC). 2015c. Human rights based approaches and gender equality: A manual. *Sustainable Artisanal Mining Project and Center for Human Rights Development*. SDC, Mongolia

Swiss Agency for Development and Cooperation (SDC). 2016a. Artisanal and small-scale mining—Organisation in Mongolia. *Sustainable Artisanal Mining Project*. SDC, Mongolia.

Swiss Agency for Development and Cooperation (SDC). 2016b. Social insurance Mongolia—Security for your life. *Sustainable Artisanal Mining Project and Social Insurance General Office*. SDC, Mongolia.

Swiss Agency for Development and Cooperation (SDC). 2017. Revised regulation improves conducive environment for artisanal and small-scale miners. *Sustainable Artisanal Mining Project*. SDC, Mongolia.

The Asia Foundation. 2015. Sustainable development: Responsible resource use—Multi-stakeholder engagement training handbook for stakeholders. *The Asia Foundation (ESEC II) and Mongolian Cooperatives Training and Information Center*. The Asia Foundation, Mongolia.

Villegas, C. 2013. Ninja miners and rural change in Mongolia. ASM-PACE Programme, Estelle Levin. 1st November.

World Bank. 2009. BirdLife Asia. Safeguarding important areas of natural habitat alongside economic development. *Mongolia Discussion Papers*. East Asia and Pacific Region Sustainable Development Department, World Bank, Washington, DC.

Conclusion

Robert K. Perrons

This book has shed a useful amount of light on several innovations and technological forces that are materially changing the contours of the resource sector, and explored the processes and organizational structures that underpin the innovation—and research and development (R&D)—related processes behind these changes. Also, it has examined the many ways that these innovations are fundamentally reshaping relationships between resource companies and their many stakeholders, and put forward evidence of how they are bringing about significant upheavals in the value networks and constituent firms within the industry. Upon looking at the totality of the changes described in this book, one cannot help but be taken aback both by how much has happened and by how quickly. And when considered together, its various themes and messages also point to an important question: what lies ahead?

Many of the innovations and trends described throughout the book are at least partially the result of high-level, macroscopic influences that are unfolding over time horizons of decades, and that are affecting the entire marketplace. As demonstrated within the preceding chapters, several of these forces have already had a noticeable and significant impact on the resource industry—but their sheer magnitude is such that their transformational effects will be felt for some time to come. These high-level effects will in turn lead to changes both in the nature of the technologies introduced into the mining and energy sectors and to the organizations that usher them in.

Front and center among these macroscopic forces is the increasingly digital nature of the world around us. From the emergence of Uber in the mobility sector to Airbnb's disruption of the hospitality market, digital technologies have upended traditional business models within an impressively diverse range of industries (Westerman, Bonnet, and McAfee 2014). By contrast, the digital transformation of the resource sector has been less pronounced so far. Companies in the mining and oil and gas industries have been gradually embracing many of the information technology (IT)-related breakthroughs behind this revolution, including Big Data (Holdaway 2014; Perrons and Jensen 2015) and cloud computing (Perrons and Hems 2013), but the effects of digital disruption within the resource sector will continue to deliver significant changes in four important ways.

First, the mix of organizations within the mining and energy industries will be materially transformed. The digitization of other industries has frequently carried with it fundamental shifts in profitability and power, and the appearance of new entrants into market segments where they simply had not previously been. And sure enough, while this book was being written, the market saw early evidence of this same dynamic playing out in the oil and gas industry when General Electric (GE) acquired Baker Hughes (Clough 2016), one of the largest service companies in that sector. GE has been a particularly vocal champion of digitization within asset-intensive industries (Evans and Annunziata 2012), and the Baker Hughes deal could quite reasonably be interpreted as a way for GE to quickly and aggressively expand the digital horizons of that industry—and, presumably, to shepherd the sector toward GE's own proprietary "industrial internet" technology platform. In tandem to its efforts in the oil and gas domain, GE has also been actively calling for more digitization within the mining sector. One therefore cannot help but wonder if GE or another company like it that is more digitally savvy than traditional mining companies might be preparing for a dramatic entrance into that industry, too.

Second, the nature of how organizations work and innovate together will change. As recently as a few decades ago, much of the R&D and innovation-related activity within the oil and gas and mining industries was led by small numbers of large companies at the top of their respective value chains. In the petroleum industry, more than 80% of R&D spending in exploration and production from the early 1970s until the early 1990s was done by just 11 companies from Big Oil (Economides and Oligney 2000). Innovation processes may have been somewhat less concentrated within the mining sector at around this same time, but anecdotal evidence from that period (e.g., Lynch 2002; Peterson, LaTourrette, and Bartis 2001) leaves little doubt that Big Mining led the way with several important breakthroughs in that industry in much the same way.

But the era of innovation occurring under only one company's roof is mostly coming to an end. In its place are newer models of collaboration that are both more open (Chesbrough 2003a, 2003b) and democratic (von Hippel 2005). The increasingly digital nature of work and communication has created an environment in which people and organizations are increasingly able to share ideas and data across boundaries with near-zero transaction costs. Under these new conditions, the solution to many of tomorrow's resource sector problems will likely come more and more from industries and knowledge bases that have previously had little direct exposure to those industries (Robertson and Smith 2008). In other words, the good ideas that will make big breakthroughs in the mining and oil and gas sectors will increasingly be imported from outside industries and knowledge bases. Competitive advantage in this kind of business environment will be conferred not to individuals or companies that can do everything themselves, but instead to people and organizations that become adept at identifying promising new ideas happening elsewhere in the marketplace (Chesbrough 2003a), and then figuring out how best to apply them within the context of the resource industry. There will still be a place for exhibiting technological leadership in this new reality, but it will frequently come in the form of platform leadership (Cusumano and Gawe, 2002; Gawer and Cusumano 2002) rather than more unilateral approaches to innovation.[1] Market dominance in a world of platform leaders usually cannot be accomplished by having all the good ideas yourself, but by knowing how to lead a broad ecosystem of collaborators toward a shared technology architecture and vision (Baldwin and Woodard 2009).

Third, we will probably start to notice a significant acceleration in the rate at which technological change happens within the resource sector—that is, the industry clockspeed of the mining and oil and gas industries will become faster. Although the resource industry's past has been punctuated by brief periods of bullishness where innovation is concerned (e.g., Priest 2007; Yergin 2011), much of the mining and oil and gas sectors' respective histories can be fairly characterized as "slow clockspeed" (e.g., Perrons 2015). There are some obvious and logical reasons why these industries have historically behaved this way. This phenomenon has been explained within energy industry by drawing attention to the fact that the sector is "unusually large, diverse, and complex" (Newell 2011, 27), and because it often attracts an uncommon amount of political sensitivity when changes are proposed (Deutch 2011). These factors work together to create business conditions where it "typically takes at least several years to move a new energy technology from 'proof of concept' in the laboratory to full-scale demonstration, and many more years or even decades before that technology can achieve significant market penetration" (Lester and Hart 2012, 26). Similar structural forces have caused the mining sector to exhibit many of these same behaviors throughout much of its history (Lynch 2002; Peterson et al. 2001).

Both industries are changing, however. As noted earlier in this book, several members of Big Oil have pointed to technology as an increasingly important strategic priority (e.g., Chazan 2013; Kulkarni 2011; Parshall 2011), and Big Mining also seems to be putting significantly more emphasis on technology than it used to (e.g., Bellamy and Pravica 2011; Perrons and McAuley 2015). But here, too, the increasingly digital nature of these sectors is unleashing structural changes that go above and beyond these stated priorities. Salim Ismail, the Silicon Valley strategist and former head of Yahoo!'s technology incubator division, once quipped that "Once you turn something digital, it hops on that Moore's Law[2] pace of change and accelerates." And sure enough, this prediction seems to have come true in several industries which, like taxis prior to the arrival of Uber, had business models

---

[1] In fact, the "industrial internet" strategy discussed earlier appears to be GE's attempt at this kind of platform leadership.
[2] Established in 1965 by Gordon Moore of Intel, Moore's Law originally stated that microprocessors tended to double in power and transistor density about every 2 years (an exponential growth rate). The timeframe for this doubling has since been shortened to 18 months.

that remained relatively unchanged for several decades prior to being disrupted. It therefore seems quite reasonable to guess that the rate of technological change within the energy and mining sectors will probably also gather pace as digital technologies become more and more widespread and mission-critical to these sectors' business models.

Fourth, the digital revolution and social media have fundamentally changed many aspects of how society is connected (Fuchs 2017). The rise of platforms such as Facebook and LinkedIn has led to a reshaping of how people are linked to other people, how companies connect to their customers and collaborators, and how governments engage their citizens. It therefore follows that these technologies will also have a game-changing impact on how resource companies manage their "license to operate"—that is, the approval that these companies require to be able to do business. Whereas traditional approaches to managing a firm's license to operate several years ago might have included a healthy amount of trust and goodwill between stakeholders, the ubiquity of smart phones and the internet have resulted in a situation where today's resource companies are being monitored in real-time absolutely everywhere. This clearly results in a much higher degree of transparency between mining and oil and gas firms and their many stakeholders. Almost any misstep by these companies, no matter how unintentional, will quite likely be captured by someone nearby with a smartphone. This information can then spread virally all over the internet, sometimes with the resource company at the center of the story having little control over the situation (Gaines-Ross 2010). It has been hard for resource companies to be perceived as trustworthy corporate citizens even at the best of times (Hofmeister 2010), but the rise of smartphones and social media have made this vitally important job even harder.

Finally, there is also one particularly noteworthy technological development unfolding in the resource sector that is not fundamentally digital in nature. Underpinning all of the contributions in this book is an unspoken assumption that the innovations would be applied on Earth. However, the resource sector and the space and satellite industry have made significant inroads that make this assumption less certain in the long-term future. Comparisons between the two industries are not new—one petroleum industry veteran noted, for example, that installing subsea oil production equipment into place on the ocean floor is like "landing a ship on Mars but with more extreme temperatures and pressures" (Gold and Campoy 2007, B6)—but those discussions have for the most part explored how space and satellite technologies could be usefully applied by the resource industry on Earth (Perrons and Richards 2013). What has changed in the past few years, however, is that the literature and discussions have reached a critical mass of people talking about how resources can be extracted from extraterrestrial locations like asteroids, other planets, and faraway moons (e.g., Dula and Zhenjun 2015; Jakhu, Pelton, and Nyampong, 2017; Lewicki et al. 2013;

Mueller et al. 2014). Far beyond its use as an inspiring metaphor for previous generations, tomorrow's moon shot projects in the resource industry might involve actual moons.

While it is always fun to speculate about the future, it is obviously impossible to say with any certainty how it will actually unfold. As the baseball legend Yogi Berra rightly observed, "It's tough to make predictions, especially about the future." What this book does highlight, however, is that tomorrow's resource industry will probably look decidedly different from the one we see today. And while mining and oil and gas companies have been derided over the years as being "low- and medium-tech" (von Tunzelmann and Acha 2006, 408), "technologically timid" (Lashinsky 2010, 88), and "low R&D intensity" (Moncada-Paternò-Castello et al. 2010; von Tunzelmann and Acha 2006), the evidence presented in this book strongly suggests that these labels will probably be extremely ill-fitting a few decades from now.

References

Baldwin, C. Y., and C. J. Woodard. 2009. The Architecture of Platforms: A Unified View. In *Platforms, Markets and Innovation*, ed. A. Gawer, 19–44). Cheltenham, United Kingdom: Edward Elgar.

Bellamy, D., and L. Pravica. 2011. Assessing the Impact of Driverless Haul Trucks in Australian Surface Mining. *Resources Policy* 36(2):149–158.

Chazan, G. 2013. Cutting-Edge Technology Plays Key Role for Repsol in Hunt for Oil. *Financial Times*, (May 5). http://www.ft.com/intl/cms/s/0/a20c0066-b420-11e2-b5a5-00144feabdc0.html#axzz2SSDP6umg.

Chesbrough, H. W. 2003a. The Era of Open Innovation. *MIT Sloan Management Review*, 44(Spring), 35–41.

Chesbrough, H. W. 2003b. *Open Innovation: The New Imperative for Creating and Profiting from Technology*. Cambridge, MA: Harvard Business School Publishing.

Clough, R. 2016. GE Creates $32 Billion Oil-Services Giant With Baker Hughes Deal. https://www.bloomberg.com/news/articles/2016-10-31/ge-to-merge-oil-division-with-baker-hughes-in-bet-on-energy.

Cusumano, M. A., and A. Gawer. 2002. The Elements of Platform Leadership. *MIT Sloan Management Review* 43(Spring), 51–58.

Deutch, J. M. 2011. *The Crisis in Energy Policy*. Cambridge, MA: Harvard University Press.

Dula, A. M., and Z. Zhenjun. 2015. Space Mineral Resources: A Global Assessment of the Challenges and Opportunities. *Space Mineral Resources: A Global Assessment of the Challenges and Opportunities*. Study conducted under the auspices of the International Academy of Astronautics (IAA).

Economides, M., and R. Oligney. 2000. *The Color of Oil: The History, the Money, and the Politics of the World's Biggest Business*. Katy, TX: Round Oak Publishing Company.

Evans, P. C., and M. Annunziata. 2012. *Industrial Internet: Pushing the Boundaries of Minds and Machines.* General Electric report. https://www.ge.com/docs/chapters/Industrial_Internet.pdf.

Fuchs, C. 2017. *Social Media: A Critical Introduction.* Thousand Oaks, CSA: Sage.

Gaines-Ross, L. 2010. Reputation warfare. *Harvard Business Review* 88(12):70–76.

Gawer, A., and M. A. Cusumano. 2002. *Platform Leadership: How Intel, Microsoft, and Cisco Drive Industry Innovation.* Boston, MA: Harvard Business School Press.

Hofmeister, J. 2010. *Why We Hate the Oil Companies.* New York: Palgrave MacMillan.

Gold, R., and A. Campoy. 2007. Wells Take Voyage to Bottom of the Sea. *Wall Street Journal,* July 26, B1, B6.

Holdaway, K. R. 2014. *Harness Oil and Gas Big Data with Analytics.* Hoboken, NJ: John Wiley & Sons.

Jakhu, R. S., J. N. Pelton, and Y. O. M. Nyampong. 2017. Private Sector Space Mining Initiatives and Policies in the United States. In *Space Mining and Its Regulation,* 59–71). New York: Springer.

Kulkarni, P. 2011. Organizing for Innovation. *World Oil* 232(3):69–71.

Lashinsky, A. 2010. There Will Be Oil. *Fortune* 161(5):86–94.

Lester, R. K., and D. M. Hart. 2012. *Unlocking Energy Innovation: How America Can Build a Low-Cost, Low-Carbon Energy System.* Cambridge, MA: MIT Press.

Lewicki, C., P. Diamandis, E. Anderson, C. Voorhees, and F. Mycroft. 2013. Planetary Resources—The Asteroid Mining Company. *New Space* 1(2):105–108.

Lynch, M. 2002. *Mining in World History.* London,: Reaktion Books.

Moncada-Paternò-Castello, P., C. Ciupagea, K. Smith, A. Tübke, and M. Tubbs. 2010. Does Europe Perform Too Little Corporate R&D? A Comparison of EU and Non-EU Corporate R&D Performance. *Research Policy* 39(4):523–536.

Mueller, R. P., L. Sibille, G. B. Sanders, and C. A. Jones. 2014. Concepts of Operations for Asteroid Rendezvous Missions Focused on Resources Utilization. *Earth and Space* 2014, 468–477.

Newell, R. G. 2011. The Energy Innovation System: A Historical Perspective. In *Accelerating Energy Innovation: Insights from Multiple Sectors,* ed. R. M. Henderson and R. G. Newell, 25–47). Chicago: University of Chicago Press.

Parshall, J. 2011. Shell: Leadership Built on Innovation and Technology. *Journal of Petroleum Technology* 63(1):32–38.

Perrons, R. K. 2015. How the Energy Sector Could Get It Wrong with Cloud Computing. *Energy Exploration & Exploitation* 33(2):217–226.

Perrons, R. K., and A. Hems. 2013. Cloud Computing in the Upstream Oil & Gas Industry: A Proposed Way Forward. *Energy Policy* 56:732–737.

Perrons, R. K., and J. E. Jensen. 2015. Data as an Asset: What the Oil and Gas Sector can Learn from Other Industries about "Big Data." *Energy Policy* 81:117–121.

Perrons, R. K., and D. McAuley. 2015. The Case for "n << all": Why the Big Data Revolution will Probably Happen Differently in the Mining Sector. *Resources Policy* 46(2):234–238.

Perrons, R. K., and M. G. Richards. 2013. Applying Maintenance Strategies from the Space and Satellite Sector to the Upstream Oil and Gas Industry: A Research Agenda. *Energy Policy* 61:60–64.

Peterson, D. J., T. LaTourrette, and J. T. Bartis. 2001. *New Forces at Work in Mining: Industry Views of Critical Technologies.* Santa Monica: RAND.

Priest, T. 2007. *The Offshore Imperative: Shell Oil's Search for Petroleum in Post-War America.* College Station, TX: Texas A&M University Press.

Robertson, P. L., and K. Smith. 2008. Distributed Knowledge Bases in Low- and Medium-Technology Industries. In *Innovation in Low-Tech Firms and Industries,* ed. H. Hirsch-Kreinsen and D. Jacobson, 93–117. Cheltenham, UK: Edward Elgar.

von Hippel, E. 2005. *Democratizing Innovation.* Cambridge, MA: MIT Press.

von Tunzelmann, N., and V. Acha. 2006. Innovation in "Low-Tech" Industries. In *The Oxford Handbook of Innovation,* ed. J. Fagerberg, D. C. Mowery, and R. R. Nelson, 407–432. Oxford, UK: Oxford University Press.

Westerman, G., D. Bonnet, and A. McAfee. 2014. *Leading Digital: Turning Technology into Business Transformation.* Cambridge, MA: Harvard Business Press.

Yergin, D. 2011. *The Quest: Energy, Security, and the Remaking of the Modern World.* New York: Penguin Press.

Index

Note: Page numbers followed by f, n, and t refer to figures, notes, and tables respectively.